Advances in Intelligent Systems and Computing

Volume 670

Series editor

Janusz Kacprzyk, Polish Academy of Sciences, Warsaw, Poland
e-mail: kacprzyk@ibspan.waw.pl

The series "Advances in Intelligent Systems and Computing" contains publications on theory, applications, and design methods of Intelligent Systems and Intelligent Computing. Virtually all disciplines such as engineering, natural sciences, computer and information science, ICT, economics, business, e-commerce, environment, healthcare, life science are covered. The list of topics spans all the areas of modern intelligent systems and computing such as: computational intelligence, soft computing including neural networks, fuzzy systems, evolutionary computing and the fusion of these paradigms, social intelligence, ambient intelligence, computational neuroscience, artificial life, virtual worlds and society, cognitive science and systems, Perception and Vision, DNA and immune based systems, self-organizing and adaptive systems, e-Learning and teaching, human-centered and human-centric computing, recommender systems, intelligent control, robotics and mechatronics including human-machine teaming, knowledge-based paradigms, learning paradigms, machine ethics, intelligent data analysis, knowledge management, intelligent agents, intelligent decision making and support, intelligent network security, trust management, interactive entertainment, Web intelligence and multimedia.

The publications within "Advances in Intelligent Systems and Computing" are primarily proceedings of important conferences, symposia and congresses. They cover significant recent developments in the field, both of a foundational and applicable character. An important characteristic feature of the series is the short publication time and world-wide distribution. This permits a rapid and broad dissemination of research results.

More information about this series at http://www.springer.com/series/11156

Bijaya Ketan Panigrahi · Munesh C. Trivedi
Krishn K. Mishra · Shailesh Tiwari
Pradeep Kumar Singh
Editors

Smart Innovations in Communication and Computational Sciences

Proceedings of ICSICCS 2017, Volume 2

Springer

Editors
Bijaya Ketan Panigrahi
Department of Electrical Engineering
Indian Institute of Technology Delhi
New Delhi
India

Munesh C. Trivedi
Department of Computer Science
    and Engineering
ABES Engineering College
Ghaziabad, Uttar Pradesh
India

Krishn K. Mishra
Department of Computer Science
    and Engineering
Motilal Nehru National Institute
    of Technology Allahabad
Allahabad, Uttar Pradesh
India

Shailesh Tiwari
Department of Computer Science
    and Engineering
ABES Engineering College
Ghaziabad, Uttar Pradesh
India

Pradeep Kumar Singh
Department of Computer Science
    and Engineering
Jaypee University of Information
    Technology
Waknaghat, Solan, Himachal Pradesh
India

ISSN 2194-5357   ISSN 2194-5365 (electronic)
Advances in Intelligent Systems and Computing
ISBN 978-981-10-8970-1   ISBN 978-981-10-8971-8 (eBook)
https://doi.org/10.1007/978-981-10-8971-8

Library of Congress Control Number: 2018937328

Printed on acid-free paper

This Springer imprint is published by the registered company Springer Nature Singapore Pte Ltd.
The registered company address is: 152 Beach Road, #21-01/04 Gateway East, Singapore 189721, Singapore

Preface

The International Conference on *Smart Innovations in Communications and Computational Sciences (ICSICCS 2017)* has been held at Moga, Punjab, India, during June 23–24, 2017. ICSICCS 2017 has been organized and supported by the "North West Group of Institutions, Moga, Punjab, India."

The main purpose of ICSICCS 2017 is to provide a forum for researchers, educators, engineers, and government officials involved in the general areas of communication, computational sciences, and technology to disseminate their latest research results and exchange views on the future research directions of these fields.

The field of communications and computational sciences always deals with finding the innovative solutions to problems by proposing different techniques, methods, and tools. Generally, innovation refers to find new ways of doing usual things or doing new things in different manner but due to increasingly growing technological advances with speedy pace *Smart Innovations* are needed. Smart refers to "how intelligent the innovation is?" Nowadays, there is massive need to develop new "intelligent" "ideas, methods, techniques, devices, tools." The proceedings cover those systems, paradigms, techniques, technical reviews that employ knowledge and Intelligence in a broad spectrum.

ICSICCS 2017 received around 350 submissions from around 603 authors of 9 different countries such as Taiwan, Sweden, Italy, Saudi Arabia, China, and Bangladesh. Each submission has gone through the plagiarism check. On the basis of plagiarism report, each submission was rigorously reviewed by at least two reviewers. Even some submissions have more than two reviews. On the basis of these reviews, 73 high-quality papers were selected for publication in proceedings volumes, with an acceptance rate of 20.8%.

This proceedings volume comprises 33 high-quality research papers in the form of chapters. These chapters are further subdivided into different tracks named as "Smart Computing Technologies," "Web and Informatics," and "Smart Hardware and Software Design."

We are thankful to the speakers: Prof. B. K. Panigrahi, IIT Delhi; Dr. Dhanajay Singh, Hankuk (Korea) University of Foreign Studies (HUFS), Seoul, South Korea; and Dr. T. V. Vijay Kumar, JNU Delhi; delegates, and the authors for their participation and their interest in ICSICCS as a platform to share their ideas and innovation. We are also thankful to Prof. Dr. Janusz Kacprzyk, Series Editor, AISC, Springer, and Mr. Aninda Bose, Senior Editor, Hard Sciences, Springer, India, for providing continuous guidance and support. Also, we extend our heartfelt gratitude and thanks to the reviewers and Technical Program Committee Members for showing their concern and efforts in the review process. We are indeed thankful to everyone directly or indirectly associated with the conference organizing team leading it toward the success.

We hope you enjoy the conference proceedings and wish you all the best!

Organizing Committee
ICSICCS 2017

Organizing Committee

Chief Patron

S. Lakhbir Singh Gill (Chairman)

Patron

S. Prabhpreet Singh Gill (Managing Director)
S. Dilpreet Singh Gill (Executive Member)

Advisory Committee

Prof. Dr. J. S. Hundal, MRSPTU, Punjab, India
Prof. Dr. A. K. Goel, MRSPTU, Punjab, India
Prof. Gursharan Singh, MRSPTU, Punjab, India
Dr. Buta Singh, IKGPTU, Punjab, India
Dr. B. S. Bhatia, SGGSWU, Punjab, India
Dr. D. S. Bawa, Rayat & Bahra Group of Institutes, Hoshiarpur, Punjab, India
Prof. R. S. Salaria, Rayat & Bahra Group of Institutes, Hoshiarpur, Punjab, India

Principal General Chair

Dr. N. K. Maheshwary

Conference Co-Chair

Dr. R. K. Maheshwary
Dr. Mohita

Finance Chair

Mr. Rishideep Singh (HoD CSE)

Publicity Chair

Prof. Surjit Arora

Publication Chair

Prof. K. S. Panesar (Mechanical)

Registration Chair

Ms. Navjot Jyoti (AP CSE)

Organizing Chair

Dr. R. K. Maheshwary (Dean)

Technical Program Committee

Prof. Ajay Gupta, Western Michigan University, USA
Prof. Babita Gupta, California State University, USA
Prof. Amit K. R. Chowdhury, University of California, USA
Prof. David M. Harvey, G.E.R.I., UK
Prof. Ajith Abraham, Director, MIR Labs
Prof. Madjid Merabti, Liverpool John Moores University, UK
Dr. Nesimi Ertugrual, University of Adelaide, Australia
Prof. Ian L. Freeston, University of Sheffield, UK
Prof. Witold Kinsner, University of Manitoba, Canada
Prof. Anup Kumar, M.I.N.D.S., University of Louisville
Prof. Sanjiv Kumar Bhatia, University of Missouri, St. Louis
Prof. Prabhat Kumar Mahanti, University of New Brunswick, Canada
Prof. Ashok De, Director, NIT Patna
Prof. Kuldip Singh, IIT Roorkee
Prof. A. K. Tiwari, IIT (BHU) Varanasi
Mr. Suryabhan, ACERC, Ajmer, India
Dr. Vivek Singh, IIT (BHU), India
Prof. Abdul Quaiyum Ansari, Jamia Millia Islamia, New Delhi, India
Prof. Aditya Trivedi, ABV-IIITM Gwalior
Prof. Ajay Kakkar, Thapar University, Patiala, India
Prof. Bharat Bhaskar, IIM Lucknow, India
Prof. Edward David Moreno, Federal University of Sergipe, Brazil
Prof. Evangelos Kranakis, Carleton University
Prof. Filipe Miguel Lopes Meneses, University of Minho, Portugal
Prof. Giovanni Manassero Junior, Universidade de São Paulo
Prof. Gregorio Martinez, University of Murcia, Spain
Prof. Pabitra Mitra, Indian Institute of Technology Kharagpur, India
Prof. Joberto Martins, Salvador University (UNIFACS)
Prof. K. Mustafa, Jamia Millia Islamia, New Delhi, India
Prof. M. M. Sufyan Beg, Jamia Millia Islamia, New Delhi, India

Prof. Jitendra Agrawal, Rajiv Gandhi Proudyogiki Vishwavidyalaya, Bhopal, MP, India

Prof. Rajesh Baliram Ingle, PICT, University of Pune, India

Prof. Romulo Alexander Ellery de Alencar, University of Fortaliza, Brazil

Prof. Youssef Fakhri, Faculté des Sciences, Université Ibn Tofail

Dr. Abanish Singh, Bioinformatics Scientist, USA

Dr. Abbas Cheddad, UCMM, Umeå Universitet, Umeå, Sweden

Dr. Abraham T. Mathew, NIT Calicut, Kerala, India

Dr. Adam Scmidit, Poznan University of Technology, Poland

Dr. Agostinho L. S. Castro, Federal University of Para, Brazil

Prof. Goo-Rak Kwon Chosun University, Republic of Korea

Dr. Alberto Yúfera, Seville Microelectronics Institute, IMSE-CNM, NIT Calicut, Kerala, India

Dr. Adam Scmidit, Poznan University of Technology, Poland

Prof. Nishant Doshi, S V National Institute of Technology, Surat, India

Prof. Gautam Sanyal, NIT Durgapur, India

Dr. Agostinho L. S. Castro, Federal University of Para, Brazil

Dr. Alberto Yúfera, Seville Microelectronics Institute, IMSE-CNM

Dr. Alok Chakrabarty, IIIT Bhubaneswar, India

Dr. Anastasios Tefas, Aristotle University of Thessaloniki

Dr. Anirban Sarkar, NIT Durgapur, India

Dr. Anjali Sardana, IIIT Roorkee, Uttarakhand, India

Dr. Ariffin Abdul Mutalib, Universiti Utara Malaysia

Dr. Ashok Kumar Das, IIIT Hyderabad

Dr. Ashutosh Saxena, Infosys Technologies Ltd., India

Dr. Balasubramanian Raman, IIT Roorkee, India

Dr. Benahmed Khelifa, Liverpool John Moores University, UK

Dr. Björn Schuller, Technical University of Munich, Germany

Dr. Chao Ma, Hong Kong Polytechnic University

Dr. Chi-Un Lei, University of Hong Kong

Dr. Ching-Hao Lai, Institute for Information Industry

Dr. Ching-Hao Mao, Institute for Information Industry, Taiwan

Dr. Chung-Hua Chu, National Taichung Institute of Technology, Taiwan

Dr. Chunye Gong, National University of Defense Technology

Dr. Cristina Olaverri Monreal, Instituto de Telecomunicacoes, Portugal

Dr. Chittaranjan Hota, BITS Hyderabad, India

Dr. D. Juan Carlos González Moreno, University of Vigo

Dr. Danda B. Rawat, Old Dominion University

Dr. Davide Ariu, University of Cagliari, Italy

Dr. Dimiter G. Velev, University of National and World Economy, Europe

Dr. D. S. Yadav, South Asian University, New Delhi

Dr. Darius M. Dziuda, Central Connecticut State University

Dr. Dimitrios Koukopoulos, University of Western Greece, Greece

Dr. Durga Prasad Mohapatra, NIT Rourkela, India

Dr. Eric Renault, Institut Telecom, France

Dr. Olga C. Santos, aDeNu Research Group, UNED, Spain
Dr. Pramod Kumar Singh, ABV-IIITM Gwalior, India
Dr. Prasanta K. Jana, IIT, Dhanbad, India
Dr. Preetam Ghosh, Virginia Commonwealth University, USA
Dr. Rabeb Mizouni, KUSTAR, Abu Dhabi, UAE
Dr. Rahul Khanna, Intel Corporation, USA
Dr. Rajeev Srivastava, CSE, IIT (BHU), India
Dr. Rajesh Kumar, MNIT Jaipur, India
Dr. Rajesh Bodade, MCT, Mhow, India
Dr. Rajesh Bodade, Military College of Telecommunication Engineering, Mhow, India
Dr. Ranjit Roy, SVNIT, Surat, Gujarat, India
Dr. Robert Koch, Bundeswehr University München, Germany
Dr. Ricardo J. Rodriguez, Nova Southeastern University, USA
Dr. Ruggero Donida Labati, Università degli Studi di Milano, Italy
Dr. Rustem Popa, "Dunarea de Jos" University of Galati, Romania
Dr. Shailesh Ramchandra Sathe, VNIT Nagpur, India
Dr. Sanjiv K. Bhatia, University of Missouri, St. Louis, USA
Dr. Sanjeev Gupta, DA-IICT, Gujarat, India
Dr. S. Selvakumar, National Institute of Technology, Tamil Nadu, India
Dr. Saurabh Chaudhury, NIT Silchar, Assam, India
Dr. Shijo M. Joseph, Kannur University, Kerala
Dr. Sim Hiew Moi, University of Technology, Malaysia
Dr. Syed Mohammed Shamsul Islam, University of Western Australia
Dr. Trapti Jain, IIT Mandi, India
Dr. Tilak Thakur, PED, Chandigarh, India
Dr. Vikram Goyal, IIIT Delhi, India
Dr. Vinaya Mahesh Sawant, D. J. Sanghvi College of Engineering, India
Dr. Vanitha Rani Rentapalli, VITS Andhra Pradesh, India
Dr. Victor Govindaswamy, Texas A&M University, Texarkana, USA
Dr. Victor Hinostroza, Universidad Autónoma de Ciudad Juárez
Dr. Vidyasagar Potdar, Curtin University of Technology, Australia
Dr. Vijaykumar Chakka, DA-IICT, Gandhinagar, India
Dr. Yong Wang, School of IS & E, Central South University, China
Dr. Yu Yuan, Samsung Information Systems America, San Jose, CA, USA
Eng. Angelos Lazaris, University of Southern California, USA
Mr. Hrvoje Belani, University of Zagreb, Croatia
Mr. Huan Song, Super Micro Computer, Inc., San Jose, USA
Mr. K. K. Patnaik, IIITM Gwalior, India
Dr. S. S. Sarangdevot, Vice Chancellor, JRN Rajasthan Vidyapeeth University, Udaipur
Dr. N. N. Jani, KSV University Gandhinagar
Dr. Ashok K. Patel, North Gujarat University, Patan, Gujarat
Dr. Awadhesh Gupta, IMS, Ghaziabad
Dr. Dilip Sharma, GLA University, Mathura, India

Dr. Li Jiyun, Donghua University, Shanghai, China
Dr. Lingfeng Wang, University of Toledo, USA
Dr. Valentina E. Balas, Aurel Vlaicu University of Arad, Romania
Dr. Vinay Rishiwal, MJP Rohilkhand University, Bareilly, India
Dr. Vishal Bhatnagar, Ambedkar Institute of Technology, New Delhi, India
Dr. Tarun Shrimali, Sunrise Group of Institutions, Udaipur
Dr. Atul Patel, C.U. Shah University, Wadhwan, Gujarat
Dr. P. V. Virparia, Sardar Patel University, VV Nagar
Dr. D. B. Choksi, Sardar Patel University, VV Nagar
Dr. Ashish N. Jani, KSV University, Gandhinagar
Dr. Sanjay M. Shah, KSV University, Gandhinagar
Dr. Vijay M. Chavda, KSV University, Gandhinagar
Dr. B. S. Agarwal, KIT, Kalol
Dr. Apurv Desai, South Gujarat University, Surat
Dr. Chitra Dhawale, Nagpur
Dr. Bikas Kumar, Pune
Dr. Nidhi Divecha, Gandhinagar
Dr. Jay Kumar Patel, Gandhinagar
Dr. Jatin Shah, Gandhinagar
Dr. Kamaljit I. Lakhtaria, AURO University, Surat
Dr. B. S. Deovra, B.N. College, Udaipur
Dr. Ashok Jain, Maharaja College of Engineering, Udaipur
Dr. Bharat Singh, JRN Rajasthan Vidyapeeth University, Udaipur
Dr. S. K. Sharma, Pacific University, Udaipur
Dr. Naresh Trivedi, Ideal Institute of Technology, Ghaziabad
Dr. Akheela Khanum, Integral University, Lucknow
Dr. R. S. Bajpai, Shri Ramswaroop Memorial University, Lucknow
Dr. Manish Shrimali, JRN Rajasthan Vidyapeeth University, Udaipur
Dr. Ravi Gulati, South Gujarat University, Surat
Dr. Atul Gosai, Saurashtra University, Rajkot
Dr. Digvijai sinh Rathore, BBA Open University, Ahmedabad
Dr. Vishal Goar, Government Engineering College, Bikaner
Dr. Neeraj Bhargava, MDS University, Ajmer
Dr. Ritu Bhargava, Government Women Engineering College, Ajmer
Dr. Rajender Singh Chhillar, MDU, Rohtak
Dr. Dhaval R. Kathiriya, Saurashtra University, Rajkot
Dr. Vineet Sharma, KIET, Ghaziabad
Dr. A. P. Shukla, KIET, Ghaziabad
Dr. R. K. Manocha, Ghaziabad
Dr. Nandita Mishra, IMS Ghaziabad
Dr. Manisha Agarwal, IMS Ghaziabad
Dr. Deepika Garg, IGNOU, New Delhi
Dr. Goutam Chakraborty, Iwate Prefectural University, Iwate Ken, Takizawa, Japan
Dr. Amit Manocha, Maharaja Agrasen University, HP, India
Prof. Enrique Chirivella-Perez, University of the West of Scotland, UK

Prof. Pablo Salva Garcia, University of the West of Scotland, UK
Prof. Ricardo Marco Alaez, University of the West of Scotland, UK
Prof. Nitin Rakesh, Amity University, Noida, India
Prof. Mamta Mittal, G. B. Pant Government Engineering College, Delhi, India
Dr. Shashank Srivastava, MNNIT Allahabad, India
Prof. Lalit Goyal, JMI, Delhi, India
Dr. Sanjay Maurya, GLA University, Mathura, India
Prof. Alexandros Iosifidis, Tampere University of Technology, Finland
Prof. Shanthi Makka, JRE Engineering College, Greater Noida, India
Dr. Deepak Gupta, Amity University, Noida, India
Dr. Manu Vardhan, NIT Raipur, India
Dr. Sarsij Tripathi, NIT Raipur, India
Prof. Wg Edison, HeFei University of Technology, China
Dr. Atul Bansal, GLA University, Mathura, India
Dr. Alimul Haque, V.K.S. University, Bihar, India
Prof. Simhiew Moi, Universiti Teknologi Malaysia
Prof. Rustem Popa, "Dunarea de Jos" University of Galati, Romania
Prof. Vinod Kumar, IIT Roorkee, India
Prof. Christos Bouras, University of Patras and RACTI, Greece
Prof. Devesh Jinwala, SVNIT Surat, India
Prof. Germano Lambert-Torres, PS Solutions, Brazil
Prof. Byoungho Kim, Broadcom Corp., USA

Contents

Part III Smart Hardware and Software Design

About the Editors

Dr. Bijaya Ketan Panigrahi is working as a Professor in the Electrical Engineering Department, IIT Delhi, India. Prior to joining IIT Delhi in 2005, he has served as a Faculty in Electrical Engineering Department, UCE Burla, Odisha, India, from 1992 to 2005. He is a Senior Member of IEEE and Fellow of INAE, India. His research interest includes application of soft computing and evolutionary computing techniques to power system planning, operation, and control. He has also worked in the field of biomedical signal processing and image processing. He has served as the editorial board member, associate editor, and special issue guest editor of different international journals. He is also associated with various international conferences in various capacities. He has published more than 100 research papers in various international and national journals.

Dr. Munesh C. Trivedi is currently working as a Professor in the Computer Science and Engineering Department, ABES Engineering College, Ghaziabad, India. He has rich experience in teaching the undergraduate and postgraduate classes. He has published 20 textbooks and 80 research publications in different international journals and proceedings of international conferences of repute. He has received Young Scientist Visiting Fellowship and numerous awards from different national as well international forums. He has organized several international conferences technically sponsored by IEEE, ACM, and Springer. He has delivered numerous invited and plenary conference talks throughout the country and chaired technical sessions in international and national conferences in India. He is on the review panel of IEEE Computer Society, International Journal of Network Security, Pattern Recognition Letter and Computer & Education (Elsevier's Journal). He is an Executive Committee Member of IEEE UP Section, IEEE India Council, and also IEEE Asia Pacific Region 10. He is an Active Member of IEEE Computer Society, International Association of Computer Science and Information

Technology, Computer Society of India, International Association of Engineers, and a Life Member of ISTE.

Dr. Krishn K. Mishra is currently working as a Visiting Faculty, Department of Mathematics and Computer Science, University of Missouri, St. Louis, USA. He is an alumnus of Motilal Nehru National Institute of Technology Allahabad, India, which is also his base working institute. His primary areas of research include evolutionary algorithms, optimization techniques, and design and analysis of algorithms. He has also published more than 50 publications in international journals and in proceedings of international conferences of repute. He is serving as a program committee member of several conferences and also editing few Scopus and SCI-indexed journals. He has 15 years of teaching and research experience during which he made all his efforts to bridge the gaps between teaching and research.

Dr. Shailesh Tiwari is currently working as a Professor in the Computer Science and Engineering Department, ABES Engineering College, Ghaziabad, India. He is also administratively heading the department. He is an alumnus of Motilal Nehru National Institute of Technology Allahabad, India. He has more than 16 years of experience in teaching, research, and academic administration. His primary areas of research are software testing, implementation of optimization algorithms, and machine learning techniques in software engineering. He has also published more than 50 publications in international journals and in proceedings of international conferences of repute. He is also serving as a program committee member of several conferences and also editing few Scopus and E-SCI-indexed journals. He has organized several international conferences under the sponsorship of IEEE and Springer. He is a Fellow of Institution of Engineers (FIE), Senior Member of IEEE, Member of IEEE Computer Society, and Former Executive Committee Member of IEEE Uttar Pradesh Section. He is also a member of reviewer and editorial board of several international journals and conferences. He has also edited several books published under various book series of Springer.

Dr. Pradeep Kumar Singh is currently working as an Assistant Professor (Senior Grade) in the Department of Computer Science and Engineering, Jaypee University of Information Technology (JUIT), Waknaghat, India. He has 10 years of vast experience in academics at reputed colleges and universities of India. He has completed his Ph.D. in Computer Science and Engineering from Gautam Buddha University (State Government University), Greater Noida, UP, India. He received his M.Tech. (CSE) with distinction from Guru Gobind Singh Indraprastha University, New Delhi, India. He has obtained his B.Tech. (CSE) from Uttar Pradesh Technical University (UPTU), Lucknow, India. He is having Life Membership of Computer Society of India (CSI) and promoted to Senior Member Grade from CSI. He is Member of ACM, IACSIT, Singapore, and IAENG, Hong Kong. He is associated with many IEEE International Conferences as TPC member, reviewer, and session chair. He is an Associate Editor of International Journal of Information Security and Cybercrime (IJISC) a scientific peer-reviewed journal from Romania. He has published nearly 50 research papers in various international journals and conferences of repute. He has organized various theme-based special sessions during the international conferences also.

Part I
Smart Computing Technologies

Classification of the Shoulder Movements for Intelligent Frozen Shoulder Rehabilitation

Shweta, Padmavati Khandnor, Neelesh Kumar and Ratan Das

Abstract Frozen shoulder is a medical condition leading to stiffness in the shoulder joint and also restricting the range of motion of the shoulder joint. The paper compiles the details about the four basic movements of the shoulder joint, namely the flexion/extension, abduction/adduction, internal rotation and external rotation movements. Shoulder movements of 150 subjects were recorded, and the data was further analyzed and classified using the K-nearest neighbor algorithm, support vector machine, and also using logistic regression algorithm. The data is recorded using a module consisting of a triaxial accelerometer, a HC-05 Bluetooth module and triaxial gyroscope. SVM shows an accuracy of approximately 99.99% over the classification of the four shoulder movements and is proved to be better than other classifiers. Classification of the shoulder movements can be further used to classify an individual as either a patient suffering from frozen shoulder or a normal individual.

Keywords Signal processing · Pattern recognition · Machine learning
Bioinformatics · Frozen shoulder

1 Introduction

Recognition of the human body movements has proved to be utterly important in various sectors of life like sports activities, bioinformatics [1], medical diagnosis, rehabilitation purposes. [2]. The tracking of movements of spheroidal joint like

Shweta · P. Khandnor (✉)
PEC University of Technology, Chandigarh, India
e-mail: padmavati@pec.ac.in

N. Kumar · R. Das
CSIR-CSIO, Chandigarh, India

© Springer Nature Singapore Pte Ltd. 2019
B. K. Panigrahi et al. (eds.), *Smart Innovations in Communication and Computational Sciences*, Advances in Intelligent Systems and Computing 670,
https://doi.org/10.1007/978-981-10-8971-8_1

shoulder joint is of great significance in the field of medical diagnosis and rehabilitation for the problem of adhesive capsulitis. Adhesive capsulitis or frozen
shoulder is a medical condition leading to the stiffness of the shoulder joint as well
as leading to reduction in the active and passive range of motion [3]. The shoulder
movements are majorly divided into four types, namely shoulder flexion, shoulder
abduction, internal rotation, and external rotation. Figures 1, 2, 3, and 4 show the
various shoulder movements. Table 1 shows the details of the movements of the
shoulder joint.

Fig. 1 Abduction movement [4]

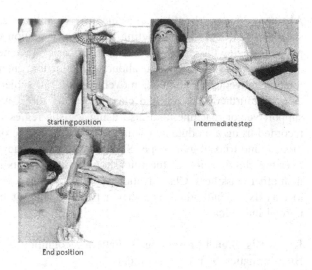

Fig. 2 Flexion movement [4]

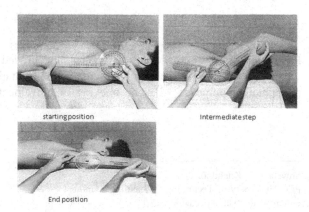

Fig. 3 Internal rotation [4]

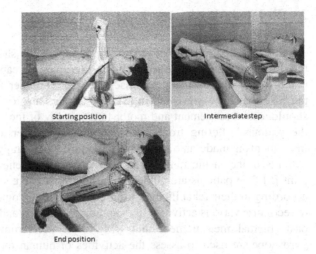

Starting position   Intermediate step

End position

Fig. 4 External rotation [4]

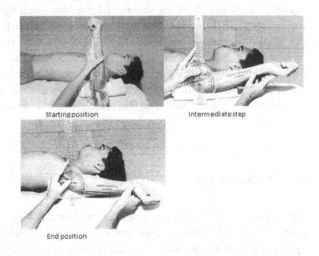

Starting position   Intermediate step

End position

Table 1 Details about the shoulder joint movements

S. No.	Shoulder movement	Mean angle (°)
1	Abduction	180
2	Flexion	180
3	Internal rotation	70
4	External rotation	90

2   Related Work

A self-controlled rehabilitation system to monitor the shoulder joint is the need of the hour for managing the problem of adhesive capsulitis [3, 4]. A virtual reality-based approach for the assessment of shoulder joint for its rehabilitation purposes has been presented in [3]. The authors also detail the analysis about the shoulder joint movement and motor functionality of the shoulder joint. To monitor the patients suffering from frozen shoulder or upper arm problems, a track of movements is made and accordingly the physical therapy is provided to assist the patients in increasing range of motion (ROM) and relieving the pain in shoulder joint [5]. The patients are guided with the proper exercises like stretching, rotation according to their rehabilitation progress. Activity recognition has been performed to recognize various activities like running, walking, standing, climbing stairs in the past. Inertial measurement unit (IMU) like accelerometer, magnetometer, and gyroscope are used to assess the activities of human motion following steps like extracting the features, selection of the higher accuracy features, and the classification of activities. Various supervised machine learning algorithms such as Naive Bayes (NB) [6], hidden Markov model (HMM) [6], support vector machine (SVM) [7], K-nearest neighbor (K-NN) [8], decision trees, or the combination of one or more algorithms have been used for classification in the past. Many algorithms which are semi-supervised or unsupervised are also used for the purpose of classification of activities. Likewise, these classifiers can also be used to classify the shoulder movements and help in the vision of smart rehabilitation of frozen shoulder.

3   Methodology

This work is done using a wireless inertial sensor-based wristband module which comprises of a triaxial accelerometer [9], a HC-05 Bluetooth chip and a triaxial gyroscope. The data is sent using the Bluetooth module to the central system using the serial port communication. The frequency of the sensor module is 20 Hz. For the experiments, we have used just the triaxial accelerometer and the biaxial gyroscope to measure the gravitational acceleration (m/s^2) in x, y, and z directions as well as the angular velocity in the x and y directions. For the analysis of the shoulder joint motion, we focus on the four elementary motions as mentioned below.

1. Action A: Keeping the arm straight in the vertical plane and then moving the arm in the sagittal plane along the socket joint (flexion and extension of the shoulder joint);
2. Action B: Keeping the arm straight in the vertical plane and then moving the arm in the coronal plane along the socket joint (abduction and adduction of the shoulder joint);

3. Action C: Keeping the angle between the elbow joint as 90° and moving the forearm toward the central axis of the body (internal rotation);
4. Action D: Keeping the angle between the elbow joint as 90° and moving the forearm away from the central axis of the body (external rotation).

The experiments were performed using the normal subjects having age in the range of 15–70 years. The ROM of the subjects is measured using the wristband of all the right dominant subjects. The training database is generated from 100 normalized subjects by making them perform all the above-mentioned actions (Action A, Action B, Action C, and Action D). The testing of the data is done using one-tenth of the database.

3.1 Data Processing

The elementary steps involved for the processing of data of an individual are specified in the flowchart in Fig. 5.

Data Gathering and Data Pre-processing Phase
The inertial sensor-based wristband is used to gather the data from the shoulder joint movement of the normal subject. The data communication is established with the central system via a HC-05 Bluetooth chip in the sensor module. The raw data from the band is filtered by applying a band-pass filter having a frequency of 0.1 Hz to remove the noise. In feature extraction phase, various feature combinations or sets like the frequency domain or time domain have been used in [10]. Application of the accelerometer and gyroscope signals for the identification of upper arm movements comes into focus because of the variability and uniqueness of the two signals [11]. For our research work, we had considered over a mix of frequency- and time-domain features. Each triaxial accelerometer signal and each triaxial gyroscope signal is considered to derive a feature set comprising of following:

1. Weighted mean for the duration of the shoulder movement;
2. Standard deviation for the duration of the shoulder movement;

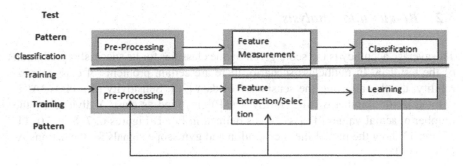

Fig. 5 Flowchart showing the process of classification

3. Kurtosis—to predict the 'peakedness' of the signal;
4. Skewness—to determine the symmetry of the signal [12];
5. Root mean square (RMS value);
6. Entropy of the signal—to measure the randomness of the signal;
7. Absolute difference between maximum and minimum values of the signal;
8. Index of dispersion.

Along with the following above derived features, the module was programmed to calculate the angle in x and y directions. Using the angle values, maximum and minimum angle values in x and y directions are calculated, respectively. ROM is measured by taking the difference between maximum and minimum angle values. For the ROM, the mean angular velocities along the x, y, and z directions are measured. Kurtosis and skewness are the variables to know the probability distribution of the data, whereas maximum amplitude is used to assess the motional capacity of the shoulder joint. Hence, a set of eight features on each accelerometer signal (accx, accy, accz) and each gyroscope signal (gyrox, gyroy, gyroz) are computed.

Feature Selection Phase

This phase of the entire analysis reflects on selecting the perfect group of features to get the desired accuracy. Clamping algorithm [13], principal component analysis (PCA) [14], RELIEF algorithm [15] are the most widely and most used algorithms for feature selection. In our experiments and analysis of data, we have used PCA to give the best results.

Classification

Some of the classification algorithms constitute of generalized linear models like logistic regression [16], linear discriminant analysis (LDA), SVM [17], Naives Bayes [18], K-NN [19]. In our analysis, we have implemented the K-NN, logistic regression, and the SVM to classify the four shoulder movements. The principle of working of K-NN is to find the 'K' number of nearest training points to the tested sample of data and therefore predict the label of the tested data. All the implementation of the programming code is done using Python in Enthought Canopy.

3.2   Results and Analysis

The average accuracy is the score given by the classifier for its successful prediction of the test data. In multiclass scenario, there are certain problems of class inseparability, thereby affecting the sensitivity of data [20]. Sensitivity of a class 'A' is defined as the number of true positives (TP) classified as class A divided by the number of actual values of class 'A' as defined in Eq. 1. Figures 6, 7, 8, 9, 10, 11, 12 and 13 show the plot of the acceleration and gyroscope signals for the actions A, B, C, and D.

Fig. 6 Action A acceleration
signal

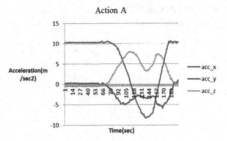

Fig. 7 Action A gyroscope
signal

Fig. 8 Action B acceleration
signal

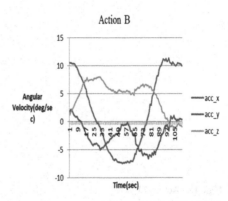

Fig. 9 Action B gyroscope
signal

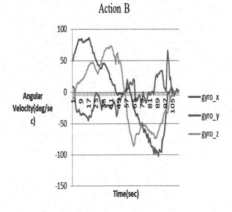

Fig. 10 Action C
acceleration signal

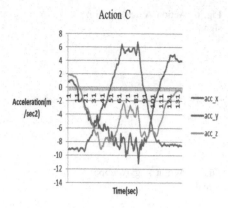

Fig. 11 Action C gyroscope
signal

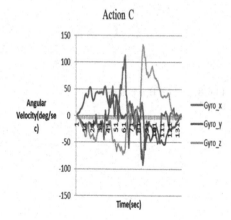

Fig. 12 Action D
acceleration signal

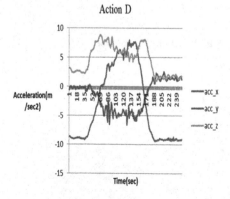

Fig. 13 Action D gyroscope
signal

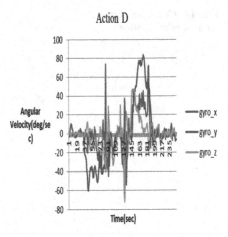

$$\text{Sensitivity of a class} = TP/(TP + TN) \tag{1}$$

Sensitivity is same as the true positive rate (TPR) and recall value of a class. Other performance evaluators used for the evaluation of a classifier are precision, specificity, F-score. Precision, specificity, and F-score are computed using Eqs. 2, 3, and 4 [21].

$$\text{Precision of a class} = \frac{TP}{TP + FP} \tag{2}$$

$$\text{Specificity of a class} = \frac{TN}{FP + TN} \tag{3}$$

$$F - \text{Score of a class} = \frac{2*TP}{2*TP + FP + FN} \tag{4}$$

Table 2 specifies the accuracies of the three classifiers K-NN, SVM, and logistic regression. We had 150 movements (62 for class A, 62 for class B, 14 for class C, and 12 for class D) to be classified by the classifiers. The SVM [22] classifier gives an accuracy of 99.99% for the classification of the shoulder movements (multiclass classification) and is best as compared to other classifiers. Figures 14 and 15 show the confusion matrices designed for SVM classifier with and without normalization. Using the confusion matrix, the performance metrics calculated for evaluation of all

Table 2 Comparative
analysis of various machine
learning algorithms used

S. No.	Machine learning algorithm	Accuracy
1	K-NN	84.6154
2	Logistic regression	92.33
3	SVM	99.99

Fig. 14 Confusion matrix
without normalization

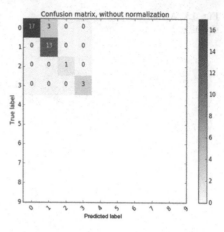

Fig. 15 Confusion matrix
with normalization

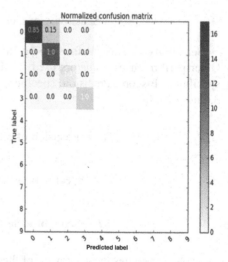

the four different shoulder movements representing respective four different classes is shown in Figs. 16, 17, 18, and 19. Accuracy of class A and class B is 91.89%, but class A has 85% sensitivity as compared to other classes having sensitivity of 99.99%. Figure 20 shows the precision–recall curve for the classification of the four shoulder movements by implementing one-vs-rest classifier, and the classes C and D have a precision–recall value of 1 which is due to the reason that the training vectors for the classes C and D were a few. Precision–recall values of each class (A, B, C, and D) are calculated and plotted. Finally, a cross-validation curve is plotted for the SVM classifier, to tell how accurately the classifier predicts for any individual general set of data (Fig. 21). The study is to identify accurately the shoulder movements using a certain classifier (in our case SVM was best) since when a user performs a specific movement wearing the sensor band the feature set has features

Fig. 16 Performance measures for class A

Class A	Confusion Matrix without normalization	Confusion Matrix with normalization
Accuracy	91.89%	96.25%
True Positive Rate/Recall/Sensitivity	85%	85%
Precision	99.99%	99.99%
Specificity	99.99%	99.99%

Fig. 17 Performance measures for class B

Class B	Confusion Matrix without normalization	Confusion Matrix with normalization
Accuracy	91.89%	96.25%
True Positive Rate/Recall/Sensitivity	99.99%	99.99%
Precision	81.25%	86.95%
Specificity	87.5%	99.99%

Fig. 18 Performance measures for class C

Class C	Confusion Matrix without normalization	Confusion Matrix with normalization
Accuracy	99.99%	99.99%
True Positive Rate/Recall/Sensitivity	99.99%	99.99%
Precision	99.99%	99.99%
Specificity	infinite	99.99%

Fig. 19 Performance measures for class D

Class D	Confusion Matrix without normalization	Confusion Matrix with normalization
Accuracy	99.99%	99.99%
True Positive Rate/Recall/Sensitivity	99.99%	99.99%
Precision	99.99%	99.99%
Specificity	infinite	99.99%

Fig. 20 Precision–recall
curve for one-vs-rest SVM
classifier

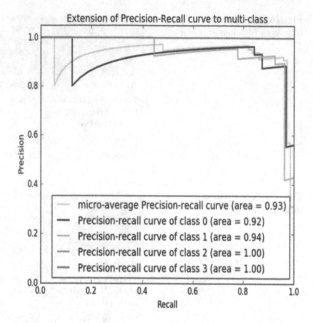

Fig. 21 The cross-validation
curve

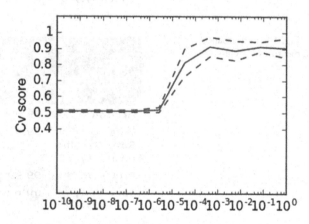

like gyrox, gyroy, gyroz; therefore, the statistical features as mentioned earlier in
data pre-processing phase are derived to know about the shape, peakedness,
maximum value, minimum value, etc., of the signal. This way the activities are
classified by the classifier.

4   Conclusion and Future Scope

In our paper, we have classified the four shoulder movements (extension/flexion, abduction/adduction, internal rotation, external rotation) using three classifiers SVM, K-NN, and logistic regression.

SVM proved to be the best in terms of accuracy as compared to K-NN and logistic regression. In the cross-validation phase, we have applied tenfold cross-validation on data. SVM gives an accuracy of 99.99% for the classification. Every class (A, B, C, and D) is checked on the parameters of specificity, TPR, precision, and F-score. Class C and D are having 99.99% accuracy as compared to class A and class B. Class B has a high precision when confusion matrix is normalized. Class A as compared to class B has more precision and specificity. Classification of the shoulder movements can be further used to classify an individual as either a patient suffering from frozen shoulder or a normal individual and to work toward the idea of self-home rehabilitation.

Statements for Ethical Approval
The paper was ethically supported by Central Scientific Institute of Research (CSIR), Central Scientific Instruments Organization (CSIO). The study is concerned with the use of a wireless inertial sensor band to monitor the shoulder joint movement so that in future it can be used to provide rehabilitation to the frozen shoulder patients.

Proper consent from the participants was taken, and then, only the further study and analysis are performed by the authors. All the individuals who participated in the data collection process were either 18 years of age or above the age of 18 years. The individuals were physically monitored during the data collection phase. The data was collected using a wireless wrist inertial band which measured the four movements of the shoulder joint.

The images and diagrams used in the paper have been taken from the books and from the research papers surveyed during the study. Proper citations have been given already for all the images and diagrams which are referenced from the research papers and books. For the remaining figures, the analysis was done on the dataset for classification of four activities using the machine learning algorithms (support vector machines, K-nearest neighbor, and logistic regression)

References

1. San-Segundo, R., Montero, J.M., Barra-Chicote, R., Fernandez, F., Pardo, J.M.: Feature extraction from smartphone inertial signals for human activity segmentation. Signal Processing 120, 359–372 (2016)
2. Patel, S., Hughes, R., Hester, T., Stein, J., Akay, M., Dy, J.G., Bonato, P.: A novel approach to monitor rehabilitation outcomes in stroke survivors using wearable technology. Proceedings of the IEEE 98(3), 450–461 (2010)

3. Huang, M.C., Lee, S.H., Yeh, S.C., Chan, R.C., Rizzo, A., Xu, W., Han-Lin, W., Shan-Hui, L.: Intelligent frozen shoulder rehabilitation. IEEE Intelligent Systems 29(3), 22–28 (2014)
4. Norkin, C.C., White, D.J.: Measurement of joint motion: a guide to goniometry. FA Davis (2016)
5. Kelley, M.J., Mcclure, P.W., Leggin, B.G.: Frozen shoulder: evidence and a propose model guiding rehabilitation. Journal of orthopaedic & sports physical therapy 39(2), 135–148 (2009)
6. Mantone, J.K., Burkhead, W.Z., Noonan, J.: Nonoperative treatment of rotator cuff tears. Orthopedic Clinics of North America 31(2), 295{311 (2000)
7. Gayathri, K., Elias, S., Ravindran, B.: Hierarchical activity recognition for dementia care using markov logic network. Personal and Ubiquitous Computing 19(2), 271–285 (2015)
8. Reyes-Ortiz, J.L., Oneto, L., Sama, A., Parra, X., Anguita, D.: Transition-aware human activity recognition using smartphones. Neurocomputing 171, 754–767 (2016)
9. Biswas, D., Cranny, A., Gupta, N., Maharatna, K., Achner, J., Klemke, J., Jobges, M., Ortmann, S.: Recognizing upper limb movements with wrist worn inertial sensors using k-means clustering classification. Human movement science 40, 59–76 (2015)
10. Mannini, A., Sabatini, A.M.: Machine learning methods for classifying human physical activity from on-body accelerometers. Sensors 10(2), 1154–1175 (2010)
11. Altun, K., Barshan, B., Tuncel, O.: Comparative study on classifying human activities with miniature inertial and magnetic sensors. Pattern Recognition 43(10), 3605–3620 (2010)
12. Banos, O., Damas, M., Pomares, H., Prieto, A., Rojas, I.: Daily living activity recognition based on statistical feature quality group selection. Expert Systems with Applications 39(9), 8013–8021 (2012)
13. Lowe, S.A., OLaighin, G.: Monitoring human health behaviour in one's living environment: a technological review. Medical engineering & physics 36(2), 147–168 (2014)
14. Chernbumroong, S., Cang, S., Atkins, A., Yu, H.: Elderly activities recognition and classification for applications in assisted living. Expert Systems with Applications 40(5), 1662 {1674 (2013)
15. Uguz, H.: A two-stage feature selection method for text categorization by using information gain, principal component analysis and genetic algorithm. Knowledge-Based Systems 24(7), 1024{1032 (2011)
16. Gupta, P., Dallas, T.: Feature selection and activity recognition system using a single triaxial accelerometer. IEEE Transactions on Biomedical Engineering 61(6), 1780{1786 (2014)
17. Liu, W., Liu, H., Tao, D., Wang, Y., Lu, K.: Multiview hessian regularized logistic regression for action recognition. Signal Processing 110, 101{107 (2015)
18. Khan, A.M., Tufail, A., Khattak, A.M., Laine, T.H.: Activity recognition on smartphone via sensor-fusion and kda-based svms. International Journal of Distributed Sensor Networks (2014)
19. Shoaib, M., Bosch, S., Incel, O.D., Scholten, H., Havinga, P.J.: A survey of online activity recognition using mobile phones. Sensors 15(1), 2059{2085 (2015)
20. Trost, S.G., Zheng, Y., Wong, W.K.: Machine learning for activity recognition: hip versus wrist data. Physiological measurement 35(11), 2183 (2014)
21. Zhu, W., Zeng, N., Wang, N., et al.: Sensitivity, specificity, accuracy, associated confidence interval and roc analysis with practical sas implementations
22. Caballero, J.C.F., Martinez, F.J., Hervas, C., Gutierrez, P.A.: Sensitivity versus accuracy in multiclass problems using memetic pareto evolutionary neural networks. IEEE Transactions on Neural Networks 21(5), 750{770 (2010)

Markov Feature Extraction Using Enhanced Threshold Method for Image Splicing Forgery Detection

Avinash Kumar, Choudhary Shyam Prakash, Sushila Maheshkar and Vikas Maheshkar

Abstract Use of sophisticated image editing tools and computer graphics makes easy to edit, transform, or eliminate the significant features of an image without leaving any prominent proof of tampering. One of the most commonly used tampering techniques is image splicing. In image splicing, a portion of image is cut and paste it on the same image or different image to generate a new tampered image, which is hardly noticeable by naked eyes. In the proposed method, enhanced Markov model is applied in the block discrete cosine transform (BDCT) domain as well as in discrete Meyer wavelet transform (DMWT) domain. To classify the spliced image from an authentic image, the cross-domain features play the role of final discriminative features for support vector machine (SVM) classifier. The performance of the proposed method through experiments is estimated on the publicly available dataset (Columbia dataset) for image splicing. The experimental results show that the proposed method performs better than some of the existing state of the art.

Keywords Image forensics · Image splicing · Markov feature · DMWT · SVM

A. Kumar · C. S. Prakash (✉) · V. Maheshkar
Department of Computer Science and Engineering,
Indian Institute of Technology (Indian School of Mines), Dhanbad, India
e-mail: shyamprakash2008@yahoo.com
URL: http://www.iitism.ac.in/

S. Maheshkar
Department of Computer Science and Engineering,
National Institute of Technology Delhi, Delhi, India

V. Maheshkar
Division of Information Technology,
Netaji Subhas Institute of Technology, New Delhi, Delhi, India
URL: http://www.nsit.ac.in/

© Springer Nature Singapore Pte Ltd. 2019
B. K. Panigrahi et al. (eds.), *Smart Innovations in Communication and Computational Sciences*, Advances in Intelligent Systems and Computing 670,
https://doi.org/10.1007/978-981-10-8971-8_2

(a)   (b)   (c)

Fig. 1 Examples of image forgery **a** copy-move, **b** image splicing, **c** image resampling

1 Introduction

In recent years, due to immediacy and the easily understandable image content, images have become the prime source of information exchange. It is being used as evidence in legal matters, proof of an experiment, media, real-world events etc. At the same time, availability of sophisticated image manipulation software and pervasive imaging devices gave rise to the need for forensic toolboxes which can access authenticity of images without knowing the original source information. Hence, numerous forensic methods are proposed which focuses on detection of such malicious post-processing of images. On the basis of method used for manipulation of an image, image forgery is divided into three categories: copy-move, image splicing, and image resampling. Copy-move forgery is also called as image cloning, where a sub-part of an image is copied and pasted to the other part of the same image, to hide important information, whereas image splicing uses cut and paste technique, in which part of one or more images are pasted to different or same image. Image resampling is done on an image by geometric transformation like scaling, stretching, rotation, skewing, flipping. Figure 1 depicts few example of image manipulation which is downloaded from Internet.

In the course of this paper, we shall focus on image splicing, which is one of the most used techniques for image tempering. It involves combining or composition of two or more images to produce a forged image. Splicing detection uses passive approach where no prior information of image is known. In recent years, researchers have proposed several methods on image splicing forgery detection. Shi et al. [1] proposed a image model, which reduce statistical moments by treating the neighboring differences of block discrete cosine transform of an image as 1D signal and the dependencies between neighboring nodes along certain directions have sculpted as Markov model. These features are considered as discriminative feature for support vector machine classifier. Xuefang Li et al. [2] proposed an approach based on Hilbert–Huang transform (HHT) and moment of wavelet transform characteristic function. They used SVM as a classifier for spliced image classification with an accuracy of 85.86%.

Method proposed by Zhao et al. [3] is based on gray level run length (RLRN) feature and chroma channel. Features are extracted using gray level RLRN vectors along four different directions from decorrelated chroma channel. Extracted features are introduced to SVM for classification.

Pevny et al. [4] proposed a method which is based on SPAM feature and modeled it as second-order Markov matrix along certain directions, which is treated as discriminative feature for SVM classifier. Later, Kirchner and Fridrich [5] used SPAM and extended it to detect median filter of JPEG compressed image which is supportive for image tempering detection. Other than above proposed method Markov model-based approach which utilizes local transition feature has shown promising splicing detection accuracy. He et al. [6] introduced Markov model in DCT domain as well as DWT domain. The difference coefficient array and transition probability matrix are modeled as feature vector and cross-domain Markov feature are considered as discriminative feature for SVM classifier. However, the proposed approach requires up to 7290 features. An enhanced state selection method is proposed by B. Su et al. [7]. In this approach, author considers some already proposed function model and maps the large number of coefficients extracted from transform domain to specific states. However, by reducing the number of features, this method sacrifices the detection performance. X. Zhao et al. [8] proposed a model in which an image is modeled as a 2D non-casual signal and captures the dependencies between the current node and its neighbors. This model is applied on BDCT and DMWT, and combined extracted features are used for classification. It is found that their method has better detection rate with the cost of higher dimension of 14,240.

As per the above discussion, it is concluded that Markov model-based approach suffers from information loss and higher feature dimension, which is directly proportional to the threshold election. Larger threshold value can minimize information loss, but it will increase the feature dimension too, which can create overfitting problem, and detection capability will get reduced. Therefore, the choice of threshold becomes a trade-off between the detection performance and computational cost. In this paper, an enhanced threshold method is proposed which gives much lesser dimension of features even with large threshold value, which improves the computational cost as well as the detection rate as discussed in step 3 in proposed work.

The rest of the paper is organized as follows. Section 2 shows proposed work algorithm framework. The experimental results and the comparison with other methods are depicted in Sect. 3, followed by conclusion in Sect. 4.

2 Proposed Method

In this paper, we proposed a model in which features are extracted from discrete cosine transform (DCT) and discrete Meyer wavelet transform (DMWT) domain and an enhanced threshold method is used to reduce the information loss as well as the computational cost, which results in improved detection capability. After all the related features are generated, SVM is used as classifier to distinguish the authentic and spliced image. The proposed algorithm framework is shown in Fig. 2.

2.1 Algorithm Flow

- Divide the input image into 8 × 8 non-overlapping blocks.
- DCT is applied on each sub-block.

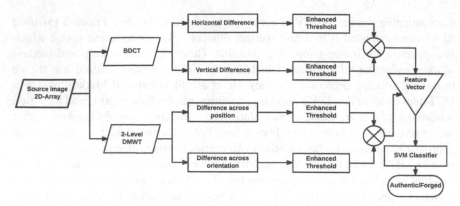

Fig. 2 Flow diagram of proposed model

- Round the coefficient and difference array is obtained in horizontal and vertical direction.
- Enhanced threshold method is applied to calculate Markov matrix.
- Above process is applied in DMWT domain also; considering dependency among wavelet coefficient, more Markov features are extracted.
- Combine all the features, extracted from DCT and DWT domain.
- SVM classifier is used to distinguish authentic and spliced image.

2.2 Extracting Splicing Artifacts

Feature Extraction in DCT Domain: The Markov feature in DCT and DWT is proposed in [6], in which correlation of neighboring coefficients is considered to differentiate authentic and spliced images. The process involved in calculation of difference arrays followed by transition probability matrix. Threshold value T introduced in [6] is to minimize the computational cost, which achieved a feature dimension of $(2T + 1) \times (2T + 1) \times 4$, but it is still on higher side. To minimize the dimension of feature vector and limit the overfitting problem, we introduced an enhanced threshold method which achieves much lesser feature dimension of $(T + 1) \times (T + 1) \times 4$. Proposed approach is explained in step 3 of this section. Markov features in DCT domain are computed as follows:

Step 1: In the first step, DCT coefficient is obtained by applying non-overlapping 8×8 block discrete cosine transform (BDCT) on the input image and denoted as S. We used BDCT in our proposed model due to its energy compaction and decorrelation capability.

Step 2: In the second step, round the DCT coefficient to the nearest integer value. Then, horizontal (F_h) and vertical (F_v) difference array is calculated using the following equations:

$$F_h (i,j) = S (i,j) - S (i + 1,j) \tag{1}$$

$$F_v (i,j) = S (i,j) - S (i,j + 1) \tag{2}$$

where $i \in [1, S_m - 1]$, $j \in [1, S_n - 1]$, and S_m and S_n is the dimension of input source image.

Step 3: Enhanced threshold method: Considering threshold $T(T \in N_+)$, it is replaced with T or $-T$, if the value of an element in difference array is either $>T$ or $<-T$, respectively, and the range of threshold we considered is $(u, v) \in \{-T, -T + 2, \ldots, T + 2, T\}$. Under given range, we calculate the horizontal and vertical Markov matrices using Eqs. (3), (4), (5), (6), which minimize the feature dimension to $4 \times (T + 1) \times (T + 1)$.

$$p\{F_h(i+1,j) = v \mid F_h(i,j) = u\} =$$
$$\frac{\sum_{i=1}^{S_m-2} \sum_{j=1}^{S_n-1} \delta\left((F_h(i,j) = u \parallel F_h(i,j) = u - 1), (F_h(i+1,j) = v \parallel F_h(i+1,j) = v - 1)\right)}{\sum_{j=1}^{S_n-2} \sum_{i=1}^{S_m-1} \left(\delta\left(F_h(i,j) = u\right) \parallel \delta\left(F_h(i,j) = u - 1\right)\right)} \quad (3)$$

$$p\{F_h(i,j+1) = v \mid F_h(i,j) = u\} =$$
$$\frac{\sum_{i=1}^{S_m-1} \sum_{j=1}^{S_n-2} \delta\left((F_h(i,j) = u \parallel F_h(i,j) = u - 1), (F_h(i,j+1) = v \parallel F_h(i,j+1) = v - 1)\right)}{\sum_{j=1}^{S_n-1} \sum_{i=1}^{S_m-2} \left(\delta\left(F_h(i,j) = u\right) \parallel \delta\left(F_h(i,j) = u - 1\right)\right)} \quad (4)$$

$$p\{F_v(i+1,j) = v \mid F_v(i,j) = u\} =$$
$$\frac{\sum_{i=1}^{S_m-2} \sum_{j=1}^{S_n-1} \delta\left((F_v(i,j) = u \parallel F_v(i,j) = u - 1), (F_v(i+1,j) = v \parallel F_v(i+1,j) = v - 1)\right)}{\sum_{j=1}^{S_n-2} \sum_{i=1}^{S_m-1} \left(\delta\left(F_v(i,j) = u\right) \parallel \delta\left(F_v(i,j) = u - 1\right)\right)} \quad (5)$$

$$p\{F_v(i,j+1) = v \mid F_v(i,j) = u\} =$$
$$\frac{\sum_{i=1}^{S_m-1} \sum_{j=1}^{S_n-2} \delta\left((F_v(i,j) = u \parallel F_v(i,j) = u - 1), (F_v(i,j+1) = v \parallel F_v(i,j+1) = v - 1)\right)}{\sum_{j=1}^{S_n-1} \sum_{i=1}^{S_m-2} \left(\delta\left(F_v(i,j) = u\right) \parallel \delta\left(F_v(i,j) = u - 1\right)\right)} \quad (6)$$

where $(u, v) \in \{-T, -T + 2, -T + 4, \ldots T - 4, T - 2, T\}$, and S_m and S_n denote the dimension of original source image and

$$\delta(A = u, B = v) = \begin{cases} 1 & \text{if } A = u, B = v. \\ 0 & \text{otherwise} \end{cases} \quad (7)$$

Finally, all the captured elements of the Markov matrix can be used as features for image splicing detection.

Similarly, inter-block correlation is considered to extract more Markov features. Here, inter-block difference 2D array is calculated using Eqs. (8) and (9).

$$E_h(i,j) = S(i,j) - S(i+8,j) \quad (8)$$

$$E_v(i,j) = S(i,j) - S(i,j+8) \quad (9)$$

where $i \in [1, S_m - 1]$, $j \in [1, S_n - 1]$ and, S_m and S_n is the dimension of original input image.

Now, enhanced threshold method is applied to the inter-block difference array $E_h(i,j)$ and $E_h(i,j)$ as explained in step 3 where $S_m - 1$, $S_m - 2$, $S_n - 1$, and $S_n - 2$ are replaced with $S_m - 8$, $S_m - 16$, $S_n - 8$, and $S_n - 16$, respectively. Hence, by considering inter-block correlation $4 \times (T + 1) \times (T + 1)$, more features have been extracted. Thus, a total of $2 \times 4 \times (T + 1) \times (T + 1)$ features are extracted from DCT domain which can be used to distinguish the authentic image from spliced one.

Feature Extraction in DMWT Domain: Most of the previously proposed approach based on DWT [9, 10] deals with all the sub-bands independently after wavelet decomposition, but [6] shows that there is dependency among wavelet components across position, scales, and orientation. However, it is observed that among the three dependencies contribution of position and orientation is more than scale in splicing detection. So, in this paper, we only consider dependency across position and orientation. Hence, Markov features with different dependencies are extracted as follows.

Step 1: We apply two-level discrete Meyer wavelet transform on the input image and round the coefficient of eight sub-bands to absolute value. Processed sub-bands are denoted as $\{W_a^b, W_h^b, W_v^b, W_d^b\}$, where b = $\{1, 2\}$.

Step 2: Consider dependency across position in DMWT domain, which is similar to characterize correlation between neighboring coefficients in DCT domain. Hence, by replacing F in Eqs. (1) and (2) with each of the eight sub-bands of DMWT domain followed by using Eqs. (3), (4), (5), and (6), we captured a total of $(T + 1) \times (T + 1) \times 32$ more Markov features.

Step 3: Now, considering the dependency among orientation, more features can be extracted using the following difference arrays.

$$W_h W_v^b (I, J) = W_h^b (i,j) - W_v^b (i,j) \tag{10}$$

$$W_v W_d^b (I, J) = W_v^b (i,j) - W_d^b (i,j) \tag{11}$$

$$W_d W_h^b (I, J) = W_d^b (i,j) - W_h^b (i,j) \tag{12}$$

where b = $\{1, 2\}$ and $W_a^b, W_h^b, W_v^b, W_d^b$ denote bth level approximation, horizontal, vertical, and diagonal sub-bands, respectively.

Now, F_h in Eqs. (3) and (4) is replaced by each of the difference arrays obtained in (10), (11), and (12) to capture more Markov matrix. Hence, $(T + 1) \times (T + 1) \times 12$ more Markov features are obtained.

By combining $(T + 1) \times (T + 1) \times 8$ Markov features captured in DCT domain and $(T + 1) \times (T + 1) \times 44$ Markov features captured in DMWT domain, resultant feature vector is used to differentiate spliced image from an authentic one. We choose threshold T = 6. So, we got a total of 2548 features.

3 Experimental Results and Performance Analysis

3.1 Dataset and Classifier

We use Columbia image dataset [11] provided by DVMM. It consists of 933 authentic and 912 spliced images without any post-processing enhancement. All the forged images are spliced image. This dataset is designed to test the blind image splicing detection method. Some images from the DVMM dataset are shown in Fig. 3, in which first row shows the set of authentic images and second row shows the set of spliced images.

To classify the images, support vector machine (SVM) is used in our experiments. In this experiment, SVM classifier is trained to solve the binary decision problem (classification of authentic and spliced images).

To evaluate the performance, all the experiments are performed on Columbia image splicing dataset [11] using same classifier. In each experiment, 80% randomly selected images are used to train the SVM classifier and remaining 20% images are used for testing.

3.2 Performance Analysis of the Proposed Model

Some experiments are carried out to verify and compare the detection accuracy of the proposed approach. T is set to 6 in these experiments. Feature vectors from DCT and DMWT domain are captured and effect on the detection performance of the proposed method with Z. He et al. [6] is evaluated in both the domain. The obtained results are shown in Table 1 and Table 2, respectively. In Table 2, level 1 and level 2 represent the first-level DMWT and second-level DMWT, respectively. It can be observed

Fig. 3 Some sample of authentic and spliced images from Columbia image splicing evaluation dataset [11]

Table 1 Comparison and detection rate of original and proposed method in BDCT domain with T = 6

BDCT	H-Difference (%)	V-Difference (%)
Z. He et al. [6]	85.30	85.00
Proposed	87.12	87.40

Table 2 Comparison and detection rate of original and proposed method in DMWT domain with T = 6

DMWT	Level 1 (%)	Level 2 (%)
Z. He et al. [6]	81.60	70.00
Proposed	84.12	72.35

Table 3 Comparison of proposed approach and existing methods

Feature vector	TP	TN	Accuracy (%)
Z. He et al. [12]	82.3	78.9	80.6
B. Su et al. [7]	87.5	87.6	87.5
Proposed method	85.42	91.44	88.43

from Table 1 and Table 2 that our method has improved the detection rate by approximately 1.0%–3.1% and 2.1%–2.3 % in BDCT and DMWT domains, respectively. Further, it is observed that by combining feature vectors from DCT and DMWT domains, we are getting much better accuracy.

Table 3 shows the comparison and detection rate of proposed work and some previous splicing detection methods [7, 12]. The complete implementation of the proposed method has achieved an accuracy of 88.17%, which makes a significant progress in splicing detection. In Table 3, true positive (TP) and true negative (TN) are calculated as:

$$TP = \frac{N_{ca}}{N_a}, \quad TN = \frac{N_{cs}}{N_s} \tag{13}$$

where N_{ca} = number of correct authentic classification, N_{cs} = number of correct spliced classification, N_a = total number of authentic images, and N_s = total number of spliced images.

The experimental results of proposed and other methods are shown in Table 3. It can be observed that our proposed method performs best out of the three presented splicing detection scheme in Table 3.

3.3 Recognizing Real Images

In Fig. 4, publicly available on Internet, we have given three original images (b), (c), and (e) and their associated altered images (a) and (d). To test these five images (three authentic and two spliced), we trained the classifier using experiments mentioned in Sect. 3.1. The test has been performed 20 times, and the results are shown in Table 4. It can be observed that there are only four cases in which images are wrongly classified.

3.4 Threshold Selection

Selecting a threshold is an issue because in general for a smaller T value, information loss will be higher; in that case, Markov matrices may be insufficient to distinguish authentic and forged images, whereas a larger T value can reduce information loss but a larger number of features can generate an overfitting problem, which results in low detection performance. Therefore, the choice of T and size of Markov matrix have an important impact on detection performance and computational cost.

The performance analysis of proposed approach for different thresholds ($T = 4, 6,$ and 8) is shown in Table 5. From Table 5, it can be observed that $T = 6$ is the best choice which balances the detection rate and computational cost with the accuracy of 88.43%.

(a)   (b)   (c) (d)   (e)

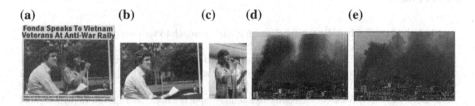

Fig. 4 Test images. **a** Spliced image. **b** Authentic image. **c** Authentic image. **d** Altered image. **e** Authentic image

Table 4 Splicing detection on real images (✓ correct, ✗ wrong)

Test rounds	1	2	3	4	5	6	7	8	9	10	11	12	13	14	15	16	17	18	19	20
Figure 4a	✓	✓	✓	✓	✓	✓	✓	✓	✓	✓	✓	✓	✓	✓	✓	✓	✓	✓	✓	✓
Figure 4b	✓	✓	✓	✓	✓	✓	✗	✓	✓	✓	✓	✓	✓	✓	✓	✓	✓	✓	✓	✓
Figure 4c	✓	✓	✓	✓	✓	✗	✓	✓	✓	✓	✓	✓	✗	✓	✓	✓	✓	✓	✓	✓
Figure 4d	✓	✓	✓	✓	✓	✓	✓	✓	✓	✗	✓	✓	✓	✓	✓	✓	✓	✓	✓	✓
Figure 4e	✓	✓	✓	✓	✓	✓	✓	✓	✓	✓	✓	✓	✓	✓	✓	✓	✓	✓	✓	✓

Table 5 Performance analysis for different thresholds

Feature sets	TN	TP	Accuracy
T = 4	83.85	90.90	87.33
T = 6	**85.41**	**91.44**	**88.43**
T = 8	82.30	93.58	87.86

4 Conclusion

In this paper, an enhanced threshold method is proposed to extract Markov feature which generates reduced feature set without any feature loss which improves the detection rate. Reduced feature sets are extracted from DCT and DMWT domains by performing difference operation followed by enhanced threshold method. Features extracted from DCT domain consider the correlation between the DCT coefficients, while DMWT domain distinguishes the dependency among coefficients across orientations and positions. Finally, the combined reduced feature vector from both the domain is considered as distinguished feature for classification. SVM is used as a classifier in our experiments. Our experimental results are encouraging, yielding the accuracy of over 88.43% correct classification which outperforms some state-of-the-art methods.

References

1. Shi YQ, Chen C, Chen W. A natural image model approach to splicing detection. In Proceedings of the 9th workshop on Multimedia & security 2007 Sep 20, ACM, 51–62.
2. Li X, Jing T, Li X.: Image splicing detection based on moment features and Hilbert-Huang Transform. In Information Theory and Information Security (ICITIS), 2010 IEEE International Conference on 2010 Dec 17, IEEE, 1127–1130.
3. Zhao X, Li J, Li S, Wang S.: Detecting digital image splicing in chroma spaces. In International Workshop on Digital Watermarking 2010 Oct 1, Springer Berlin Heidelberg, 12–22.
4. Pevny T, Bas P, Fridrich J.: Steganalysis by subtractive pixel adjacency matrix. IEEE Transactions on information Forensics and Security. 2010 Jun, 5(2):215–224.
5. Kirchner M, Fridrich J.: On detection of median filtering in digital images. In IS&T/SPIE Electronic Imaging, International Society for Optics and Photonics, 2010 Feb 4, 754110–754110.
6. He Z, Lu W, Sun W, Huang J.: Digital image splicing detection based on Markov features in DCT and DWT domain. Pattern Recognition. 2012 Dec 31, 45(12):4292–4299.
7. Su B, Yuan Q, Wang S, Zhao C, Li S.: Enhanced state selection Markov model for image splicing detection. EURASIP Journal on Wireless Communications and Networking. 2014 Dec 1, 2014(1):1–10.
8. Zhao X, Wang S, Li S, Li J.: Passive image-splicing detection by a 2-D noncausal Markov model. IEEE Transactions on Circuits and Systems for Video Technology. 2015 Feb;25(2):185–199.
9. Chen W, Shi YQ, Su W.: Image splicing detection using 2-d phase congruency and statistical moments of characteristic function. In Society of photo-optical instrumentation engineers (SPIE) conference series 2007 Feb 15, (6505), 26.

10. Lu W, Sun W, Chung FL, Lu H.: Revealing digital fakery using multi resolution decomposition and higher order statistics. Engineering Applications of Artificial Intelligence. 2011 Jun 30;24(4):666–672.
11. Ng TT, Chang SF, Sun Q.: A data set of authentic and spliced image blocks. Columbia University, ADVENT Technical Report. 2004 Jun:203–204.
12. He Z, Sun W, Lu W, Lu H.: Digital image splicing detection based on approximate run length. Pattern Recognition Letters. 2011 Sep 1;32(12):1591–1597.

An Adaptive Algorithm
for User-Oriented Software Engineering

Anisha, Gurpreet Singh Saini and Vivek Kumar

Abstract Efficient resource allocation process is a necessity in varied environments such as software project management, operating system, construction models for reducing the risk of failure by utilizing only those many resources which are required. This paper presents an algorithm which uses fuzzy methodology of soft computing with the concept of dynamic graph theory for generating the graphs which helps in allocating the resources efficiently. With proper observation of requisites, this algorithm presents the importance of formulating a model which could be invoked at the time of chaos or failure. Due to the chaotic behavior of the software engineering environment, the resource allocation graph continuously evolves after its initial design, which is a unique factor of dynamicity signified by the algorithm. This dynamicity contains a reasoning perspective that can be validated by appending more information, and it is eliminated through the inference mechanism of fuzzy logic. The calculative process is effective and has ability to change according to the environment; therefore, it is much more effective in reducing the failures, answering the allocation of resources and specifying the work initiated by using those resources. The development of the algorithm is specifically focused on product and will be done with respect to the perspective of the developer during development process accompanied by the views of the customer regarding the needed functionalities.

Keywords Dynamic graph theory · Resource allocation · Fuzzy logic
Soft computing · Chaos theory

Anisha (✉) · G. S. Saini · V. Kumar
Delhi College of Technology and Management, New Delhi, India
e-mail: anishanagpal@outlook.com

G. S. Saini
e-mail: g.saini4888@live.com

V. Kumar
e-mail: drvivekkumar@rediffmail.com

© Springer Nature Singapore Pte Ltd. 2019
B. K. Panigrahi et al. (eds.), *Smart Innovations in Communication and
Computational Sciences*, Advances in Intelligent Systems and Computing 670,
https://doi.org/10.1007/978-981-10-8971-8_3

29

1   Introduction

In basic terminology, the word "dynamic graph" [1] is related to the field of mathematics, and in the world of computing, dynamic graph theory [2, 3] is a new concept. It is an old technique in the field of engineering [4]. With the study of publications in well-known journals, it can be seen that the concept of implementation of dynamic graph theory in the area of resource allocation is relatively new.

There are multiple instances in the development cycle where paths do occur with varied man power, time, skill, and other attributes but none of those lead to success. In such situations, dynamic graph theory is applied and a developer tries to reduce the damages by assessing his present situation using his natural conscience or intuition. However, if a developer fails to evaluate the current status of work, he can never ascertain necessary steps that should be taken in the future to minimize the damages. The ideology behind this is the chaos theory of physics [5] which states that "when the present determines the future, but the approximate present does not approximately determine the future." The most recent outcome of this chaos theory is "dynamic graph theory" which has evolved due to the increase in knowledge and accommodates the changes which occur at various levels. It facilitates construction of networks consisting path discovery based on more than one factor from links within the network [6].

Dynamic graph theory has been used in various fields [7, 8], and there have been few instances of its use in the computer science. However, the field of software engineering is still not explored to its full extent and its application is confined to the area of testing [4, 6, 9, 10]. Software engineering is already incorporated with some of the models of resource allocation and is mainly associated with the terms development. The software quality can be defined as perceptual, conditional, and subjective in nature which can be interpreted distinctly by different people. It is one of the factors that determine the nature of the system developed by the developer, but it relies on the end user and the vendor for whom it is developed. End-user expectations with the developed system are associated with the quality of the system, but all the models are developers oriented in their way of working which sometimes lead to non-satisfaction among the users. However, due to new innovations using chaos theory there has been significant improvement in the development of software [11].

As per the reports by Richard Schmidt, Sirrush Corporation in October 2012 [12], around two-third of the software projects developed in 2009 either "failed" or were "critically challenged" leading to their extensions in timeline by many years and hence incurring huge losses. Taking this into consideration, we present an efficient algorithm for allocation of resources.

The idea of dynamic graph for allocating the resource is stable, and based on the requirements stated during the initial phase of software cycle, a dynamic network will be formed. Once gathering of requirements is done, they are separated in terms of critical and most required functionalities (based on the decision made through dynamic network). Developer should then develop a system based on critical requirements, and remaining ones should be implemented into the system as an update.

1.1 Literature Background

Uncertainty and predictability are the important aspects of a complex system, and in order to answer the future state of activities in these nonlinear systems, chaos theory is widely used. The formal definition of chaos theory is "the definite present determines the future, but the approximate present does not approximately determine the future." As described earlier, dynamic graph theory [11] was developed out of "chaos theory" after multiple evolutions to answer those situations which are related to the future and are going to happen, by evaluating the current status.

Similarly, in order to achieve the objectives stated at the initial stage of the software development life cycle, the field of software engineering completely relies on uncertainty and predictability. A simple strategy can be derived from the initial literature review, i.e., dividing the problem state into smaller problems which are not contradictory and have their own state definition. The problem can be divided as follows:

1. A module having different connected parts is a partially completed task having larger number of dependencies.
2. Division of modules depends upon certain issues such as relationship between previous tasks and the succeeding tasks, functionalities which are expected to be delivered, and time taken to complete a particular task.

    (a) The functionalities of software which are the core of a development in order to give software or project a specific structure fall in category of large issues.
    (b) Timely issues are critical issues which require continuous observation such that immediate actions can be taken in order to avoid delay in other tasks as this can lead to the overall delay in development which increases cost.
    (c) Dependent issues are easy to impart as they are tried and tested modules taken from past projects and can be combined with other module interfaces as per the need of the project.

3. At the end, if the state of stability is achieved then only output is presented.

Fuzzy algorithm for generating stability in this approach through resource allocation is required as it will enable the developer to determine at a given point of time what kind of requirement and functionality needs to be designed and which all resources we need for the same? It could further be used to calculate how much time is required along with the strategy that should be used for testing. However, the chaos model and the graph model alone have failed to answer the questions related to "how it is to be done?" as they are unable to determine how we can divide the system into smaller subsystems to get a prototype and to what level we can minimize the risk of failure.

2 Algorithm

The purpose of designing the fuzzy algorithm-based solution for resource allocation is to eliminate the problems occurring in chaos model of engineering in such a way that proper allocation of resources could be done in different modules so as to generate a failure-free system. Thus, for allocating the resources using dynamic graph we need to prioritize the requirements according to the weights allocated to them by developers which is further dependent upon the nature of requirement, i.e., large, timely, or dependent. The decision regarding weights is generated by considering the probabilistic data, and it is improved throughout the fuzzy process so as to achieve precise outputs. After the completion of the fuzzy process, the priorities are given to the requirements, while keeping into consideration that for predicting the future state, the algorithm requires values for the current state. After completion of each module, iterations are made to the model created at the very first step. The important steps involved in formulation of algorithm are:

1. The bigger problems should be divided into smaller subproblems, and then a search is to be done to convert them into a question.
2. Conduct an intensive search through various methods for answering the above-generated questions.
3. Carry out a test for correctness on all the answerable questions so that we can find out the effect of every question on the project along with its domain applicability.
4. Transform those findings into realistic facts for supporting decisions, and re-evaluate as per the views of the customer.
5. Performances are evaluated over the set parameters, and do continuous refinement as per the modules integrated into the project.

2.1 Advantages of Fuzzy Algorithm

The need of efficient resource allocation mechanism based on predicting the future events by evaluating the current status of the system came into existence due to the fact that there are none of the models designed to answer the basic questions stated in the text below. Also, the algorithms which tried answering the questions deviated from the main agenda of development cycle in that process and finished the aspects of dynamicity. Each model was clear in its approach, but inadequate in answering the following questions:

1. What kind of requirement and functionality needs to be designed at a given moment? Also, for same what resources we need, and how much time is required along with the strategy that should be used for testing.
2. How can we divide the system into subsystem and produce an early working model out of it; such that, it becomes easy for the end user to understand the system at the very initial phase reducing the risk of loss or failure at both developer and user end.
3. How to reduce the risk of "failures" by avoiding those critical situations which may lead to permanent shutdown or system crash.

On the basis of the available strategies and some traditional approaches of the software development methods, the software development team imposes a damage recovery procedure which is essential in some of the situations of dynamicity.

The fuzzy algorithm provides solution to these questions and also which are based on the perception of a person. The problem lies with the non-predictive models of the development cycles used earlier; however, it is the strategy of dynamic graph which works to avoid chaos generated during the development cycle and accomplishes the task within the deadlines. This strategy is used at every step of the software development life cycle. On the basis of the literature survey and observations, it can be stated that "how one can approve that 'someone' who prioritizes the tasks has done that efficiently as the software quality states 'quality is a perceptual, conditional, and somewhat subjective'". Therefore, the algorithm has successfully answered the major backdrops which are human perspective-based and gives the approach an area for improvement.

However, at the stage of critically challenged project, a user expects system which can fulfill its basic needs rather than the standard working. A developer can only provide such system by prioritizing the work in an efficient way within a specified amount of time.

2.2 Design

The ideology behind the design of the algorithm is simple and can be described as, "generate a dynamic network through the stable marriage resource allocation based on the requirements stated at initial phase of software cycle. Now, gather the requirements that provide critical and most required functionalities based on the decision taken through this dynamic network, and start developing a system around them. The requirements that are left should be imparted to the system as an update."

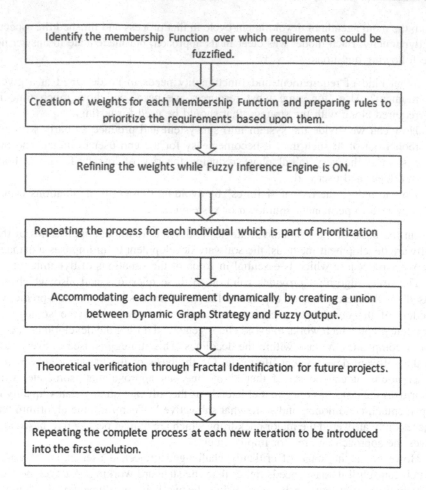

It can be understood by the following steps:

1. All requirements of the user should be retrieved first and then divided on the basis of the priorities given. These priorities are based on the collective weights given by the end users and the experienced members of the team, not decided by the individual.

2. Formulate a heuristic that can assess the weights or priorities of the modules.

3. For allocating the resources, stable marriage approach must be applied for every requirement having priorities and then predict the order of the tasks from starting to the end. The iteration of the modules that are supposed to be processed after the initial phase should be given order using the methodology of fuzzy logic [13–15].

4. Now start processing the modules, construct prototype of the initial working module, and then add new functionality to the existing or the older version of the module with each successive iteration.

This helps in providing advantages in certain issues such as

1. Prioritizing the requirements which was earlier done by an individual only, but now it is done with the inputs from experienced team and group of end users.
2. Extension in timeline for the complete development of the system.
3. Early working system, which is now available as a prototype with initial module.
4. Clarity in project development cycle for the end user.
5. Independence to explore new techniques or ways for developing a solution.
6. Interchangeable components that can be reused in terms of software as they are already tested at individual levels and can be combined easily.

2.3 Backdrops

The designing process of the algorithm contains some key points which are difficult to neglect and may improve to some extent:

1. History cannot always give adequate structure to build future upon: Projects are dynamic in nature that generates complex equations as one can only estimate how to proceed in a particular direction by using historical data but it never informs about how to use resources efficiently.
2. Values which are not predicted easily and are an issue of concern: Due to the dynamic nature of the problem, one cannot define all the constraints and cannot initialize all the variables while planning to construct a complex system. This changing environment always consists of some instances where non-predictive values appear.
3. When an algorithm is used deliberately, it causes failure: After algorithm has been brought into regular practice, people recognize the intended meaning of the algorithm and its protocols; they can implement same algorithm with alterations as per their personal needs which violate its significance.

2.4 Algorithmic Outcomes

The research will also consider some factors such as time period of the iteration, man power required, testing required, addition of more functionalities into the system, integrating new updates into the older modules, and training of the end user for better understanding of the developed system.

We can use this approach at any phase of the project development life cycle or whenever a development team faces situations which are challenging to handle.

With the efficient deployment of fuzzy logic for computation of appropriate number of resources and their allocation, the algorithm has given the solution to all the backdrops. This will result in efficient planning of projects, and it is not compulsory to implement it within the IT projects as long as there can be alterations and customizations with respect to relevant domain are done.

3 Conclusion

The proposed work will be beneficial in answering the allocation of resources and how the work has to be done using those resources which helps in reducing the failure. It will give importance in formulating a model with proper observation of consequences so that it could be invoked at the time of chaos and failure. The evolution of the model will be done by considering the perspective of the developer during development process accompanied by the views of the customer regarding the needed functionalities.

4 Future Scope

A framework is developed from the proposed algorithm which could be used in various environments such as software project management, operating system, construction models. The future actions may consist of developing a framework for specific environment by improving this generalized framework so that it could become the real-time project management solution to the particular problem.

References

1. M. D. Konig, "Dynamic R&D Networks - The Efficiency and Evolution of Interfirm Collaboration Networks", Dissertation–18182, ETH ZURICH, 2010.
2. M. D. Konig, S. Battiston, M. Napoletano and F. Schweitzer, "On Algebraic Graph Theory and the Dynamics Of Innovation Networks", Networks And Heterogeneous Media, Volume 3, Number 2, June 2008.
3. O. Shai, K. Preiss, "Graph theory representations of engineering systems and their embedded knowledge", Artificial Intelligence in Engineering, Elsevier, Vol 13, 1999.
4. G. Toffetti, M. Pezze, "Graph transformations and software engineering: Success stories and lost chances", Journal of Visual Languages and Computing, Elsevier, Vol 24, 2013.
5. S. Boccaletti, C. Grebogi, Y.C. Lai, H. Mancini, D. Maza, "The Control Of Chaos: Theory and Applications", S. Boccaletti et al./Physics Reports 329, Elsevier, 2000.
6. C. Y. Chong, S. P. Lee, "Analyzing maintainability and reliability of object oriented software using weighted complex network", The Journal of Systems and Software, Elsevier, Vol 110, 2015.

7. R. Attri, S. Grover, N. Dev, "A graph theoretic approach to evaluate the intensity of barriers in the implementation of total productive maintenance (TPM)", International Journal of Production Research, Taylor and Francis, 2013.
8. F. Schweitzer, G. Fagiolo, D. Sornette, F. V. Redondo, D. R. White, "Economic Networks: What do we know and what do we need to know?", ACS - Advances in Complex Systems, vol 12, no 4, 2009.
9. H. Robinson, "Graph Theory Techniques in Model-Based Testing", International Conference on Testing Computer Software, 1999.
10. Peter C.R. Lane, F. Gobet, "A theory-driven testing methodology for developing scientific software", Journal of Experimental & Theoretical Artificial Intelligence, Taylor & Francis, 2012.
11. D. K. Saini, M. Ahmad, "Software Failures and Chaos Theory", WCE-London, Vol II, 2012.
12. R. Schmidt, "Software engineering: Architecture-driven Development", NDIA 15th Annual Systems Engineering Conference, October 2012.
13. G. Kumar, P. K. Bhatia, "Neuro-Fuzzy Model to Estimate & Optimize Quality and Performance of Component Based Software Engineering", ACM SIGSOFT Software Engineering Notes, Vol 40, 2015.
14. S. Mishra, A. Sharma, "Maintainability Prediction of Object Oriented Software by using Adaptive Network based Fuzzy System Technique", International Journal of Computer Applications, Vol 119, 2015.
15. K. Tyagi, A. Sharma, "A rule-based approach for estimating the reliability of component based systems", Advances in Engineering Software, Elsevier, Vol 54, 2012.
16. H. Jifeng, X. Li, Z. Liu, "Component-Based Software Engineering-The Need to Link Methods and their Theories", 973 project 2002CB312001 of the Ministry of Science and Technology of China.
17. B. Boehm, "Value-Based Software Engineering: Overview and Agenda", USC-CSE-2005-504, February 2005.
18. T. Dybå, B. A. Kitchenham, M. Jørgensen, "Evidence-Based Software Engineering for Practitioners", IEEE Software, Published by the IEEE Computer Society. January-February 2005.
19. D. Kelly, "Scientific software development viewed as knowledge acquisition: Towards understanding the development of risk-averse scientific software" The Journal of Systems and Software, Elsevier, Vol 109, 2015.

A Systematic Review on Scheduling Public Transport Using IoT as Tool

Dharti Patel, Zunnun Narmawala, Sudeep Tanwar
and Pradeep Kumar Singh

Abstract Public transport can play an important role in reducing usage of private vehicles by individuals which can, in turn, reduce traffic congestion, pollution, and usage of fossil fuel. But, for that public transport needs to be reliable. People should not have to wait for the bus for a long time without having any idea when the bus will come. Further, people should get a seat in the bus. To ensure this, efficiently and accurately scheduling and provisioning of buses is of paramount importance. In fact, nowadays buses are scheduled as per the need. But these scheduling is being done manually in India. Our survey shows that there are many algorithms proposed in the literature for scheduling and provisioning of buses. There is a need to tailor these algorithms for Indian scenario. We present a brief overview of these algorithms in this paper. We also identify open issues which need to be addressed.

Keywords Scheduling · Provisioning · Public transport · IoT

1 Introduction

With an increase in population, the demand for public transport service also increases. This situation pushes the public transport infrastructure to its limits in peak hours. This has lead to the late arrivals of public transport at their scheduled

D. Patel (✉) · Z. Narmawala · S. Tanwar
Department of CE, Institute of Technology, Nirma University, Ahmedabad, Gujarat, India
e-mail: 15mcen18@nirmauni.ac.in

Z. Narmawala
e-mail: zunnun80@gmail.com

S. Tanwar
e-mail: sudeep149@rediffmail.com

P. K. Singh
Department of CSE, Jaypee University of Information Technology,
Waknaghat, Himachal Pradesh, India
e-mail: pradeep_84cs@yahoo.com

© Springer Nature Singapore Pte Ltd. 2019
B. K. Panigrahi et al. (eds.), *Smart Innovations in Communication and Computational Sciences*, Advances in Intelligent Systems and Computing 670,
https://doi.org/10.1007/978-981-10-8971-8_4

stops, inefficient use of resources, and a decrease in operating cost. And again this situation leads to switching of other modes of transportation by people which is undesirable. The traditional scheduling operation in India is not efficient for large-scale running routes [1]; i.e., public transport still lags behind in India. The intelligent public transportation scheduling becomes the main research area. Hence, by using the emerging technologies, public transportation can be scheduled in an optimized way to meet the public demand. Hence, to provide the better transportation services, the scheduling algorithms have been proposed which focuses on the many factors such as a number of passengers, the capacity of buses, waiting time of passengers, arrival time, and departure time. The best scheduling algorithm is one which leads to the decrease in waiting time of passengers while maintaining the capacity of the passengers in public transport. Hence, IoT is the most emerging technology which has been used by public transport services.

As per the survey of researchers, it has been concluded that all the present systems which are based on transportation system provide the service of either reservation or displays bus status. In transportation system, it is also important to maintain the schedule of buses. But due to traffic and many more parameters, schedules get disturbed. Usually, passengers have to wait for the buses because of bad scheduling, overcrowding, traffic congestion, and breakdowns. Overcrowding occurs because of not determining the correct frequency for demand and improper scheduling. Frequency relates to the number of trips which are needed to cope up with the passenger demand during a peak or fixed period. Hence, it is important to have exact traffic information to notify the exact arrival and departure timings of buses. This information can be available by using GPS [2]. Along with these, it is also important to schedule buses according to the requirements of the public. The solution is to develop efficient scheduling algorithm which can schedule the buses automatically by analyzing the historical data.

There are many scheduling algorithms available such as priority, round robin, first come first serve, last come first serve, and much more. But these algorithms are developed for operating systems. Algorithms like genetic algorithm [3], greedy randomized adaptive search procedure (GRASP) [4], trip frequency scheduling, and vehicle routing problem are such scheduling algorithms which either schedule the vehicles or finds proper routes for buses. These algorithms have the drawback in terms of a number of jobs or vehicles. There is a need of scheduling algorithm which can schedule buses as per the requirement of passengers on proper time and proper routes. Some algorithms use heuristics to calculate the parameters and schedule it accordingly. But these heuristics require same number of jobs with the same number of resources. Hence, if we use these algorithms for scheduling of buses, it will require the same number of buses with the same number of routes. It becomes compulsory to have the same number of parameters. Hence, it is necessary to develop such algorithm which can even schedule one bus. We can use assignment problem to assign buses for required route with heuristics to schedule the buses as per the requirement and reschedule the buses which are not required for that particular route.

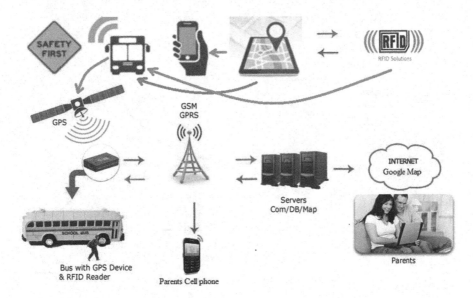

Fig. 1 A future scenario of public transport

Figure 1 shows the present scenario of the public transportation system which uses RFID tags to track the public transport to have live data about the stops covered by the buses. It also uses GSM/GPRS to notify the passengers about the current status of buses. Servers are used from which data are fetched to notify the passengers. In more general, this system shows the school bus which is tracked using RFID tags and GSM/GPRS, and its status is notified to the parents so that their waiting time at the stops decreases waiting unnecessarily at stops.

According to the Planning Commission (2013) [5] survey, it has been concluded that there is an increase in traffic of public transport as people refer it much for traveling. According to the survey, people prefer railways and roadways increasingly and it is expected to grow up to 11 percent and 7 percent per year, respectively. By this, it is expected that traffic could increase up to 16 factors by next 20 years. Since it was about 7 to 8 factors in last 10 years, it is growing very fast. Hence, this proves the need for public transport which should be scheduled according to the need. Figure 2 shows an increase in traffic in roadways and railways in billion passenger kilometer (bpkm).

1.1 Contribution of This Paper

The main contribution of this paper is as follows:

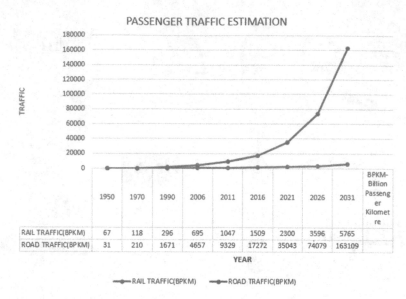

Fig. 2 Passenger traffic estimation

- A complete taxonomy of public transport, which can be further classified into general and selective approach, is provided using IoT as a tool.
- Approaches used for scheduling public transport are analyzed with respect to parameters, such as positioning, time.
- Finally, a systematic comparison of the various public transport system with pros and cons of each is provided in the text.

1.2 Organization

Rest of the paper is organized as follows. Section 2 presents the similar work done by various researchers in this domain with tabular comparison of each approach. Section 3 provides the challenges and open research problems in this domain, and finally, Sect. 4 concludes the article with future scope.

2 Literature Survey

This section provides the detailed description of the work done under this domain by various researchers. We have divided this description into two categories: generalized and selective. Next subsection explains each category in detail with pros and cons of each.

2.1 Generalized Approach

Since last few years, research has been carried out to develop the optimization models which will increase the convenience of passengers, and on the bus management side, bus operations are reduced. Depending on the approach for determining optimal solutions, the bus scheduling models studied so far can be classified into a number of types. These models use heuristics to calculate total traffic cost.

Fu et al. [6] proposed a new operating strategy in which service vehicles is followed by the lead vehicle with all stop service and also by providing the facility to skip some stops as an express service. Chen [7] measures bus service reliability, vehicle load capacity, by considering the headway adherence and average waiting time. Yan proposed a network flow problem using a mathematical model which uses Lagrangian relaxation. Kim et al. [8] constructed a schedule based on the starting point and stops by using travel time response model for critical scheduling areas.

2.2 Selective Approach

Vehicle Routing Problem In vehicle routing problem with time window, a number of vehicles are allocated to route and each with given capacity which is located at a single depot which is serving passengers dispersed geographically. In this problem, each passenger has been given the demand and they must be served in specific time window. The main objective of this problem is to minimize the total cost of traveling while serving the customers with minimum cost. Figure 3 shows the vehicle routing problem (VRP), and how a route is set up from base depot to each stop reducing the traveling cost. It reduces the path cost by setting the optimized route between stops from one station to other from depot. The main objective of VRP is as follows:

1. It minimizes the global transportation cost based on global distance.
2. Minimizes the number of resources needed to serve all customers.
3. Variation in travel time and vehicle load is least.

Online Dial-A-Ride Problem with Time Window (DARPTW) Routing and scheduling of buses can be referred as a On-line-Dial-A-Ride problem [7]. It takes care of the available set of resources and constraints. All the services related to the transportation is Web-based and handled by mobile phones. Here, request is generated one day prior to the beginning of service. Due to a high number of variables involved in it, the solutions available are based on heuristics. In the Online Dial-A-Ride Problem with time windows (ODARPTW) [9] requests are exhibited

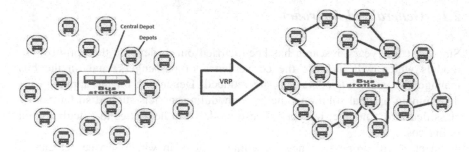

Fig. 3 Vehicle routing problem

after some time, requiring the server to convey the articles from sources to desti-nation, while the server is en-route for serving the different request. On the off chance that a demand is to be served, the server must achieve the time between the demand's landing and its due date. The objective here is to plan techniques for the server to fill in whatever number of approaching requests as could reasonably be expected by their due dates in an on-line way. The system for ODARPTW neither has data about the discharge time of the last demand nor has the aggregate number of request. It must decide the conduct of the server at a specific minute of time as an element of the considerable number of request discharged up to time t (and the present time t). Interestingly, a disconnected technique has data about all requests in the entire succession as of now at time 0.

Assignment algorithm is operation research method to optimize the cost. It deals with the transportation problem and the assignment of jobs to the workers or assignment of resources to the workers. The ultimate objective is how to assign the jobs efficiently to the workers, and which worker should be assigned which type of job. In this problem, at least 3×3 matrix is required and also the same number of parameters. Assignment problem uses heuristics, in which the operating cost of the system can be maximized or minimized. If we are adding any constant in each and every element of columns and rows of the matrix, then it will generate a matrix, which can minimize the total effectiveness. A situation also exists, where we need to add a dummy column or a row when there are no exact number of resources, and jobs. The main disadvantage of this algorithm is a number of jobs must be equal to the number of workers or else allocation of jobs will not be efficient.

The data flow diagram of assignment algorithm is discussed below in Fig. 4. This diagram shows how assignment algorithm works with its two conditions. In assignment problem, basically there are two situations: First, when the number of cutting lines is equal to the number of rows and columns, and second, when the number of cutting lines is not equal to the number of rows and columns.

Fig. 4 Assignment algorithm

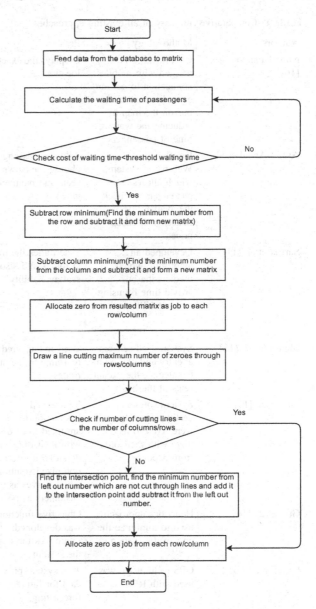

Table 1 represents the summary of the whole survey. In the table, there are four columns. The first column is the name of author, second is the algorithm/method they have used for implementation, third is what are the pros of paper, and the fourth one is what can be added/improved or cons in paper.

Table 1 Comparative summary of all existing approaches

Authors	Methodology	Pros	Cons
Nagadapally et al. [10]	GPS tracking is merged with mobile application to get the exact location of the bus on the map by reducing the waiting time of passengers	Displays the exact bus location	In situation there is lack of data in mobile, this system would not give an accurate result and it works only for selected versions of android and iOS
Asaad Abdul et al.	Developed a Web-based system, which allows passengers to check the availability of tickets with cheap prices.	Allowed passenger to choose their own seats and payment gateway	The passengers must be registered for searching and reservation
Sarma et al. [11]	Developed an algorithm, which predicts the bus arrival time by using existing GPS system	Determines the traffic at routes and also the seat availability	User receive the SMS using USSD technique and such messages cannot be saved so user does not have any other backup
Menon et al. [12]	System used sensors at entry and exit point to calculate the vacant seats of the bus	Used IoT to predict real-time traffic at routes	Cost of system increase installation of costly IoT-enabled components
Lau et al. [13]	System source and destinations timings were fixed to feed them with their order from depot	Source and destination timings are fixed to feed them with their order from depot so it returns an optimized solution which serves as many customers as possible	Restricted number of vehicle is used.
Guy et al. [14]	Heuristics approach used to minimize the cost of the system	Objective function was developed, which minimizes the cost of the system	How to order the nodes is the main problem of this system
Sulaima Le bbe [15]	GPS technology was used with RFID	This system provides data for future forecasting, congestion identification, travel route planning, and modeling	System is handled manually
Singh [16]	Focused on how assignment is done using different methods	Methods are used to optimize the cost of system	Some methods lack in optimization due to improper assignment

(continued)

Table 1 (continued)

Authors	Methodology	Pros	Cons
Olarthichachart et al. [17]	Used to find optimal trip frequency	Reduces cost of operation passenger service quality by efficient scheduling	Applicable to finite set only
Sulaima et al. [18]	Used wireless sensor networks to provide visual information for real-time system	Gives real-time information about the buses using GPS/GSM and Arduino	Only registered passengers can have real-time information

3 Challenges and Open Research Problems in Scheduling Public Transport

In existing public transportation systems, passengers need to register compulsory for traveling. Scheduling of transportation service is limited for the limited resources. Scheduling of public transport is not much efficient as it requires accurate data on traffic. Due to this, the waiting time of passengers increases. Passengers used to travel in public transport mostly in peak hours. Therefore, in peak hours due to nonexistence scheduling mechanism, passengers increase beyond the capacity of public transport. As capacity of public transport increases, the counting of passengers in the existing system is not efficient. Waiting time of passengers increases at stops in peak hours that is in morning and evening which increases traffic in buses. Hence, there is a need to develop an efficient and accurate system for scheduling and provisioning of buses dynamically based on current demand.

Public transport should provide safe traveling, cheap fare, and less travel time. In order to achieve these attributes, govt must focus on road network, optimal routing, and minimum delay. Public transport efficiency depends upon all other related factors, i.e. optimization of route, transfer optimization, and coordination among feeder bus service.

4 Conclusion and Future Scope

In India, traffic congestion problem is increasing day by day in urban centers. It is because of increase in population in urban areas with very high rate, and people prefer private vehicles over public transport, as they need to wait for a long amount of time to get a bus and often buses are crowded too. Hence, it is important to develop an efficient scheduling and provisioning of public transports so that waiting time of travelers reduces and everyone gets seat in the bus. Many algorithms have been proposed by various researchers as discussed in Sect. 2. These proposals do have their drawbacks as highlighted in Table 1. We need to improve these algorithms in terms of scheduling and provisioning of public transport for Indian scenario.

In future, we can develop a system which can allow seat reservation in real time. Sensors can be used to track seat occupancy and allow reservation for unoccupied seats.

References

1. K. Shankar, "Applications of advanced technologies to transportation systems in indian context," *International Journal of Earth Sciences and Engineering*, vol. 4(6), pp. 394–397, 2011.
2. T. Halonen, J. Romero, and J. Melero, "Gsm, gprs and edge performance: evolution towards 3 g/umts," 2004.
3. F. A. Kidwai, B. R. Marwah, K. Deb, and M. R. Karim, "A genetic algorithm based bus scheduling model for transit network," vol. 5, pp. 477–489, 2005.
4. R. M. Aiex, M. G. Resende, P. M. Pardalos, and G. Toraldo, "Grasp with path relinking for three-index assignment," *INFORMS Journal on Computing*, vol. 17(2), pp. 224–247, 2005.
5. "School bus tracking." http://schoolpixa.com/images/GPS-School-Bus-Tracking-System.png.
6. J. Zhou and W. Lam, "Models for optimizing transit fares. in: Advanced modeling for transit operations and service planning," *Publication of: Elsevier Science Publishers BV*, 2003.
7. W. Chen, C. Yang, F. Feng, and Z. Chen, "An improved model for headway-based bus service unreliability prevention with vehicle load capacity constraint at bus stops," *Discrete Dynamics in Nature and Society*, vol. 2012, 2012.
8. W. Kim, B. Son, J.-H. Chung, and E. Kim, "Development of real-time optimal bus scheduling and headway control models," *Transportation Research Record: Journal of the Transportation Research Board*, no. 2111, pp. 33–41, 2009.
9. F. Yi, Y. Song, C. Xin, and K. Ivan, "Online dial-a-ride problem with unequallength time-windows," pp. 1–5, 2009.
10. A. S. R. Nagadapally, B. B. Krishnaswamy, C. A. R. Shettap, D. D. Prathika, and E. R. Lee, "A step ahead to connect your bus-with gps," p. 171, 2015.
11. S. Sarma, "Bus tracking and ticketing using ussd," vol. 1, no. 7 (December-2014), 2014.
12. A. Menon and R. Sinha, "Implementation of internet of things in bus transport system of Singapore," 2013.
13. H. C. Lau, M. Sim, and K. M. Teo, "Vehicle routing problem with time windows and a limited number of vehicles," *European journal of operational research*, vol. 148, no. 3, pp. 559–569, 2003.
14. S. N. Parragh, K. F. Doerner, and R. F. Hartl, "A survey on pickup and delivery problems," *Journal für Betriebswirtschaft*, vol. 58, no. 1, pp. 21–51, 2008.
15. A. Haleem, S. Lebbe, and S. S. Nawaz, "Real time bus tracking and scheduling system using wireless sensor and mobile technology," 2016.
16. S. Singh, G. Dubey, R. Shrivastava, J. P. Siregar, E. B. K. Antariksa, S. Ravindra, V. V. Reddy, S. Sivanagaraju, N. N. Parandkar, and B. HIMABINDU, "A comparative analysis of assignment problem," *OSR Journal of Engineering (IOSRJEN)*, vol. 2, pp. 1–15, 2012.
17. P. Olarthichachart, S. Kaitwanidvilai, and S. Karnprachar, "Trip frequency scheduling for traffic transportation management based on compact genetic algorithm," vol. 2, 2010.
18. A. Haleem, S. Lebbe, and S. S. Nawaz, "Real time bus tracking and scheduling system using wireless sensor and mobile technology," 2016.

Blood Vessel Detection in Fundus Images Using Frangi Filter Technique

Adityan Jothi and Shrinivas Jayaram

Abstract Blood vessels are an important factor in identification of various retinal vascular defects such as hypertensive retinopathy, retinal vein occlusion, central retinal artery occlusion, diabetic retinopathy. In this paper, we will develop an algorithm that makes use of Frangi filter technique to detect and segment blood vessels in fundus images for further diagnosis. We make use of database from Friedrich-Alexander-Universitat that comprises of fundus having healthy images, diabetic retinopathy images, glaucoma images. We also make use of the DIARETDB which comprises of 89 other fundus images and also 400 images from the STARE project. The algorithm makes use of Frangi filter technique which would produce more accurate results.

Keywords Image processing · Machine learning · Blood vessel detection
Fundus imaging · Frangi filter

1 Introduction

The eye is the central organ of the sensory system which we use in day-to-day life to perceive the environment around us. The retina is the inner coat of the eye which acts as a screen which sends impulses to the brain via the optic nerve when light strikes it which is later processed by the brain to form images. The optic nerve fibers are alongside the retina which helps in the traversal of these electrical and chemical impulses to the brain. Retinal vascular disorders refer to a wide range of diseases that occur due to blood vessels in the eye such as hypertensive retinopathy, retinal vein occlusion, diabetic retinopathy.

A. Jothi (✉)
Amity School of Engineering, Noida, India
e-mail: research.adityan@gmail.com

S. Jayaram
Ramakrishna Mission Vivekananda College, Chennai, India
e-mail: jayaramshrinivas@gmail.com

© Springer Nature Singapore Pte Ltd. 2019
B. K. Panigrahi et al. (eds.), *Smart Innovations in Communication and Computational Sciences*, Advances in Intelligent Systems and Computing 670,
https://doi.org/10.1007/978-981-10-8971-8_5

Retinal vascular disorders are critical and should be diagnosed and detected at an early stage in order to prevent vision loss and further complications. Retinal vein occlusion which is a type of retinal vascular disorders is a condition where the vein becomes narrow or obstructed. It is one of the primary contributors to vision loss after diabetic retinopathy. Another disorder due to blood vessels is central retinal artery occlusion; this is a blockage of the central retinal artery that carries blood and oxygen to the eye. This is a serious condition that requires emergency treatment as it can cause irreparable damage to the eye. Due to the seriousness of retinal vascular diseases, the identification and isolation of blood vessels in fundus images is of utmost importance for early diagnosis. We will develop an algorithm that does this with increased accuracy and reduced computation time based on color differences; the algorithm would be able to detect and isolate blood vessels in retina effectively using the Frangi filter technique.

1.1 Retinal Vascular Disorders

The disorders that occur due to defects or damages in the retinal blood vessel network are known as retinal vascular disorders. We have presented certain major disorders due to retinal vascular problems below.

Diabetic retinopathy is a complication that arises in patients who have diabetes. This disorder is quite common and chronic that leads to loss of vision among adults aged between 20 and 60 years. In this disorder, the retina is damaged due to leakage in blood vessels into the retina. It occurs usually due to blockage in retinal veins, rupture of the blood vessels in eye and hypertension. It leads to further complications in the eye such as microaneurysm, haemorrhages, which leads to vision loss in the eyes of patients (Fig. 1).

Fig. 1 Fundus image of an eye having diabetic retinopathy

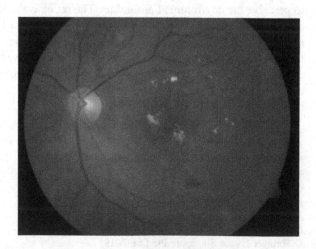

Hypertensive retinopathy is generally damage that occurs in the retinal blood circulation and the retina due to high blood pressure or hypertension. Patients having hypertensive retinopathy do not exhibit any symptoms generally; however, some cases might report blurred or reduced clarity in vision and episodic headaches. This disorder can be characterized by abnormalities at points where arterioles and venules cross such as arteriolar narrowing, nicking and changes in arteriolar wall.

Retinal vein occlusion is the disorder that causes blockage of small veins that transport blood away from the retina. It is usually caused by the hardening of the arteries and due to a blood clot. The thickened veins cause pressure on the retinal vein and lead to blockage of smaller veins. This may lead to blurred vision or vision loss in eye.

The above-mentioned retinal vascular disorders underline the importance for using better tools to diagnose and identify retinal vascular diseases at an earlier stage to prevent any complications. Therefore, we require an algorithm that can effectively trace the retinal nerves accurately and at a faster rate.

2 Previous Works

Research work has been extensive in the field of blood vessel detection in retina for diagnosing glaucoma or other diseases. One such paper makes use of morphological operations mainly morphological opening followed by morphological dilation in order to identify blood vessels [1]. Another paper discusses an algorithm that adopts an approach based on integral channels and random forests [2]. Different approaches have been taken by researchers in order to obtain better and more accurate results over the years such as mathematical morphology [3, 4], model-based approaches [5, 6], pattern recognition techniques [7, 8], matched filtering [9], and various other techniques.

The problem with the existing algorithms is that the algorithms that make use of supervision or neural networks do achieve accuracy; however, they consume a lot of time and resource for processing and training the network. On the other hand, simpler algorithms do not achieve the desired level of accuracy and detect false veins and lead to incorrect diagnosis most of the time. Therefore, there is a need to develop an algorithm that is faster in computation as well as it achieves a desirable level of detection of blood vessels in the retina.

We have proposed an algorithm that makes use of Frangi filter technique; this method would yield better results than other techniques because it segments the image based on the likelihood of the region containing a vessel. Therefore, for our application of extracting blood vessels, Frangi filter technique would be apt.

3   Proposed Methodology

See Fig. 2.

The algorithm comprises of standard phases such as image acquisition, image denoising where the image acquisition process makes use of images obtained from fundus cameras and the denoising is carried out through means of median filtering in order to remove noises if any from the fundus images.

The frangi filter technique's principle follows the claim that vessels in an image constitute a small area of the total area being examined in the images and are either lighter than their background or darker than their background; this feature is the main principle behind the development of the Frangi filter technique. A Frangi filter function finds the likelihood that a pixel is a part of a vessel because this contrast of intensity values. The Hessian, a 2×2 matrix that consists of the partial derivatives of a function, is calculated at each pixel as follows:

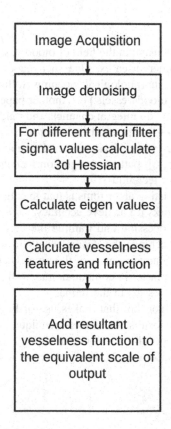

Fig. 2 Flowchart of proposed algorithm

$$H(x,y) = \begin{bmatrix} \frac{\partial^2 I}{\partial x^2} & \frac{\partial^2 I}{\partial x \partial y} \\ \frac{\partial^2 I}{\partial y \partial x} & \frac{\partial^2 I}{\partial y^2} \end{bmatrix} \tag{1}$$

The Hessian comprises of all the second-order information required by the image for each pixel. This implies that the Hessian is a discrete function. However, the Hessian can be approximated to a continuous function by making use of the Gaussian filter and the differentiation property of convolution in the following way:

$$H(x,y) \approx G * \begin{pmatrix} \frac{\partial^2 I}{\partial x^2} & \frac{\partial^2 I}{\partial x \partial y} \\ \frac{\partial^2 I}{\partial y \partial x} & \frac{\partial^2 I}{\partial y^2} \end{pmatrix} = \begin{pmatrix} \frac{\partial^2 G}{\partial x^2} & \frac{\partial^2 G}{\partial x \partial y} \\ \frac{\partial^2 G}{\partial y \partial x} & \frac{\partial^2 G}{\partial y^2} \end{pmatrix} * I(x,y) \tag{2}$$

Now, to extract obtain information from this matrix regarding the contrast and direction, we calculate the eigenvalues. Let us consider $|\lambda_1| \leq |\lambda_2|$ to be the two eigenvalues corresponding to the Hessian matrix and v_1 and v_2 be the corresponding eigenvectors. Since λ_1 is the eigenvalue having the smallest magnitude, it corresponds to the eigenvector v_1 that indicates the direction of the smallest curvature and λ_2 corresponds to the eigenvector v_2 that indicates the direction of largest curvature. In the perspective of the vessel pixels, v_1 points to the direction that the vessel is traveling and v_2 indicates the edge of the vessel.

We make use of the two parameters, namely anisotropy and contrast of the pixel to detect the vesselness of the pixels. The anisotropy will be low for vessel pixels, which means the lower anisotropy is more likely to be a vessel pixel. Contrast value will be low if both eigenvalues are small for the lack of contrast so that the larger contrast value implies to more probability of the pixel belonging to a vessel.

$$A = |\lambda_1|/|\lambda_2|$$
$$C = \sqrt{\lambda_1^2 + \lambda_2^2} \tag{3}$$

In images where the vessels' intensity is darker than their background, vessels are valleys and the curvature will be negative such that $\lambda_2 < 0$. Therefore, the vesselness equation is defined as

$$F = \begin{cases} 0 & \text{if } \lambda_2 > 0 \\ e^{\left(\frac{-A^2}{(2\alpha^2)}\right)} * \left(1 - e^{\left(\frac{C^2}{(2\beta^2)}\right)}\right) & \text{otherwise} \end{cases} \tag{4}$$

where α and β are variables to adjust anisotropy and contrast, respectively. If the pixel that is under consideration has a high probability of being a part of a vessel, then a high response value is designated to that pixel. This process is iterated for every pixel, and response values are the output.

4 Results

See Figs. 3, 4, 5, and 6.

After testing the algorithm using the Diaret DB and other databases [8], the above results are a representative for the set of results obtained. For the representative image, we obtained the image histogram, contour plot, and surface plots. We process the image using hessian matrix to get the partial derivatives of the image (Fig. 5). These derivatives are then processed using the vesselness equation to compute lambda similarities which would provide us an accurate representation of the segmented vessels (Fig. 7).

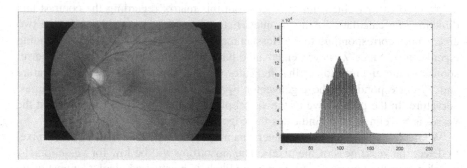

Fig. 3 Original fundus image (left) and image histogram (right)

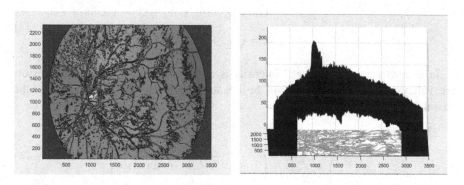

Fig. 4 Contour plot (left) and surf plot (right) of the given fundus image

(a)

(b)

(c)

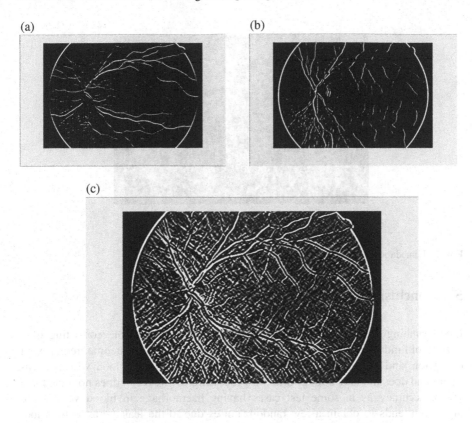

Fig. 5 Second-order derivatives of fundus image xx (**a**), xy (**b**), and yy (**c**)

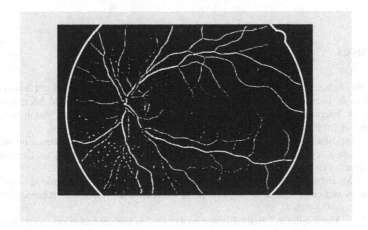

Fig. 6 Lambda similarity measure of image using λ_1 (*produces accurate detection of vessels*)

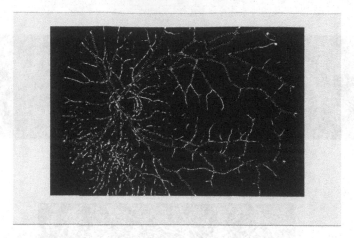

Fig. 7 Lambda similarity measure of the fundus image using λ_2

5 Conclusions

Upon running the algorithm against various databases available consisting of a variety of fundus images infected with diabetic retinopathy, glaucoma, retinal vein occlusion, and healthy eye images, the algorithm detects the blood vessels accurately and does not take a lot of time for processing. However, it does not guarantee 100% accuracy as in some test cases having haemorhage in blood vessels the algorithm tends to obtain a few random noises due to the leakage as vessels too. This can be fixed in future by making use of $L * a * b *$ color space conversion and color-based segmentation algorithms integrated into the Frangi filter for better detection of blood vessels.

References

1. N. Kaur, J. Kaur, M. Accharya and S. Gupta,: Automated detection of red lesions in the presence of blood vessels in retinal fundus images using morphological operations, 2016 IEEE 1st International Conference on Power Electronics, Intelligent Control and Energy Systems (ICPEICES), Delhi, 2016, pp. 1–4.
2. Zhun Fan, Jiewei Lu and Yibiao Rong,: Automated blood vessel segmentation of fundus images using region features of vessels, 2016 IEEE Symposium Series on Computational Intelligence (SSCI), Athens, 2016, pp. 1–6.
3. Fraz M M, Barman S A, Remagnino P.: An approach to localize retinal blood vessels using bit planes and centerline detection, Computer Methods and Programs in Biomedicine, 2012, 108(2), pp. 600–616.
4. Miri M S, Mahloojifar A,: Retinal image analysis using curvlet transform and multistructure elements morphology by reconstruction, IEEE Transactions on Biomedical Engineering, 2011, 58(5), pp. 1183–1192.

5. Al-Diri B, Hunter A, Steel D,: An active contour model for segmenting and measuring retinal vessels, IEEE Transactions on Medical Imaging, 2009, 28(9), pp. 1488–1497.
6. Lam B S Y, Gao Y, Liew A W C,: General retinal vessel segmentation using regularization-based multiconcavity modeling, IEEE Transactions on Medical Imaging, 2010, 29(7), pp. 1369–1381.
7. Soares J V B, Leandro J J G, Cesar Jr R M,: Retinal vessel segmentation using the 2-D gabor wavelet and supervised classification, IEEE Transactions on Medical Imaging, 2006, 25(9), pp. 1214–1222.
8. Staal J, Abrmoff M D, Niemeijer M,: Ridge-based vessel segmentation in color images of the retina, IEEE Transactions on Medical Imaging, 2004, 23(4), pp. 501–509.
9. Zhang B, Zhang L, Zhang L,: Retinal vessel extraction by matched filter with first-order derivative of Gaussian, Computers in biology and medicine, 2010, 40(4), pp. 438–445.
10. T. Balasubramanian, S. Krishnan, M. Mohanakrishnan, K. R. Rao, C. V. Kumar and K. Nirmala,: HOG feature based SVM classification of glaucomatous fundus image with extraction of blood vessels, 2016 IEEE Annual India Conference (INDICON), Bangalore, 2016, pp. 1–4.4
11. L. Shyam and G. S. Kumar,: Blood vessel segmentation in fundus images and detection of Glaucoma, 2016 International Conference on Communication Systems and Networks (ComNet), Thiruvananthapuram, 2016, pp. 34–38.
12. J. I. Orlando, E. Prokofyeva and M. B. Blaschko,: A Discriminatively Trained Fully Connected Conditional Random Field Model for Blood Vessel Segmentation in Fundus Images, in IEEE Transactions on Biomedical Engineering, vol. 64, no. 1, pp. 16–27, Jan. 2017.
13. J. Minar, M. Pinkava, K. Riha, M. K. Dutta, A. Singh and H. Tong,: Automatic extraction of blood vessels and veins using adaptive filters in Fundus image, 2016 39th International Conference on Telecommunications and Signal Processing (TSP), Vienna, 2016, pp. 546–549.
14. Z. Fan et al.,: Automated blood vessel segmentation in fundus image based on integral channel features and random forests, 2016 12th World Congress on Intelligent Control and Automation (WCICA), Guilin, 2016, pp. 2063–2068.
15. A. Budai, J. Hornegger, G. Michelson: Multiscale Approach for Blood Vessel Segmentation on Retinal Fundus Images. In Invest Ophthalmol Vis Sci 2009;50: E-Abstract 325, 2009.
16. A. Budai, J. Odstricilik, R. Kollar, J. Jan, T. Kubena, G. Michelson: A Public Database for the Evaluation of Fundus Image Segmentation Algorithms Poster at Fort Lauderdale Convention Center, The Association of Research in Vision and Ophthalmology (ARVO) Annual Meeting in Fort Lauderdale, FL, USA (02.05.2011)
17. J. Odstrcilik, J. Jan, R. Kolar, and J. Gazarek. Improvement of vessel segmentation by matched filtering in colour retinal images. In IFMBE Proceedings of World Congress on Medical Physics and Biomedical Engineering, pp. 327–330, 2009.
18. J. Odstrcilik, R. Kolar, A. Budai, J. Hornegger, J. Jan, J. Gazarek, T. Kubena, P. Cernosek, O. Svoboda, E. Angelopoulou, „Retinal vessel segmentation by improved matched filtering: evaluation on a new high-resolution fundus image database," IET Image Processing, Volume 7, Issue 4, June 2013, pp. 373–383.
19. Kauppi, T., Kalesnykiene, V., Kamarainen, J.-K., Lensu, L., Sorri, I., Raninen A., Voutilainen R., Uusitalo, H., Kälviäinen, H., Pietilä, J., DIARETDB1 diabetic retinopathy database and evaluation protocol

Headline and Column Segmentation in Printed Gurumukhi Script Newspapers

Rupinder Pal Kaur and Manish Kumar Jindal

Abstract Newspapers are vital source of information, and it is very much necessary to store newspapers in digital form. To search information from digital newspapers, text should be in computer processable form. To convert any newspaper into computer processable form, first step is to detect headline and segment the headline from body text. Next step is to segment columns if multiple columns are present in any article. In this paper, we have proposed a solution to segment headline from the body text and body text into columns. Experiments are carried out on Gurumukhi script newspaper article images.

Keywords Newspaper article · Headline segmentation · Column segmentation

1 Introduction

Optical character recognition (OCR) is the method to store huge volume of printed or handwritten data in digital form. A number of OCRs are developed in different languages to convert printed or handwritten data into digital and further into computer processable form. Efforts have been also started to archive old newspaper to preserve historical as well as the important information. One of the programs in this concern is National Digital Newspaper Program (NDNP) [1], which is running with collaboration of NEH and library of congress. This program provides historical newspapers since 1982 in digital form. In the last decades, newspapers were stored in only digital form but that was scanning the newspapers and storing on

R. P. Kaur (✉)
Department of Computer Science, Guru Nanak College for Girls, Sri Muktsar Sahib,
Punjab, India
e-mail: chatharupinder@yahoo.com

M. K. Jindal
Department of Computer Science and Applications, Panjab University Regional Centre,
Sri Muktsar Sahib, Punjab, India
e-mail: manishphd@rediffmail.com

© Springer Nature Singapore Pte Ltd. 2019   59
B. K. Panigrahi et al. (eds.), *Smart Innovations in Communication and
Computational Sciences*, Advances in Intelligent Systems and Computing 670,
https://doi.org/10.1007/978-981-10-8971-8_6

microfilms. But imagine the time to search a particular content from huge number of digital images. Solution to this problem is converting digital images through OCR. Digitized images of newspaper pages are analyzed with OCR software in order to produce text files of the newspaper content. At international level, many research papers are available for segmentation of newspaper images, but at national level, very few papers are available. There are around five renowned Gurumukhi script newspapers which are published daily, but only few of them are present in digital form. Till now, no OCR is developed for Gurumukhi script newspapers.

To store any newspaper in text form, it is necessary to segment newspaper article image into various blocks like headline, body text into columns, columns into lines, lines into words, words into characters and finally characters will be recognized. So the basic step is headline and column segmentation.

Much amount of work has been done to segment newspaper images of non-Indian script. Lam et al. [2] was the first author who experimented on newspaper article image. Authors used connected component analysis to segment at character level. Bottom-up approach is used for segmentation. Smearing and labeling technique is used by Gatos et al. [3] to segment newspaper image. Researchers decomposed full page of Greek newspaper with accuracy of 89.10%. Run length smearing algorithm was used by Lie et al. [4] to segment the headline. RLSA works well for segmentation into blocks, but there should be no overlapping in blocks or in lines of columns for good segmentation results. If there is overlapping of lines as commonly found in newspaper text due to poor printing quality, then this technique produces wrong segmentation results. Hadjar and Ingold [5] used combination of connected component analysis and RLSA techniques to segment the Arabic newspaper image. At national level, one technique to segment English newspaper article image is proposed by Bansal [6]. Researchers label each node (block) based on features of node like appearance and contextual features. Appearance features try to associate each block to label using characteristics of that block. This way, all blocks of image are segmented. Another paper available for segmentation of text and graphics in Indian newspapers is by authors Garg et al. [7]. Researchers have proposed techniques for segmentation of related images that are embedded into article from text of the article. Only text graphics segmentation is performed in the paper. Various problems in making OCR of newspaper articles are discussed by Kaur and Jindal [8] Researchers have studied out the problems particularly in Gurumukhi script newspapers. However, these problems can be faced in newspapers of any script. Some region separation techniques are also explained by Ray et al. [9]. A technique is also proposed by Omee et al. [10] to segment headline and columns in Bangla newspaper image. Accuracy achieved by researchers is about 70.2%. To the best of our knowledge, none of these techniques or any technique have been tried to implement on Gurumukhi script newspapers to segment headline and columns.

In this paper, we have used white pitch method to segment the Gurumukhi script newspaper article. Newspapers are scanned from old available papers. For the purpose, articles are manually cropped from scanned image of newspaper image. White pitch method has been implemented to segment the article image. Final outputs of our proposed method are segmented headline portion of articles and inter-column segmentation of body text.

2   Basic Steps in Segmentation of Newspaper Article Image

- *Data acquisition*
- *Image Preprocessing (enhancement, noise removal etc.)*
- *Page layout analysis*
- *Segmentation of images into different blocks*

A. *Data acquisition*: We have collected article images from old newspapers available from various newspaper headquarters and persons who have stored the newspapers. Scanning is done on 300 dpi flat bed scanner.

B. *Preprocessing*: Preprocessing is about removing the unwanted noise that may lead to false segmentation or enhancing the image features which may help in further recognition. Binarization is important step of preprocessing. In binarization, scanned image is converted into bitmap image. We have used radon transform (if required) to correct the skew angel. Median filter is used for noise removal.

C. *Page Layout Analysis*: Page layout analysis is segmenting textual regions from non-textual regions. This phase is important if article contains images describing the event or any other graphics like maps. Removing extra black lines surrounding the article is also part of page layout analysis. In this paper, we have considered only textual article. Black lines surrounding the articles are removed manually.

After page layout analysis, block segmentation is performed. For segmentation, many techniques are proposed in literature, but none of the technique can be directly implemented on Gurumukhi script article image. So, we have proposed a solution for headline and column segmentation.

3   Segmentation of Headline and Columns Using Proposed Method

We have worked on the idea of gap between headline and body text and inter-column gap. It is very common that horizontal white space in headline(s) and body text is larger than column inter-line spaces. If the gap increases from certain threshold, then that point will act as candidate of segmentation point. Similarly, continuous vertical white lines exist in columns, that point will be considered as inter-column segmentation point. Figure 1 shows gap in headline and body text and column inter-line gap.

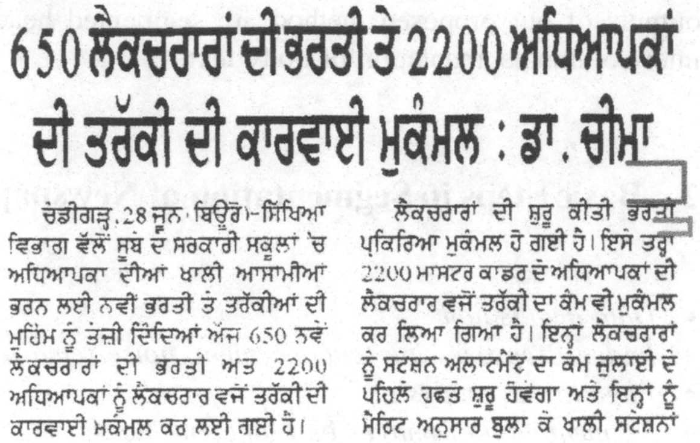

650 ਲੈਕਚਰਾਰਾਂ ਦੀ ਭਰਤੀ ਤੇ 2200 ਅਧਿਆਪਕਾਂ ਦੀ ਤਰੱਕੀ ਦੀ ਕਾਰਵਾਈ ਮੁਕੰਮਲ : ਡਾ. ਚੀਮਾ

ਚੰਡੀਗੜ੍ਹ, 28 ਜੂਨ (ਬਿਊਰੋ)-ਸਿੱਖਿਆ ਵਿਭਾਗ ਵੱਲੋਂ ਸੂਬੇ ਦੇ ਸਰਕਾਰੀ ਸਕੂਲਾਂ 'ਚ ਅਧਿਆਪਕਾਂ ਦੀਆਂ ਖਾਲੀ ਆਸਾਮੀਆਂ ਭਰਨ ਲਈ ਨਵੀਂ ਭਰਤੀ ਤੇ ਤਰੱਕੀਆਂ ਦੀ ਮੁਹਿੰਮ ਨੂੰ ਤੇਜ਼ੀ ਦਿੰਦਿਆਂ ਅੱਜ 650 ਨਵੇਂ ਲੈਕਚਰਾਰਾਂ ਦੀ ਭਰਤੀ ਅਤੇ 2200 ਅਧਿਆਪਕਾਂ ਨੂੰ ਲੈਕਚਰਾਰ ਵਜੋਂ ਤਰੱਕੀ ਦੀ ਕਾਰਵਾਈ ਮੁਕੰਮਲ ਕਰ ਲਈ ਗਈ ਹੈ।

ਲੈਕਚਰਾਰਾਂ ਦੀ ਸ਼ੁਰੂ ਕੀਤੀ ਭਰਤੀ ਪ੍ਰਕਿਰਿਆ ਮੁਕੰਮਲ ਹੋ ਗਈ ਹੈ। ਇਸੇ ਤਰ੍ਹਾਂ 2200 ਮਾਸਟਰ ਕਾਡਰ ਦੇ ਅਧਿਆਪਕਾਂ ਦੀ ਲੈਕਚਰਾਰ ਵਜੋਂ ਤਰੱਕੀ ਦਾ ਕੰਮ ਵੀ ਮੁਕੰਮਲ ਕਰ ਲਿਆ ਗਿਆ ਹੈ। ਇਨ੍ਹਾਂ ਲੈਕਚਰਾਰਾਂ ਨੂੰ ਸਟੇਸ਼ਨ ਅਲਾਟਮੈਂਟ ਦਾ ਕੰਮ ਜੁਲਾਈ ਦੇ ਪਹਿਲੇ ਹਫਤੇ ਸ਼ੁਰੂ ਹੋਵੇਗਾ ਅਤੇ ਇਨ੍ਹਾਂ ਨੂੰ ਮੈਰਿਟ ਅਨੁਸਾਰ ਬੁਲਾ ਕੇ ਖਾਲੀ ਸਟੇਸ਼ਨਾਂ

Fig. 1 Red bracket shows space in headlines and body text, and green bracket shows inter-line space

The steps of proposed algorithm are described below:

Step 1: Load the scanned image of article and binarized it (store in bitmap form in 2-D array).

Step 2: Fix a threshold value for minimum space between headline and body text and also fix a threshold value for inter-column gap. Threshold values are fixed after though experimentation on images.

Step 3: Scan the image from top to bottom to find horizontal consecutive white lines. Set starting point (y_0, x_0). y donate to row and x donate to column.

Step 3.1: Scan every row to find a horizontal line containing all white pixels. Store position of white line in a variable and call it *pos1*.

Step 3.2: Continue for scanning white lines. Where consecutive white line ends, store position of that line in a variable called pos2.

Step 3.3: If count (*pos2-pos1*) of horizontal consecutive white lines is greater than fixed threshold value, mark that position as segmentation point.

//headline segmentation is complete

Step 4: Cut the headline portion from starting point (y_0, x_0) to the point (y_1, x_1) where value increases from threshold value and consecutive horizontal white lines end.

Step 5: To segment body text into columns, set row starting point equal to ($y_1 + 1$, x_0).

Step 6: Scan image column wise from left to right to find vertical consecutive white lines to segment into columns.

Step 6.1: Store position of white line found in a variable. Call it *poisition1*. Continue the scanning to find more white lines.

Step 6.2: Where consecutive white lines ends, store position of that line into a variable. Call it *position2*.

Step 6.3: If count (*position2–position1*) of vertical consecutive white lines is greater than fixed threshold value, mark that position as column segmentation point.

Step 6.4: Cut the column portion from starting point of column that is $(y_1 + 1, x_0)$ to point $(y_{h-1}, x_{position1})$ where value increases from threshold value and consecutive vertical white lines end. h is height of image.

Step 7: Now set starting point equal to $(y_1 + 1, x_{position2})$ for next column segmentation. Repeat step 6 for segmentation and continue until the scanning of image reached to width of image.

//end of column segmentation and algorithm

Benefits of implementing this algorithm are that we do not have to load sub-image every time for further segmentation. Proposed algorithm can bear slight skew in image if space in headline and body text is large. This algorithm works well on single column, two columns as well as on multicolumn documents.

4 Results

To test the accuracy of proposed algorithm, we scanned newspapers on flat bed 300 dpi scanner (scanning less than 300 dpi can cause more noise in scanned images which can increase the difficulties in segmentation) and manually cropped the article images. We have used visual C++ for coding the algorithm. Algorithm is tested on single column and multicolumn article images. Results are very encouraging.

Figure 2 shows bitmap image of newspaper scanned articles image containing headline and two columns. Following figures show output of proposed algorithm. Figure 3 shows the segmentations point of headline and columns. Lines in green color show segmentation points of headline from body text. Lines in red color show column segmentation. Figure 4 shows segmented headline portion of article image. Figures 5 and 6 show segmented portions of first and second column, respectively.

We have also implemented this algorithm on multicolumn documents and single-column documents. Figure 7 shows a bitmap image of multicolumn

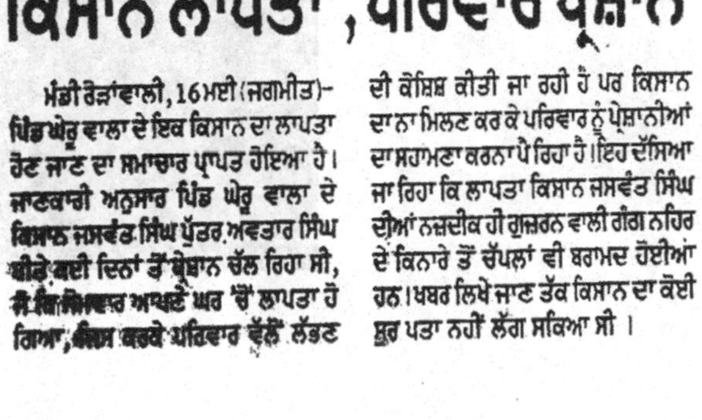

Fig. 2 Bitmap article image of scanned article

Fig. 3 Segmentation points of headline and columns (red lines showing segmentation point of headline and body text, and green lines showing segmentation points in columns)

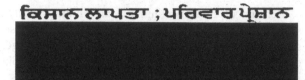

Fig. 4 Segmented headline of article image from body text shown in Fig. 3

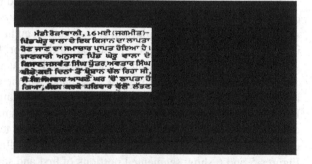

Fig. 5 First column segmented from rest of columns (article shown in Fig. 3)

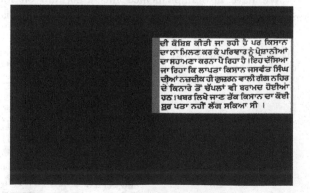

Fig. 6 Seccond (left) column segmented from rest of image (article image shown in Fig. 3)

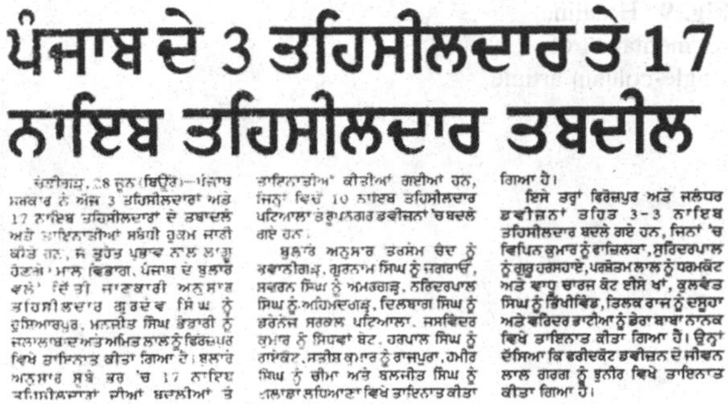
ਪੰਜਾਬ ਦੇ 3 ਤਹਿਸੀਲਦਾਰ ਤੇ 17
ਨਾਇਬ ਤਹਿਸੀਲਦਾਰ ਤਬਦੀਲ

Fig. 7 A sample of multicolumn article image

document containing headlines and three columns. Figure 8 shows segmentation points of headlines from body text and inter-column segmentation. Figure 9 shows single-column article headline segmentation.

Around 30 single and multicolumn article images are used to test the algorithm. These images are scanned from three different newspapers that are Ajit, Jagbani, and Punjabi Tribune. All the three newspapers differ slightly in formatting, but this algorithm works well on all three types of newspaper images with slight variation in threshold value that is fixed for space in headline and body text and inter-column spacing. Among the 30 images, 24 images were segmented correctly. Sometimes space in headlines and body text or inter-column spacing is less than threshold value which can cause wrong segmentation results. Figure 10 shows wrong segmentation results because space in headlines and body text is less than fixed threshold value.

ਪੰਜਾਬ ਦੇ 3 ਤਹਿਸੀਲਦਾਰ ਤੇ 17
ਨਾਇਬ ਤਹਿਸੀਲਦਾਰ ਤਬਦੀਲ

Fig. 8 Segmentation point of multicolumn document (green color showing headline segmentation points, and red color showing column segmentation)

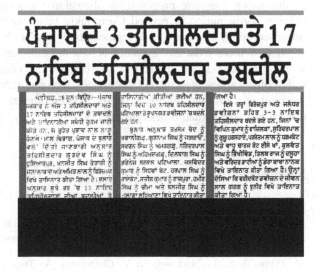

Fig. 9 Headline
segmentation of
single-column article

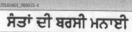

Fig. 10 Wrong segmentation
results

5   Conclusion

In this paper, we have proposed an algorithm for segmentation of headline and
column segmentation in printed Gurumukhi script newspaper articles. Although
proposed solution gives good segmentation results, but this algorithm fails to
segment if there exist any graphics across the columns in the article. Algorithm
works on fixed space threshold, so if space in headline and body text and
inter-column space is less than fixed spaces, then algorithm will not perform correct
segmentation. Our next efforts would be to overcome the limitations and to separate
textual and non-textual blocks in article.

References

1. https://www.loc.gov/ndnp/.
2. Lam, Stephen W., Dacheng Wang, and Sargur N. Srihari.: Reading newspaper text, Pattern
   Recognition: Proceedings of 10th International Conference on document analysis and
   recognition, Vol. 1, pp. 703–705 IEEE (1990).
3. Gatos, B. S. L. Mantzaris, K. V. Chandrinos, A. Tsigris, S. J. Parintonis.: Integrated
   algorithms for newspaper page decomposition and article tracking: Proceedings of the Fifth

International Conference on Document Analysis and Recognition. IEEE Computer Society, pp. 559–562. (1999).

4. Liu, Qing Hong, and Chew Lim Tan: Newspaper headlines extraction from microfilm images: International journal on Document Analysis and recognition, Vol. 6, pp 201–210 (2004).

5. Hadjar, Karim, and Rolf Ingold.: Arabic newspaper page segmentation: 12th International Conference on Document Analysis and Recognition: pp. 1186–1189, Vol. 2, IEEE Computer Society (2003).

6. Bansal, Anukriti.: Newspaper article extraction using hierarchical fixed point model: Document Analysis Systems (DAS), 11th IAPR International Workshop on IEEE, pp. 257–261 (2014).

7. Garg, Ritu, et al. "Text graphic separation in Indian newspapers." Proceedings of the 4th International Workshop on Multilingual OCR, p. 13, ACM (2013).

8. Kaur, Rupinderpal, and Manish Kumar Jindal: Problems in Making OCR of Gurumukhi Script Newspapers: International Journal of Advanced Research in Computer Science, pp. 16–22, vol. 7 issue 6 (2016).

9. Ray Chaudhuri, A., Mandal, A.K., Chaudhuri, B.B.: Page layout analyzer for multilingual Indian documents: In: Proceedings of the Language Engineering Conference, pp. 24–32 IEEE (2002),

10. Omee, Farjana Yeasmin, Md Shiam Shabbir Himel, and Md Abu Naser Bikas.: An Algorithm for headline and column separation in bangla documents: Intelligent Informatics. ASCI 182, pp. 307–315, Springer Berlin Heidelberg, (2013).

An Innovative Technique Toward the Recognition of Carcinoma Using Classification and Regression Technique

M. Sangeetha, N. K. Karthikeyan and P. Tamijeselvy

Abstract Usually, the nature of a human body is that the cells start to develop, grow, live for some time, and die after a certain period of time. This phenomenon proceeds until the life span. But instead of this normal phenomenon, if the cells grow abundantly, then they end in carcinoma. Cancer cells destroy the entire functioning of the body. If cancer is found in the preliminary stage, then the extent of life span has been guaranteed to some extent. Existing categorization techniques do not provide exact classification results. Here, this work used the standard leukemia dataset for categorizing the cancer cells and non-cancer cells. Missed values are replaced through enhanced independent component analysis method. Further best features are selected through geometric particle swarm optimization. As a last step, cancer cells are categorized through classification and regression technique.

Keywords Bi-clustering Bayesian principal component analysis (Bi-BPCA)
Decision tree · Classification and regression technique

M. Sangeetha (✉)
IT Department, Sri Krishna College of Technology, Coimbatore 641042,
Tamil Nadu, India
e-mail: godsan2003@gmail.com

N. K. Karthikeyan
CSE & IT Department, Coimbatore Institute of Technology, Coimbatore 641014,
Tamil Nadu, India
e-mail: karthiaish1966@gmail.com

P. Tamijeselvy
CSE Department, Sri Krishna College of Technology, Coimbatore 641042,
Tamil Nadu, India
e-mail: p.tamijeselvy@skct.edu.in

© Springer Nature Singapore Pte Ltd. 2019
B. K. Panigrahi et al. (eds.), *Smart Innovations in Communication and Computational Sciences*, Advances in Intelligent Systems and Computing 670,
https://doi.org/10.1007/978-981-10-8971-8_7

1 Introduction

Cancer is the outcome of the development of unusual cells. Carcinoma is also known as cancer or malignant tumor. Generally, cancer may occur due to the presence of viruses. The viruses may be classified as RNA virus or DNA virus. The patients suffering from carcinoma have to undergo the medication of chemotherapy and radiotherapy that may spoil the normal functioning of the human body gradually. In fact, carcinoma can be curable if found in the preliminary stage. In clinics, a biopsy test has been made where a specific amount of tissues are eliminated from the tumor and handed over for a laboratory test. In this work, the leukemia dataset (www.gems-system.org) reported by Golub et al. [1] has been used. Gene expression has been taken from this dataset. Microarray technique is the best method for analyzing gene expression. Initially, the dataset may have missing values. Missed values are replaced through bi-clustering Bayesian principal component analysis. Duplicate values are removed through enhanced independent component analysis. To examine the data, feature selection has been used to reduce the dimensionality of the data. To categorize the gene as either cancer gene or normal gene, different classification algorithms such as support vector machine (SVM), K-nearest neighbor (KNN), Naive Bayesian (NB) have been used. In this proposed work, the gene has been classified through classification and regression tree (CART).

2 Related Works

Data mining has the great potential to handle the healthcare systems. Thus, data mining plays a significant part in reducing the cost. This work illustrates about different categorization methods that have been utilized in the clinical field. It specifies that K-nearest neighbor is the smooth classifier, decision tree is the famous technique that aids to select the other substitute techniques, support vector machine yields the correct result, neural network has been referred as the perfect categorization algorithm; it has the leading elements as neurons, but this method is costly. For large datasets, Bayesian approach functions fast and provides the exact result [2]. Healthcare and therapeutic fields possess the huge amount of data, but they are not accurately utilized for making decisions. This paper aims at the unmined data, and they use Naive Bayesian (NB) and K-means approach for forecasting the heart attack [3]. In the current atmosphere, commonly widening disease is swine flu [4]. This work estimates the swine flu by Naïve Bayesian method through particular parameters. As the technique is flexible to interpret, NB is preferred as the choice. In the current scenario, abundant information is shrouded in the database [5]. The information has been retrieved from the database by different data extracting methods. If the mining process is exactly correct, then the clinics will yield benefit by the clear decision making. The neural network is opted for interpreting the dilemma of data mining because of its characteristics such as high degree of fault tolerance, good robustness, parallel processing, distributed storage [6]. This paper illustrates a literature survey of different classification methods. Naïve Bayesian is the smooth classifier approach that utilizes Bayesian theorem. Support vector machine is famously known for quick and exact results. K-nearest neighbor is primarily used for text categorization and pattern recognition. Neural network harvests a communication within input and output [7]. Lupus is a disorder that can modify any portion of the body. It is an incurable disease which may be due to genetic factors or the environment. In this paper, ID3 approach has been utilized to identify the disorder in the preliminary stage [8]. Data mining methods are utilized to increase the essence of prognostication of disease. Different methods of data mining have been utilized to identify the similar patterns [9]. The process of identifying the hidden information from a large amount of data has been referred as data mining. ID3 algorithm and neural network have been used to identify the presence of any general disorder [10]. In this paper, microarray genetic data have been used to categorize the cancer genes. Here, genes are ranked based on the distance between them. Then, the maximal gene subset has been chosen by applying cross-validation. In such a way, the classification model has been built [11]. In this paper, various methods have been used for reducing the dimensions. They have used continuous wavelet transform for extracting the features. In addition to that, they have used fast correlation-based filter for selecting the features.

3 Dataset

In this work, the Leukemia dataset (www.gems-system.org) reported by Golub et al. [1] has been used. The dataset consists of details of more than 40 patients with acute lymphoblastic leukemia (ALL) and more than 22 patients with acute myeloid leukemia (AML). The bone marrow samples of more than 70 patients obtained at the time of leukemia diagnosis, containing 5328 gene expressions, are used.

The rest of the paper is organized as follows: Section 4 explains the proposed method that specifies about EICA, Bi-BPCA, GPSO, and CART algorithm. Section 5 includes the discussion of experimental methods, and conclusion is given in Sect. 6.

4 Proposed Method

The primary goal of this proposed work for classifying leukemia gene expression is to develop a method for the intention of leukemia cancer cell investigation utilizing gene classification method with the process of feature selection. There are multiple approaches that identify the cancer cell investigation in an automatic way. The proposed leukemia gene classification process makes use of enhanced independent component analysis (EICA) method and Bi-PCA with geometric particle swarm optimization (GPSO)-based feature selection and classification. The proposed method has to utilize the following steps for cancer cells and non-cancer cells classification process with the use of leukemia dataset; they are preprocessing, feature selection, and classification methods. Initially, the given dataset is preprocessed for the purpose of efficient final classification result and missing value calculation and duplication elimination, which will be done with the assistance of the newly enhanced independent component analysis process. Once the preprocessing is completed, the analyzed dataset is used for feature selection process, in this stage, the most important features are selected with the assistance of geometric particle swarm optimization (GPSO), then the selected features are used for classification process, this process will be done with the assistance of classifier. The proposed classifier, CART, will be utilized to classify cancer cells and non-cancer cells based on the gene microarrays. At last, the given dataset is subjected to the proposed technique to assess the performance in classifying cancer cells or non-cancer cells from the leukemia dataset. In this proposed method, for experimentation, the dataset will be subjected to analyze the performance of the proposed approach utilizing classification accuracy, F-score sensitivity, and specificity.

4.1 Preprocessing

The standard dataset may have missing values or noisy data. These data are to be analyzed using data extracting techniques. The data have to be preprocessed to enhance the accuracy of the result. Data preprocessing is an essential step in the data mining process. In this work, the preprocessing techniques such as bi-clustering Bayesian principal component analysis (Bi-BPCA) and enhanced independent component analysis (EICA) are discussed and implemented. Initially, the dataset may have missing values and duplicate values. Bi-clustering Bayesian principal component analysis has been used to find out the missing values. The duplicate values have been eliminated with the help of enhanced independent component analysis.

4.1.1 Bi-cluster Principal Component Analysis (Bi-PCA)

Bi-clustering Bayesian principal component analysis (Bi-BPCA) is used for examining enormous amount of microarray data, eliminating unwanted component in the data, and also measuring the missing values of leukemia gene dataset. In this method, two clusters are utilized to deal with the local matrix. Gene expression is represented as a microarray. Rows consist of genes, and columns consist of experimental conditions. Rows and columns with missing values are selected to calculate the missing entry of the dataset. Based on the eigenvalues and eigenvectors, missing entries can be found.

4.1.2 Enhanced Independent Component Analysis (EICA)

Enhanced independent component analysis (EICA) is a statistical approach and is found to be very effective in data mining. It is capable of retrieving biotic-related gene features from a microarray. It has demanded much interest in the current years because of its broad applications. It is used for dimensionality reduction. Each component relates to a gene signal. Rows consist of detected matrix, and columns include some samples. By calculating the average error square of samples, the duplicate data are found.

4.2 Feature Selection Using Geometric Particle Swarm Optimization

Feature selection is also known as attribute selection or variable selection or variable sub-selection. One significant goal of feature selection is to simplify the models as they can be easily interpreted by researchers or users. The data may have large features. These features may be either similar or dissimilar. Feature selection aims to eliminate the unnecessary features that are not needed by the users. For selecting the best features, the following algorithms have been used: particle swarm optimization (PSO) and geometric particle swarm optimization (GPSO). In particle swarm optimization (PSO), the swarm consists of a finite count of particles. The particles are indicated as gene subsets. For each gene subset, the fitness value is calculated. The fitness value is updated, if the current fitness value is better than the former. This process is repeated until the best features are found. At every repetition, the particles shift in the space to identify the global optima. Each particle has space and velocity vectors for instructing the movement of the gene subsets. Geometric particle swarm optimization (GPSO) is proposed to overcome the shortcomings in PSO and also to enhance the selection of genes. The prime objective of geometric particle swarm optimization (GPSO) is to choose the best genes for attaining high prediction accuracy. For each particle, objective function takes part in the operation. For each particle, the current, the global/local best, and the previous best positions are examined. Finally, particles are shifted using three-parent crossover operators. This process is repeated until the final condition matches with a particular number.

4.3 Classification Using CART

Classification deals with allotting a class label to a group of unclassified cases. It is a data mining technique that is referred as decision tree (DT). It is used in the investigation of an enormous amount of data in an efficient manner. In classification and regression tree (CART), data are represented in the form of trees. The goal is to classify outcome based on the predictors. In this tree model, the leaves indicate the class labels and branches denote conjunctions. This model is not only easy to understand but also easy to interpret. They can handle both categorical and numerical data. This model performs well with massive datasets. Large volume of data can be analyzed in a shorter interval of time. By calculating the Gini index, healthy cells and cancer cells can be classified. In the CART investigation, let factors $u_1, u_2, u_3, \ldots, u_n$ be in the domain. It predicts the outcome of interest. It is an appropriate method, where the data space is divided into shorter portions which depict the precise data interactions. CART analysis uses this recursive partitioning to create a tree where each node indicates a partition cell. The nodes A, B, and C are the end nodes, where further splitting is not possible.

As in regression equations, the criterion for choosing "$d(u)$" will be the mean squared prediction error $\{d(u) - E(v|u)\}2$ or expected misclassification cost in the case of the classification tree. Sample mean is calculated using Eq. (1).

$$\hat{v} = \frac{1}{x} \sum_{x=1}^{x} v_1 \tag{1}$$

4.3.1 Leukemia Gene Classification Using CART

CART analysis by Loh (2011) relies on binary decision trees. For leukemia gene classification in cancer diagnosis system, it takes the selected gene features known as variables "x" from GPSO as input. This proposed CART-based method splits a single variable at each node "l". This approach can also produce classification trees, which depends on the type of the dependent variable (categorical or numerical value of microarray data).

With the use of the Gini index, splitting criteria is calculated as shown in Eq. (2)

$$\text{Gini } (G) = \left(1 - \sum_{x=1}^{X} p^2(x|l) \right) \tag{2}$$

Algorithm of CART

The algorithm for classification and regression technique is given below:

Step 1: Initialize gene dataset.
Step 2: Consider all possible values of all gene features.

Step 3: Select the gene feature $x = v_1$ that produces the greatest "separation" in the target, where $x = v_1$ is called a "split."

Step 4: If $x < v_1$, then send the data to the left node; otherwise, send data point to the right node.

Step 5: Repeat the same process until getting the classification result.

Step 6: To apply splitting rules

6.1 Select the feature value $x = v_1$ that produces the greatest "separation" in the target feature.

6.2 Perform "separation" using the Gini index as split criteria.

Step 7: Calculate the Gini index to classify pure nodes using Eq. (2).

Step 8: End.

The classification is performed in MATLAB 8.1 environment. There are two steps in classification, such as training and testing. In training process, 70% (50 samples) of the dataset is sent to the classifier, and remaining 30% (12 samples) of the dataset is sent as input to the testing phase. Based on the dataset entry and the class labels, the classifier model is created. From the testing phase, the output will be the predicted class labels. The actual and predicted class labels are compared to get the final output. The results of the existing methods KNN and NB and the proposed method CART in the presence and absence of GPSO are shown.

5 Experimental Results and Conclusion

The results are evaluated using the parameters such as sensitivity, specificity, and classification accuracy.

Sensitivity (S): Sensitivity is the percentage of predicted and actual classes which belong to the current cases as correctly identified and determined using Eq. (3).

$$\text{Sensitivity } (S) = \frac{T_1}{T_1 + T_2} \tag{3}$$

= (Number of True Positive Assessment)/(Number of all Positive Assessment)

Specificity (Sp): Specificity is defined as the percentage of predicted and actual classes which belong to negative cases that were correctly identified, as determined using Eq. (4).

$$\text{Specificity } (Sp) = \frac{F_2}{F_1 + F_2} \tag{4}$$

= (Number of True Negative Assessment)/(Number of all Negative Assessment)

Classification accuracy (**A**): Classification accuracy is the percentage of the total amount of predictions. It can be either positive or negative as correctly identified and determined using Eq. (5).

$$\text{Accuracy}\,(A) = \frac{T_1 + F_2}{T_1 + T_2 + F_1 + F_2} \tag{5}$$

= (Number of True Assessments)/Number of all Assessments)

The sensitivity rate of the methods KNN, NB, and CART in the presence and absence of GPSO is shown in Table 1. Figure 1 shows that the sensitivity rate increases when the feature selection method has been included. This enhanced sensitivity rate indicates that these classifiers perform well in the presence of geometric particle swarm optimization.

The specificity rate of KNN, NB, and CART in the presence and absence of GPSO is shown in Table 2. From Fig. 2, it is evident that the specificity rate increases when feature selection has been included. This enhanced specificity rate indicates that these classifiers perform well in the presence of geometric particle swarm optimization.

Table 1 Sensitivity of KNN, NB, and CART

Methods	Sensitivity rate (%)	
	Without GPSO	With GPSO
KNN	30	40
NB	39	66
CART	50	81

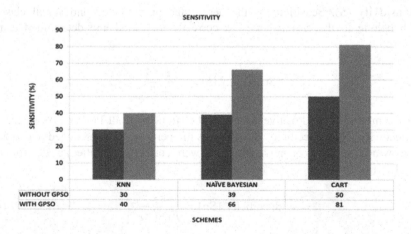

Fig. 1 Comparison of sensitivity of KNN, NB, and CART in the presence and absence of GPSO

Table 2 Specificity of KNN, NB, and CART

Methods	Specificity rate (%)	
	Without GPSO	With GPSO
KNN	45	65
Naïve Bayesian	80	88
CART	85	90

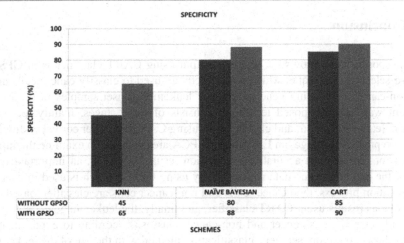

Fig. 2 Comparison of specificity of KNN, NB, and CART in the presence and absence of GPSO

Table 3 Classification accuracy of KNN, NB, and CART

Methods	Classification accuracy (%)	
	Without GPSO	With GPSO
KNN	42	81
Naïve Bayesian	45	90
CART	50	93

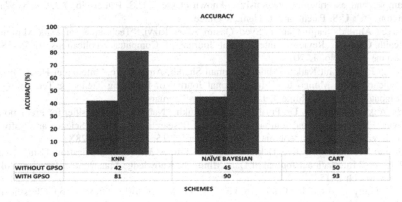

Fig. 3 Comparison of classification accuracy of KNN, NB, and CART in the presence and absence of GPSO

The classification accuracy of KNN, NB, and CART with and without GPSO is shown in Table 3. Figure 3 shows that the classification accuracy increases when feature selection has been included. This enhanced classification accuracy rate indicates that these classifiers perform well in the presence of geometric particle swarm optimization.

6 Conclusion

This proposed work shows the efficient output using CART classifier with GPSO feature selection technique, which was effectively used to classify cancer cells and the non-cancer cells using gene microarray leukemia dataset samples in the most efficient way. The proposed technique consists of three stages, mainly, preprocessing, feature selection, and classification using CART classifier correspondingly. In the preprocessing stage, an EICA and Bi-PCA are proposed to examine the input dataset for further most accurate classification result. After that, an important feature of the leukemia gene data is selected by using GPSO, which is used in cancer classification process, then cancer cells and non-cancer cells are classified based on the previous results using CART classifier, and finally, the leukemia gene dataset is efficiently classified as cancer and non-cancer classes. According to experimental results shown, the proposed gene classification method with the use of the leukemia dataset efficiently classifies the normal and abnormal classes. In future, it will be focused on an additional feature extraction process for the most accurate classification results in the proficient early cancer diagnosis.

References

1. Golub, TR, Lander, ES, Mesirov, J, Slonim, D, & Tamayo, P 2007, 'Methods for classifying samples and ascertaining previously unknown classes', U.S. Patent No. 7,239,986, Washington, DC: U.S. Patent and Trademark Office.
2. Parvez Ahmad, Saqib Qamar, Syed Qasim Afser Rizvi, "Techniques of Data Mining in Health Care—A Review", International Journal of Computer Applications (0975–8887) Volume 120—No. 1, 2015.
3. Akash Jarad, Rohit Katkar, Abdul Rehaman Shaikh, Anup Salve, "Intelligent Heart Disease Prediction With Mongo DB", International Journal of Emerging Trends & Technology in Computer Science, Volume 4, Issue 1, January–February 2015.
4. Ms. Ankita R. Borkar*, Dr. Prashant R. Deshmukh, "Naïve Bayes Classifier for Prediction of Swine Flu", International Journal of Advanced Research in Computer Science and Software Engineering, Page 120, Volume 5, Issue 4, April 2015, ISSN: 2277 128X.
5. Gaurab Tewary, "Effective Data Mining For Proper Mining Classification Using Neural Networks" International Journal of Data Mining & Knowledge Management Process, Vol. 5, No. 2, March 2015.
6. Patel Pinky, Raksha R. Patel, Antika J. Patel, Maitri Joshi, "Review on Classification Algorithms in Data Mining", International Journal of Emerging Technology and Advanced Engineering, Vol. 5, Issue 1, Jan 2015.

7. Gomathi S, Dr. V. Narayani, "A Data Mining Classification Approach to Predict Systemic Lupus Erythematosus using ID3 Algorithm" International Journal of Advanced Research in Computer Science and Software Engineering, Volume 4, Issue 3, March 2014.
8. Mohammed Abdul Khaleel, Satheesh Kumar, Pradham, G.N. Dash, "A Survey of Data Mining Techniques on Medical Data For Finding Locally Frequent Diseases", International Journal of Advanced Research in Computer Science and Software Engineering, Volume 3, Issue 8, August 2013 ISSN: 2277 128X.
9. L. Sathish. A. Padmapriya, "ID3 algorithm performance of Diagnosis for Common Disease", International Journal of Advanced Research in Computer Science and Software Engineering, Vol. 2, Issue 5, May 2012. ISSN: 2277 128X.
10. Wenyan Zhong, Xuewen Lu & Jingjing Wu, "Feature Selection for Cancer Classification using Microarray Gene Expression Data", Biostatistics & Biometrics, Vol. 1, Issue 2, April 2017.
11. Mohammed Kazem Ebrahimpour, Hamid Mirvaziri, Vahid Sattari-Naeini, "Improving Breast Cancer Classified by Dimensional Reduction on Mammograms", Computational Methods in Biomechanics & Biomedical Engineering: Imaging & Visualization, Article 6, pp. 1–11, Published online.

Model Order Reduction Using Fuzzy C-Means Clustering and Particle Swarm Optimization

Nitin Singh, Niraj K. Choudhary, Rudar K. Gautam
and Shailesh Tiwari

Abstract The hybrid method which combines the evolutionary programming technique, i.e., based on the swarm optimization algorithm and fuzzy c-means clustering method is used for reducing the model order of high-order linear time-invariant systems in the presented work. The process of clustering is used for finding the group of objects with similar nature that can be differentiated from the other dissimilar objects. The reduction of the numerator of original high-order model is done using the particle swarm optimization algorithm, and fuzzy c-means clustering technique is used for reducing the denominator of the higher-order model. The stability of the model is also verified using the pole zero stability analysis, and it was found that the obtained reduced-order model is stable. Further, the transient and steady state response of the obtained lower-order model as compared to the other existing techniques are better. The output of the obtained lower-order model is also compared with the other existing techniques in the literature in terms of ISE, ITSE, IAE, and ITΛE.

Keywords Model order reduction · Fuzzy c-means clustering
Particle swarm optimization · Pole clustering · Fuzzy logic

N. Singh (✉) · N. K. Choudhary · R. K. Gautam
Department of Electrical Engineering, MNNIT Allahabad,
Allahabad, Uttar Pradesh, India
e-mail: nitins@mnnit.ac.in

N. K. Choudhary
e-mail: niraj@mnnit.ac.in

R. K. Gautam
e-mail: gautamrudra03@gmail.com

S. Tiwari
ABES Engineering College, Ghaziabad, Uttar Pradesh, India
e-mail: shail.tiwari@yahoo.com

© Springer Nature Singapore Pte Ltd. 2019   81
B. K. Panigrahi et al. (eds.), *Smart Innovations in Communication and
Computational Sciences*, Advances in Intelligent Systems and Computing 670,
https://doi.org/10.1007/978-981-10-8971-8_8

1 Introduction

The mathematical modeling of different physical systems when estimated utilizing the hypothetical considerations results in a model of larger order. The time domain or state-space modeling technique generally results in the higher-order state-space model or TF model in the frequency domain. Therefore, in the control applications, it becomes essential to represent the higher-order system by a TF or state variable model having reduced order. Several techniques are reported in the literature for reducing the order of system, which includes both time-domain and frequency-domain analysis [1–6]. Moreover, the researchers have also proposed some hybrid models which combine the characteristics of two or more methods for model order reduction. The existing hybrid models have got their own pros and cons when used for reducing the model order of any particular system. Although there exist a large number of models/techniques, there is no approach that guarantees the best results.

In order to analyze any physical system, it has to be converted into an equivalent mathematical model which represents the physical system completely. The process of structure modeling results in comprehensive and complex depiction of the system, resulting in higher-order differential equations. It becomes difficult for control engineer to use these equations for system analysis and controller synthesis. Hence, it becomes essential to find the lower-order equations which describe the system completely and adequately reflect the prevailing attribute of the system. The high-order system model is reduced to the lower-order system model because of the following reasons:

- To understand the system in a better way.
- To reduce computational and hardware complexity.
- For making viable controller modeling.
- For reduction of the time required for simulating the model in real time.

For the desired objective, the method proposed in this article is a hybrid technique that consists of both the pole clustering technique and particle swarm optimization (PSO) algorithm. In this presented work, the desired number of pole clusters is found using fuzzy c-means (FCM) clustering [7]. FCM algorithm is used for determining the denominator of the lower-order model (ROM), and PSO is used for determining the numerator.

In the recent past, the PSO technique had gained the momentum as popular algorithm for solving the nonlinear optimization issues. The particle swarm optimization is a populace-based stochastic algorithm. The main inspiration behind the development of the PSO algorithm is the social actions of the bird flock and fish schooling [8, 9]. Both the PSO and genetic algorithm (GA) are population-based algorithm and share various similarities. The major dissimilarity between GA and PSO is that it does not have any crossover and mutation operators. The major benefit of using PSO over GA is its algorithmic ease, use of less parameters, and ease in implementation.

The particle in PSO flies through the problem space and follows the particle which has achieved the present optimal value. In the projected work, the lower-order model system is obtained using PSO by keeping the constant values of the pole centroid. The system rigidity for the model of higher-order system and model of reduced-order system is also kept constant. The integral square error (ISE) among the transient responses of original and reduced-order models using particle swarm optimization is used for finding the zeros for the unit step input. Two different case studies are considered to show the effectiveness of the proposed technique. The paper is organized into various sections: Sect. 2 defines the problem, Sect. 3 describes the different techniques used for the model order reduction, and Sect. 4 presents the obtained reduced-order model for the case studies, followed by concluding remarks in Sect. 5.

2 Problem Formulation

Equation (1) represents the transfer function of a high-order stable single-input–single-output (SISO) system of order 'n', where the poles of the original high-order system are represented using $\lambda_1, \lambda_2, \ldots \ldots \lambda_n$. The main objective of the projected work is to obtain an approximate ROM (r_{th} order) which is stable.

$$G(s) = \frac{N(s)}{D(s)} = \frac{b_0 + b_1 s + b_2 s^2 + \cdots b_{n-1} s^{n-1}}{(s+\lambda_1)(s+\lambda_2) \ldots (s+\lambda_n)} \tag{1}$$

The rth reduced-order model of the original system given by (1) can be represented by (2), where $\lambda_1', \lambda_2', \ldots \ldots \lambda_r'$ are the poles of reduced-order model.

$$G_r(s) = \frac{N_r(s)}{D_r(s)} = \frac{\alpha_0 + \alpha_1 s + \cdots \alpha_{r-1} s^{r-1}}{(s+\lambda_1')(s+\lambda_2') \ldots (s+\lambda_r')} \tag{2}$$

The main objective of the proposed work is to realize the rth reduced-order model whose TF is shown in Eq. (2) from the original higher order system whose TF is shown in Eq. (1) the reduced-order model should represent the original system well by approximating its step response as close as possible and retain all the dominant features of the original system.

3 Methodology Used

3.1 Pole Clustering

The process of grouping the large amount of data into smaller groups by lowering the dimensionality of the group for analysis is called clustering. Pattern recognition or classification problems are generally solved using the clustering techniques, but it was also extended to solve the problem of data compression and model order reduction. The existing literature of MOR contains considerable number of clustering schemes, if the cluster numbers are known beforehand than the clustering techniques such as k-means and fuzzy c-means (FCM). The model order reduction is a similar case, where the cluster numbers are known a priori, so the k-means or FCM clustering techniques can be used for reducing the model order.

For finding the cluster center in the k-means or hard c-means (HCM) clustering [11] scheme, the cost function of the divergence measure [10] is reduced. In this process, the Euclidean distance is chosen as the dissimilarity measure in most of the cases. The Euclidean space is chosen as the dissimilarity measure in most of the cases. The Euclidean space-based cost function for a vector y_k in cluster j and the equivalent cluster center d_i can be calculated using (3), where J_i represents the cost function of group i.

$$J = \sum_{i=1}^{d} J_i = \sum_{i=1}^{d} \left(\sum_{k, y_k \in G_i} \|y_k - d_i\|^2 \right)$$ (3)

The matrix U represents the $d \times n$ binary membership between the partitioned groups; the value of element u_{ij} will be equal to unity in case the jth data point (y_j) is from the cluster i, and will be 0 otherwise. After fixing the cluster centers d_i, the minimized u_{ij} for Eq. (3) can be derived as (4)

$$u_{ij} = \begin{cases} 1 & if \|y_j - d_i\|^2 \leq \|y_j - d_k\|^2, \quad for\ k \neq i \\ 0 & otherwise \end{cases}$$ (4)

After fixing the membership matrix U, the optimal center d_i that minimizes Eq. (3) which will be the mean of all vectors in group i it is shown using Eq. (5).

$$d_i = \frac{1}{|G_i|} \sum_{k, y_k \in G_i} y_k$$ (5)

The HCM clustering technique was further improved as FCM clustering [10]. The data points in the FCM clustering technique pertain to the cluster with a degree designated by the membership grade. In FCM clustering strategy, it is permitted to relate a piece of information to at least two clusters in the meantime. The components in the matrix U are permitted to have the values between 0 and 1. The summation of the considerable number of degrees of belongingness for every one of the data points in all the clusters will be equivalent to unity as represented by (6).

$$\sum_{i=1}^{d} u_{ij} = 1$$ (6)

The cost function for FCM in the generalized form can be written as (7).

$$J = \sum_{i=1}^{d} J_i = \sum_{i=1}^{n} \sum_{j=1}^{n} u_{ij}^m f_{ij}^2 \tag{7}$$

where u_{ij} lies between 0 and 1, d_{ij} represents cluster center of the fuzzy group i, $f_{ij} = \|d_i - x_j\|$ represents the Euclidean distance among the i_{th} cluster center and the j_{th} data point, and $m \in [1, \infty]$ is a weighting exponent. The essential criteria for (7) to achieve to its minimum value are specified as (8) and (9).

$$d_i = \frac{\sum_{j=1}^{n} u_{ij}^m y_j}{\sum_{j=1}^{n} u_{ij}^m} \tag{8}$$

$$u_{ij} = \frac{1}{\sum_{k=1}^{d} \left[\frac{\|d_i - y_j\|}{\|d_k - y_j\|} \right]^{\frac{2}{(m-1)}}} \tag{9}$$

The conditions specified by (8) and (9) are iteratively worked upon by the FCM algorithm until no more improvement is seen [10]. The cluster centers d_i and the membership matrix U in the batch mode operation of the FCM algorithm are determined by using the subsequent steps.

Step 1: The initialization of the relationship matrix U is completed with the arbitrary values in the vicinity of 0 and 1 fulfilling the required constraints as indicated by (6).

Step 2: The fuzzy clusters with centers $d_i = 1, \ldots, d$ are calculated using (8).

Step 3: The value of the cost function specified by (7) is evaluated. The algorithm is terminated if the value is below the tolerance limit or the value of its improvement lies below certain threshold.

Step 4: If the termination condition specified at Step 4 is not satisfied, then the new membership matrix U is calculated using (9) and the algorithm is repeated from Step 2 until it is terminated.

The steps explained above are used for getting the requisite number of clustered pole centers in the denominator poles of the high-order system (HOS) with the FCM algorithm. The steps involved in the FCM clustering technique are also shown in the form of flowchart given in Fig. 1.

Fig. 1 Fuzzy c-means
clustering algorithm

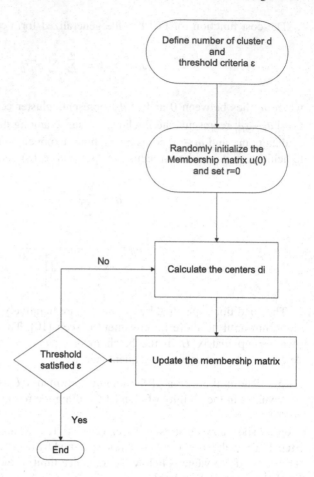

3.2 Determination of the Denominator of ROM Using Fuzzy C-Means Algorithm

In order to get better performance, FCM algorithm requires much iteration to be completed because of random initialization of the membership matrix U values. It is possible in FCM to get membership function with different degrees for each cluster in FCM algorithm. The FCM clustering scheme is used to create the desired number of clustered pole center of all the poles of the remaining system. For the lower-order model, the denominator of rth order can be found by the following method. The following cases may occur while synthesizing the reduced-order denominator polynomial using FCM clustering method.

Case 1: If all the cluster centers are real, then the rth-order denominator polynomial can be obtained using (10).

$$D_r(s) = \left(s - \lambda'_{c1}\right)\left(s - \lambda'_{c2}\right) \cdots \left(s - \lambda'_{cr}\right) \tag{10}$$

Case 2: The FCM clustering algorithm can be used for finding the cluster centers if the cluster centers are having the complex conjugate. The denominator of the new lower-order model is realized in this case by using the poles along with conjugate pairs.

Case 3: If the real poles as well as poles which are complex conjugate are present in the original system, then the synthesis of denominator polynomial of the obtained lower-order model is done using the combination of poles in the form of real as well as in the form of the complex conjugate. These cases can be used for formulating the denominator of rth order for the ROM by using (11).

$$D_r(s) = s^r + \hat{b}_1 s^{r-1} + \hat{b}_2 s^{r-2} + \hat{b}_3 s^{r-3} + \cdots + \hat{b}_r \tag{11}$$

3.3 Particle Swarm Optimization Algorithm

The objective of any optimization algorithm is to calculate or determine the optimal (i.e., best suited) solution of the given problem for a specified set of constraints. Kennedy and Eberhart in 1995 proposed an algorithm which does not need any information of gradient and rather works by following the combined behavior of bird flocking, particles, and socio-cognition [12], which was named as particle swarm optimization (PSO) [12, 13]. The PSO algorithm is based upon population in which every individual is termed as a particle which denotes a candidate solution. The particles fly in the search space with an adaptable speed which can be updated dynamically depending upon individual flying experience and also the flying experience associated with further particles. Another key feature of PSO is that every particle attempts for improving itself by replicating characters from the peers which are winning peers. Also, the memory is there in every particle due to which it is able of keeping the track of the finest location in exploration space yet attained by the particle in it. In this process, the location matching to the finest fitness is termed as pbest and the globally best among all the particles in the population is termed as gbest.

In the perspective of multivariable optimization, the swarm size is expected to be of definite or fixed size where every particle is positioned primarily at random places in the multi-dimensional design space. Every particle is correlated with two features which are location and speed. Every particle roams around the design space and memorizes the best position (with respect to the objective function value) it has determined. The particles share information or their positions among themselves and accordingly update their separate positions and velocities depending upon the information acknowledged about the good positions.

The function in a classical PSO can be represented by (12):

$$f(x_1, x_2, x_3 \ldots .x_n) = f(\vec{X}) \tag{12}$$

where $\vec{X}$ denotes the search variable vector, which signifies the set of free variables of the function mentioned. The main objective is to find $\vec{X}$ which minimizes or maximizes the value of the function $f(\vec{X})$. PSO is a search algorithm which is multi-agent and parallel in nature; in the PSO algorithm, the particles are considered as an entity which flies through the search space which is multi-dimensional. Individual particle in the swarm can be defined on the basis of its position and velocity.

If the search space is assumed to be an n-dimensional search space and the swarm is consisting of m particles, then the location and speed of jth particle can be shown using two vectors; i.e., $x_j = (x_{j1}, x_{j2}, x_{j3}, \ldots x_{jn})$ and $v_j = (v_{j1}, v_{j2}, v_{j3}, \ldots v_{j4})$, respectively, where $j = 1, 2, 3, 4, \ldots, m$. The objective function of the optimization problem determines the fitness value of each particle and determines the best earlier visited and current location of each particle.

The initial values for pxbest$_j$ and gxbest are taken as pxbest$_j(0)$ and gxbest(0) for all the particles. The velocity and new position of jth particle in the swarm can be calculated after each particle knows the best individual particle position in the swarm using Eq. (13) and (14).

$$v_j^{k+1} = w + v_j^k + c_1 \tau_1 \left(\text{pxbest}_j - x_j^k \right) + c_2 \tau_2 \left(\text{gxbest} - x_j^k \right) \tag{13}$$

$$x_j^{k+1} = x_j^k + v_j^{k+1} \tag{14}$$

where v_j^k and x_j^k are the velocity and position of the jth particle at time k, respectively. The individual best position of the jth particle and the global best position of the jth particle in the swarm are given as pxbest$_j$ and gxbest, respectively. τ_1 and τ_2 are uniformly distributed numbers within interval [0 1] that determines the impact of pxbest$_j$ and gxbest on the velocity update formula. c_1 and c_2 are two constant terms namely 'self-confidence' and 'swarm-confidence,' respectively. Initial values of c_1 and c_2 are considered as $c_1 = c_2 = 2$. w is the inertial weight and can be determined using (15).

$$w = w_{max} - (w_{max} - w_{min}) \text{Num}/\text{Num}_{max} \tag{15}$$

where w_{max} and w_{min} are the maximum and minimum values of the w, Num_{max} is the maximum iteration time of the algorithm, and Num is the current iteration of the algorithm.

Input: Initial value of position of m particles $x_1^0, x_2^0, x_3^0, \ldots\ldots, x_m^0$ and initial velocities $v_1^0, v_2^0, v_3^0, \ldots, v_m^0$ both values are generated randomly where $m = 20$.
Output: The output will be position of the approximate global optima of
x and v.
Algorithm: **Begin**
      While terminating condition
      for $1 \leq i \leq$ no.of particles (m) do
            Evaluate the fitness of the objective function;
            Update pxbest and gxbest ;
            Adapt the velocity of the particle as follows;
            $v_i^{k+1} \Leftarrow w \times v_i^k + c_1\tau_1 \left(\text{pxbest}_i - x_i^k\right) + c_2\tau_2 \left(\text{gxbest} - x_i^k\right)$
            Update the position of the particle;
            Increase the no. of iteration;
      end   for
      end

The main parameters of the PSO are inertia weight (w), constant (c_1, c_2), velocity (v_{max}), and swarm size S; the parameter values define how well the algorithm optimizes. The momentum of the particle is controlled by the w, if $w \ll 1$ gives quick changes in the direction; i.e., more exploitation of the local search space is done by choosing different directions. If $w = 0$, then the concept of velocity is lost; i.e., the next step is taken without any knowledge of the previous velocity. If $w \gg 1$, i.e., high, then it acts similar to the low values of c_1, c_2; i.e., the change in direction for the particle is difficult, and it implies that the current search direction is explored. Eberhart and Shi have referred that when V_{max} is not small (≥ 3), the value of w can be taken as 0.9 [14].

Large initial weight helps the global exploration of the algorithm, while the smaller inertial weight is used to locally explore or tune the existing search area. The maximum velocity decides the maximum change in the position coordinates of a particle in one iteration usually; the maximum value of the V_{max} is designated equal to the half of the length of the search space [15].

'E' represents the objective function which is minimized by using PSO. 'E' is basically represented by (16) which is the integral square error between the transient responses of the original higher-order system and the obtained lower-order system model.

$$E = \int\limits_0^\infty \left[\left[y(t) - y_r(t)\right]\right]^2 \mathrm{d}t \tag{16}$$

In (16), $y(t)$ and $y_r(t)$ represent the responses of the original higher-order system and lower-order system to the unit step input. The coefficients of the numerator of the obtained lower-order system model are the parameters which are required to be found out, i.e., α_i $(i = 0, 1,..., (r - 1)$. The various parameters involved in PSO algorithm for the present study are shown in Table 1. Moreover, Fig. 2 shows the structure and the flowchart of the proposed algorithm.

Table 1 Parameters of the PSO algorithm

Parameter	Value
Swarm size	15
Max. generation	85
c_1, c_2	1.7, 1.7
Starting point (wstart), end point (wend)	0.85, 0.33

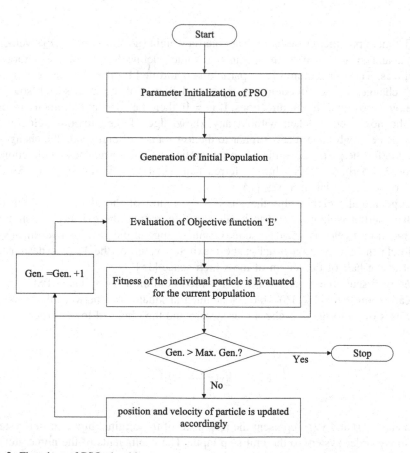

Fig. 2 Flowchart of PSO algorithm

4   Case Studies

The case study of numerical examples from the literature has been considered to show the performance of the proposed technique as compared to the other existing techniques. The main objective is to develop the second-order reduced system model from the given higher-order system model. The accuracy of the obtained lower-order system model using the projected method is checked by calculating the integral square error (ISE) [6], integral absolute error (IAE), integral time-weighted absolute error (ITAE), and integral of time multiply squared error (ITSE). The equations for the testing indices ISE, IAE, ITAE, and ITSE are shown using (17), (18), (19), and (20), where $e(t) = Y(t) - Y_r(t)$ and $Y(t)$ and $Y_r(t)$ are step responses of the original and the reduced models, respectively.

$$\text{ISE} = \int_0^\infty e^2(t)\mathrm{d}t \tag{17}$$

$$\text{IAE} = \int_0^\infty |e(t)|\mathrm{d}t \tag{18}$$

$$\text{ITSE} = \int_0^\infty te^2(t)\mathrm{d}t \tag{19}$$

$$\text{ITAE} = \int_0^\infty t|e(t)|\mathrm{d}t \tag{20}$$

4.1   Case Study I

Let us consider a stable eighth-order system given by (21). It is required to reduce the order of system and to obtain the third-order model.

$$G(s) = \frac{18s^7 + 514s^6 + 5982s^5 + 36282s^4 + 122664s^3 + 222088s^2 + 184760s + 40320}{(s+1)(s+2)(s+3)(s+4)(s+5)(s+6)(s+7)(s+8)} \tag{21}$$

From (21), it can be observed that the poles of the original higher-order system are $(-1, -2, -3, -4, -5, -6, -7, -8)$. The third-order model given by (22) is obtained using the proposed technique of FCM clustering and PSO algorithm for reducing the denominator and numerator of the original system, respectively.

$$G_3(s) = \frac{17.08s^2 + 71.56s + 21.71}{s^3 + 11s^2 + 31.63s + 21.63} \tag{22}$$

The step response of the eighth-order and the reduced-order system is plotted as well as shown in Fig. 3. It can be observed that the step reaction of the obtained model which is third-order reduced model is similar to the original system. In order to prove the effectiveness of the obtained second-order reduced model using the proposed technique, comparative analysis is done with the existing techniques shown in Table 2. The proposed method has given the least value of ISE, IAE, ITAE, and ISTE. In order to show the stability issue, the bode plot of the eighth-order and third-order reduced models is shown in Fig. 4.

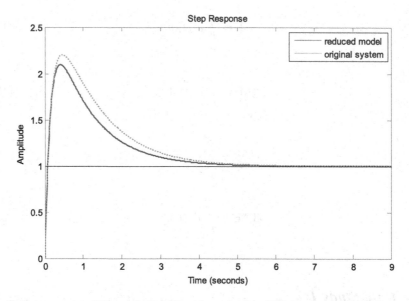

Fig. 3 Step response of the original eighth-order and reduced third-order models

Table 2 Parameter comparison of third-order model with the existing techniques

Reduction method	Reduced model	ISE	ITSE	IAE	ITAE
Proposed method	$\frac{17.08s^2 + 71.56s + 21.71}{s^3 + 11s^2 + 31.63s + 21.63}$	0.0011	0.00157	0.0588	0.0995
Padé and fuzzy c-means technique	$\frac{18s^2 + 64s + 21.63}{s^3 + 11s^2 + 31.63s + 21.63}$	0.0485	0.0652	0.3735	0.5949
IDM method	$\frac{18s^2 + 55.39s + 22.49}{s^3 + 10.52s^2 + 30.39s + 22.49}$	0.1884	0.2544	0.7363	1.1785
Impulse energy method	$\frac{11.358s^2 + 42.391s + 11.980}{s^3 + 6.274s^2 + 20.057s + 11.980}$	0.0385	0.0292	0.2781	0.3095

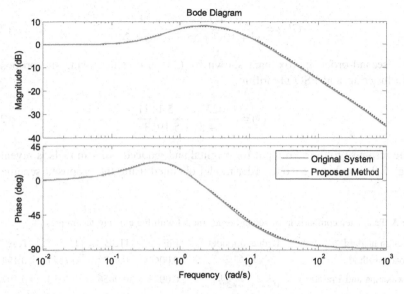

Fig. 4 Frequency response of the original eighth-order and reduced third-order models

4.2 Case Study II

Let us reflect on the fourth-order TF of a stable system given by (23). The objective is to obtain second-order transfer function by the proposed technique.

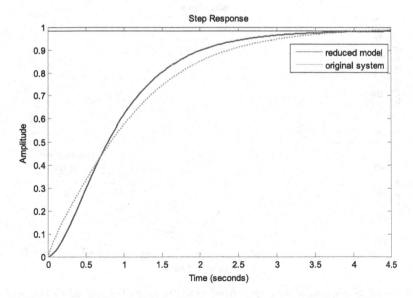

Fig. 5 Step response of the original fourth-order and reduced second-order models

$$G(s) = \frac{s^3 + 7s^2 + 24s + 24}{s^4 + 10s^3 + 35s^2 + 50s + 24} \tag{23}$$

The second-order approximant shown by (24) is obtained using the proposed FCM clustering and PSO algorithm.

$$G_2(s) = \frac{0.02578s + 5.1171}{s^2 + 5s + 5.1953} \tag{24}$$

The response to the step input for original and reduced-order models is revealed in Fig. 5. The reduced second-order model obtained using the proposed scheme is

Table 3 Parameter comparison of second-order model with the existing techniques

Reduction method	Reduced model	ISE	ITSE	IAE	ITAE
Proposed method	$\frac{0.02578s + 5.1171}{s^2 + 5s + 5.1953}$	0.0063	0.0077	0.1327	0.1941
Vishwakarma and Prasad [16]	$\frac{-0.1898s + 4.571}{s^2 + 4.762s + 4.571}$	0.0079	0.0056	0.1333	0.1978
Safonov and Chiang [17]	$\frac{s + 5.403}{s^2 + 8.431s + 4.513}$	0.0073	0.0233	0.1337	0.3635
Pal [18]	$\frac{16.0008s + 24}{30s^2 + 42s + 24}$	0.0096	0.0131	0.1726	0.2858
Prasad and Pal [19]	$\frac{s + 34.2465}{s^2 + 239.8082s + 34.2465}$	1.0682	2.5713	1.9710	4.4985
Gutmann et al. [20]	$\frac{96s + 288}{70s^2 + 300s + 288}$	0.0454	0.0522	0.3586	0.5134

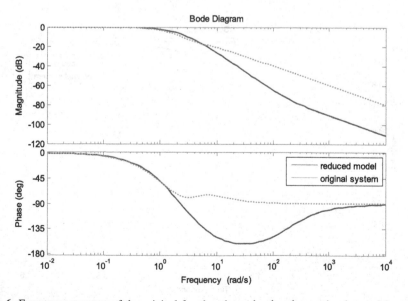

Fig. 6 Frequency response of the original fourth-order and reduced second-order models

compared with the existing methods using different performance indices. The results of comparison are shown in Table 3. The least ISE of 0.0063 is obtained for the reduced second-order model obtained using the proposed system. The bode plot of the 4th order and 2nd order reduced model is shown in Fig. 6.

5 Conclusion

A hybrid technique for order reduction of the higher-order systems has been proposed which merge the benefits of PSO algorithm and fuzzy c-means clustering techniques. In the proposed scheme, the poles and zeros are calculated using fuzzy c-means clustering and PSO algorithm, respectively. The proposed technique has been used for the reduction of the model order of the case studies with real poles. It has been observed that step response of the obtained lower-order system model is comparable with the step response of the original higher-order system model. The proposed method is simple, rugged, and fast; i.e., it takes minimal computational time. The performance comparison of the proposed technique and the other well-established techniques for reducing the model order for the case studies I and II has been presented in Tables 2 and 3, respectively. The result justifies the superiority of the proposed method as compared to the existing methods based on the four different performance indices. The projected technique also ensures the model stability by reducing the difference between the initial and final values, i.e., error of the responses of the original higher-order system and the obtained lower-order system model. The proposed technique can preserve mutually the time-domain as well as frequency-domain distinctiveness of the original system.

References

1. B. Bandyopadhyay and S. Lamba, "Time-domain Pade approximation and modal-Pade method for multivariable systems," *IEEE Transactions on Circuits and Systems*, vol. 34, no. 1, pp. 91–94, Jan. 1987.
2. W. Enright and M. Kamel, "On selecting a low-order model using the dominant mode concept," *IEEE Transactions on Automatic Control*, vol. 25, no. 5, pp. 976–978, Oct. 1980.
3. M. Hutton and B. Friedland, "Routh approximations for reducing order of linear, time-invariant systems," *IEEE Transactions on Automatic Control*, vol. 20, no. 3, pp. 329–337, Jun. 1975.
4. V. Krishnamurthy and V. Seshadri, "Model reduction using the Routh stability criterion," *IEEE Transactions on Automatic Control*, vol. 23, no. 4, pp. 729–731, Aug. 1978.
5. T. N. Lucas, "Some further observations on the differentiation method of modal reduction," *IEEE Transactions on Automatic Control*, vol. 37, no. 9, pp. 1389–1391, Sep. 1992.
6. V. Singh, D. Chandra, and H. Kar, "Improved Routh–PadÉ Approximants: A Computer-Aided Approach," *IEEE Transactions on Automatic Control*, vol. 49, no. 2, pp. 292–296, Feb. 2004.

7. T. N. Lucas, "Scaled impulse energy approximation for model reduction," *IEEE Transactions on Automatic Control*, vol. 33, no. 8, pp. 791–793, Aug. 1988.
8. N. Saxena, A. Tripathi, K. K. Mishra, and A. K. Misra, "Dynamic-PSO: An improved particle swarm optimizer," 2015, pp. 212–219.
9. S. Mukherjee, Satakshi, and R. C. Mittal, "Model order reduction using response-matching technique," *Journal of the Franklin Institute*, vol. 342, no. 5, pp. 503–519, Aug. 2005.
10. K. Hammouda and F. Karray, "A comparative study of data clustering techniques," *University of Waterloo, Ontario, Canada*, 2000.
11. D. Napoleon and S. Pavalakodi, "A new method for dimensionality reduction using K-means clustering algorithm for high dimensional data set," *International Journal of Computer Applications*, vol. 13, no. 7, pp. 41–46, 2011.
12. J. Kennedy, J. F. Kennedy, R. C. Eberhart, and Y. Shi, *Swarm intelligence*. Morgan Kaufmann, 2001.
13. T. Zeugmann *et al.*, "Particle Swarm Optimization," in *Encyclopedia of Machine Learning*, C. Sammut and G. I. Webb, Eds. Boston, MA: Springer US, 2011, pp. 760–766.
14. Y. Shi and R. Eberhart, "A modified particle swarm optimizer," 1998, pp. 69–73.
15. B. Alatas, E. Akin, and A. B. Ozer, "Chaos embedded particle swarm optimization algorithms," *Chaos, Solitons & Fractals*, vol. 40, no. 4, pp. 1715–1734, May 2009.
16. C. B. Vishwakarma and R. Prasad, "System reduction using modified pole clustering and Pade approximation," in *XXXII National systems conference, NSC*, 2008, pp. 592–596.
17. M. G. Safonov and R. Y. Chiang, "Model reduction for robust control: A schur relative error method," *International Journal of Adaptive Control and Signal Processing*, vol. 2, no. 4, pp. 259–272, Dec. 1988.
18. J. Pal, "Stable reduced-order Padé approximants using the Routh-Hurwitz array," *Electronics Letters*, vol. 15, no. 8, p. 225, 1979.
19. R. Prasad and J. Pal, "Stable reduction of linear systems by continued fractions," *Journal-Institution of Engineers India Part El Electrical Engineering Division*, vol. 72, pp. 113–113, 1991.
20. P. Gutman, C. Mannerfelt, and P. Molander, "Contributions to the model reduction problem," *IEEE Transactions on Automatic Control*, vol. 27, no. 2, pp. 454–455, Apr. 1982.

Control Chart Pattern Recognition Based on Convolution Neural Network

Zhihong Miao and Mingshun Yang

Abstract Quality control chart pattern recognition plays an extremely important role in controlling the products quality. By means of real-time monitoring control, the abnormal status of the product during production can be timely observed. A method of control pattern recognition based on convolution neural network is proposed. Firstly, the control chart patterns (CCPs) are analyzed, the statistical characteristics and shape features of the control charts are considered, and the appropriate characteristics to distinguish the different abnormal patterns are selected; secondly, deep learning convolution neural network is trained and learned; finally, the feasibility and effectiveness of the control chart pattern recognition are verified through Monte Carlo simulation.

Keywords Quality control chart · Pattern recognition · Convolution neural Network · Monte Carlo · Deep learning

1 Introduction

Quality control of the manufacturing process is the core of the total quality management, and it is also an important part of advanced manufacturing models [1].

Control chart as an effective mean of quality control, control chart pattern timely and accurately recognized, it can make the product quality production status quickly be found, and the quality of product manufacturing effectively controlled, thereby the manufacturing quality management of enterprise can be improved. The automatic recognition of the control chart patterns is realized by the analysis of the statistical characteristics and shape features of the control chart pattern, the effective

Z. Miao · M. Yang (✉)
School of Mechanical and Precision Instrument Engineering,
Xi'an University of Technology, Xi'an 710048, Shaanxi, China
e-mail: nocc@163.com

Z. Miao
e-mail: mzh910612@163.com

© Springer Nature Singapore Pte Ltd. 2019
B. K. Panigrahi et al. (eds.), *Smart Innovations in Communication and Computational Sciences*, Advances in Intelligent Systems and Computing 670,
https://doi.org/10.1007/978-981-10-8971-8_9

features between them are extracted and fused, and the control pattern through the relevant algorithm was classified. The different models of control pattern are efficiently identified [2].

Guo and Chen proposed hierarchical learning rate adaptive deep convolution neural network based on an improved algorithm, and it can diagnose bearing faults and determine their severity [3]; In order to realize the abnormal production pattern of rapid recognition and real-time monitoring of the production process, Yan et al. combined MapReduce programming model with LDA algorithm to analyse the data stored in the HDFS [4]. Hassan et al. used half-against-half multi-class support vector machine to identify the six basic control chart patterns, and the recognition speed and recognition accuracy were improved [5]. Lei and Govindaraju constructed a generalized neural network to identify the upward and downward steps of abnormal patterns [6]. For the higher dimensions of the original characteristics and other characteristics, based on principal component analysis (PCA) and support vector machine (SVM), Yang et al. proposed an intelligent identification method to identify abnormal control chart pattern [7]. Yang and Zhou proposed a hybrid approach that integrates extreme-point symmetric mode decomposition (ESMD) with extreme learning machine (ELM) to identify typical concurrent CCPs [1].

In order to realize the efficient and accurate identification of the control chart pattern, the statistical characteristics and shape features of the control charts are considered and the appropriate characteristics to distinguish the different abnormal patterns are selected. The control chart mode automatically identifies through the reasonable computational intelligence method. The convolution neural network is used to identify and classify the different patterns of the control chart, the problem mode is built, and the control chart model is simulated by Monte Carlo. The model and the results of the simulation are validated by practical example.

2   Control Chart and Its Mode

Control chart is an effective method of modern quality control management, and it is a tool for real-time quality monitoring and diagnosis of manufacturing production process. Control chart models are generally divided into seven basic modes: normal (NOR), upward/downward trends (UT/DT), cycles (CYC), upward/downward shifts (DS/US), and periodic (PRD) [4].

The control chart data can be described as follows:

$$y(t) = \mu + r(t) + s(t) \tag{1}$$

where $y(t)$ is sample values of t moments; μ is the average value of the statistical value of the quality characteristic for each process; $r(t)$ is the interference caused by accidental factors of moments; $s(t)$ is the deviation of the quality characteristic caused by the abnormal factors of "t" moment.

NOR: Manufacturing processes are controlled, $s(t) = 0$.

The trend mode: $s(t) = \pm\rho \times d \times t$, "+" represents UT, "−" represents a DT, ρ is 0 before the trend, after the trend appears to be 1; d is slope.

The shift mode: $s(t) = \pm v \times s$, "+"represents US, "−" represents DS, v is 0 before the trend, after the trend appears to be 1; s is a step amplitude, $S = 1.5\delta \sim 2.5\delta$.

CYC: $s(t) = \pm\alpha\sin 2\pi t/T$, where α is amplitude and T is a cycle.

The mixed mode: $s(t)$ is the corresponding combination of trend model, shift mode, and periodic pattern.

3 Construction of Convolution Neural Network Model

Firstly, the statistical and shape features are extracted as the feature set of control charts. Then the sample features are trained and classified by the deep learning of convolution neural network depth learned to complete the second part of the control chart pattern recognition. Finally, the effectiveness of the method and recognition accuracy is verified by the test sample.

3.1 Feature Extraction

3.1.1 Statistical Characteristics

Statistical characteristic index has the average value, the average amplitude, the root amplitude, the maximum value, the mean amplitude, the standard deviation, the skewness, the kurtosis, the peak value, the waveform, the pulse, the margin and so on. The average value, the maximum value and the standard deviation are chose as a statistical feature of the control chart. Specific indicators are expressed as follows:

$$x = \frac{1}{T}\sum_{i=1}^{n} x_i \tag{2}$$

$$x_{\max} = \max(x_i) \tag{3}$$

$$\delta = \sqrt{\frac{1}{T}\sum_{i=1}^{T}(x_i - x)^2} \tag{4}$$

3.1.2 Shape Feature

The shape feature [8] is a digital quantity which can be used to represent the geometry characteristic based on the original data of control chart pattern, and it can highlight the distinction between different models, Gauri and Chakraborty [9] summed up 32 control chart pattern shape feature components. Three typical shape features are extracted to distinguish the seven different control graphs.

Area between the pattern and the main line (APML): The normal mode has the minimum value of APML, so it can be used to distinguish between normal mode and other exception patterns.

Area between the pattern and its least square line (APSL): The periodic mode and the step mode have higher APSL values than the normal mode and the trend pattern. This feature can be used to distinguish the cycle mode, the step mode from the normal mode and the trend mode.

Area between the least square line and the line segment (ASS): The trend mode ASS value is close to zero, and the step mode has a larger value, and this feature can be used to distinguish between trend mode and step mode.

3.2 Convolution Neural Network Model Construction

The concept of deep learning is derived from the study of artificial neural networks. Depth learning creates a more abstract upper layer by combining low-level features to represent its attribute categories or characteristics, thereby discovering the distributed characteristics of the data. Its essence is to construct machine learning architecture model with multiple hidden layers. The architecture model can obtain vast and representative feature information by training large-scale data, so as to realize the classification and prediction of the samples and improve the classification and prediction accuracy of the samples.

Convolutional neural network (CNN) is a multi-layer supervised learning neural network. The core module which implements the feature extraction function of convolution neural network is the pool and convolution of hidden layer. This network model achieves reverse adjustment of weight parameters by layer through using the gradient descent method and minimizes the loss function, and the accuracy of the network is improved by frequent iterative training.

In order to achieve the classification of the control chart model, convolution neural network deep classifier is built using the deep learning ideas to build convolution neural network deep classifier as shown in Fig. 1; it has four layers: input layer, two hidden layers, and output layer. The process of pattern recognition usually consists of four steps:

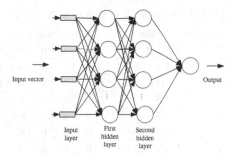

Fig. 1 Deep classification structure of convolutional neural network

1. The extraction of the statistical features and the shape features of the control chart;
2. The classification of various statistical characteristics of indicators;
3. The control chart sample of the shape characteristics of the index classification;
4. The statistical characteristics and shape features are integrated to form a more abstract high-level semantic concept, and the patterns of corresponding control charts are identified.

Based on the understanding of the control chart from low to high, from shallow to deep, a deep network classifier with two hidden layers is constructed.

Based on control chart pattern recognition of the convolution neural network, first of all, the data of each type of control chart pattern was trained and studied. The statistical characteristics and shape features of the control chart model were used as indicators to identify and classify, and the training data were preprocessed. Secondly, the constructed convolution neural network was used for training and learning the preprocessed data. According to the convergence of the output results, the training sample data was completed. Finally, the test samples were validated to realize the accuracy identify of the various abnormal modes of the control chart. The specific flowchart is shown in Fig. 2.

Fig. 2 CNN control chart pattern recognition process

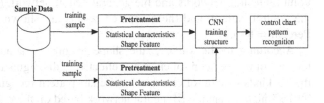

Table 1 Output value of each CNN modes

Number	Mode	Output values												
1	NOR	1	0	0	0	0	0	0	0	0	0	0	0	0	0
2	UT	0	1	0	0	0	0	0	0	0	0	0	0	0	0
3	DT	0	0	1	0	0	0	0	0	0	0	0	0	0	0
4	US	0	0	0	1	0	0	0	0	0	0	0	0	0	0
5	DS	0	0	0	0	1	0	0	0	0	0	0	0	0	0
6	CYC	0	0	0	0	0	1	0	0	0	0	0	0	0	0
7	UT and US	0	0	0	0	0	0	1	0	0	0	0	0	0	0
8	UTDS	0	0	0	0	0	0	0	1	0	0	0	0	0	0
9	UT + CYC	0	0	0	0	0	0	0	0	1	0	0	0	0	0
10	DT and US	0	0	0	0	0	0	0	0	0	1	0	0	0	0
11	DT and DS	0	0	0	0	0	0	0	0	0	0	1	0	0	0
12	DT and CYC	0	0	0	0	0	0	0	0	0	0	0	1	0	0
13	US and CYC	0	0	0	0	0	0	0	0	0	0	0	0	1	0
14	DS and CYC	0	0	0	0	0	0	0	0	0	0	0	0	0	1

4 Simulation and Verification

The input data is taken from the 40 consecutive sample data in the control chart. The output destination value is the category of the control chart, and each category is entered as shown in Table 1.

In the CNN network training, according to the input data, each model generates a total of 200 samples, and the total number of training samples produce is 2800. Each sample of the test samples produces 400 samples; the total number of samples produce is 5600. The samples are used to simulate by Monte Carlo simulation method (Monte Carlo) in MATLAB software. Through Eq. (1), 600 samples (200 training samples and 400 testing samples) are generated from each mode, $r(t) = r \times \delta$ is $N(0, \delta)$ which is a random Gauss distribution deviation of quality parameters. δ is the distributed standard deviation, $\mu = 20, \delta = 1, T = 40$. The convolution neural networks and the general neural network (BP) are used to simulate the control pattern, and the compared results are shown in Table 2 (Recognition Percent = RP).

According to the Table 2, the shape feature extraction and statistical characteristics of the key control chart can effectively distinguish the abnormal patterns of the 14 kinds of control chart; control chart pattern recognition method of CNN is 98.16% higher than the BP neural network based on the rate of 96.84. It can be seen that the depth learning CNN method has higher accuracy and better generalization ability and robustness for test sample. The recognition accuracy of CNN and BP is shown in Fig. 3. In the case of the same sample, the use of CNN for the control chart pattern recognition is better than the BP method.

Table 2 Experimental results of CNN and BP recognition methods

Number	Mode	Sample value	CNN	BP
1	NOR	400	100	100
2	UT		99.25	98.75
3	DT		99.05	99.75
4	US		97.5	96.75
5	DS		98.25	96.25
6	CYC		100	97.5
7	UT and US		96	95.25
8	UTDS		97	96.25
9	UT + CYC		97.25	96.2
10	DT and US		98	99
11	DT and DS		97.5	95.5
12	DT and CYC		98.5	92.5
13	US and CYC		96.75	96
14	DS and CYC		98.75	96

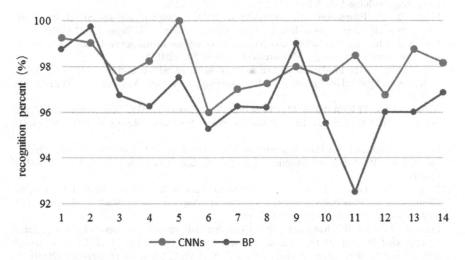

Fig. 3 Comparison between CNN and BP

5 Conclusion

From the perspective of accurate identification of control chart pattern, firstly, the characteristics of the control chart pattern are analyzed and the statistical and shape features are extracted as the feature set of control charts; secondly, the control chart pattern recognition based on convolutional neural network is proposed, and control chart pattern recognition model is constructed; finally, the model is simulated by

Monte Carlo simulation, and the correctness of the model and the effectiveness of simulation are verified by the practical example. The simulation results show that CNN method is more effective than other methods, such as BP neural network, and it has high precision and strong generalization ability. The recognition based on the convolution neural networks control graph model provides a valid basis for the subsequent quality control and has better reference and guidance for the quality control in the actual production process.

Acknowledgements This research is supported by the National Natural Science Foundation of China (Grant No: 61402361, 60903124); project supported by the scientific research project of Shaanxi Provincial Department of Education (Grant No: 14JK1521); Shaanxi province science and technology research and development project (Grant No: 2012KJXX-34).

References

1. Yang, W. A., Zhou, W.: Identification and quantification of concurrent control chart patterns using extreme-point symmetric mode decomposition and extreme learning machines [J]. Neuro computing, Volume 147, 5 January, Pages 260–270 (2015).
2. Pham, D. T.: Estimation and generation of training patterns for control chart pattern Recognition [J]. Computers & Industrial Engineering, Volume 95, Pages 72–82 (2016).
3. Guo, X. J., Chen, L.: Hierarchical adaptive deep convolution neural network and its application to bearing fault diagnosis [J]. Measurement 93, 490–502 (2016).
4. Shi, Y., Shi, Y. Q.: Production pattern recognition based on the distributed LDA algorithm under the background of big data. [J]. Manufacturing Automation, Volume 39, 3, Pages 24–28 (2016).
5. Hassan A., Shariff Nabi Baksh, M., Shaharoun, A. M.: Improved SPC chart pattern recognition using statistical features [J]. International Journal of Production Research, 41(7):1587–1603 (2010).
6. Lei, H., Govindaraju, V.: Half-Against-Half Multi-class Support Vector Machines [C]. The 6th International Workshop on Multiple. Classifier Systems, Monterrey, CA, June, 156–164 (2005).
7. Yang, Y. S., Wu, D. H., Su, H. T.: Abnormal Pattern Recognition Method for Control Chart Based on Principal Component Analysis and Support Vector Machine [J]. Journal of System Simulation. 18(5):1314–1318 (2006).
8. Lee, H., Grosse, R., Ranganath, R.: Convolutional deep belief networks for scalable unsupervised learning of hierarchical representations [C]. Proceedings of the 26th Annual International Conference on Machine Learning. New York, USA, ACM, 609–628 (2009).
9. Gauri, S. K., Chakraborty, S.: Feature-based recognition of control chart patterns [J] Computers and Industrial Engineering, 51, 726–742 (2006).

Solution to IPPS Problem Under the Condition of Uncertain Delivery Time

Jing Ma and Yan Li

Abstract Aiming at uncertain delivery time problems of process planning and job shop scheduling integration (integrated process planning and scheduling, IPPS), fuzzy number is introduced to denote the workpiece delivery time. And maximizing workpiece delivery satisfaction weighted and minimizing the maximum completion time are taken as the optimization objective to establish the mathematical model. Genetic algorithm is used to search the optimal scheduling to meet the target function. Finally, an example verifies the effectiveness and feasibility of the model and algorithm.

Keywords Process planning · Job shop scheduling · Delivery time
Fuzzy number · Genetic algorithm

1 Introduction

Process planning and job shop scheduling play an important role in the manufacturing system. Researches suggest that integrated process planning and scheduling (IPPS) can overcome the problems such as the low efficiency, goal conflict, and time difference in the process of planning and implementation when they work independently [1].

Song and Ye [2] represented delivery time with trapezoidal fuzzy number, introduced customer satisfaction index, and resolved fuzzy due date through using the quantum particle swarm algorithm under uncertain flow shop scheduling problem. Palacios and González [3] set the total time as objective and propose a genetic tabu search based on GA, TS, and heuristic seeding. Gutiérrez et al [4] set

J. Ma · Y. Li (✉)
School of Mechanical and Precision Instrument Engineering,
Xi'an University of Technology, Xi'an 710048, Shaanxi, China
e-mail: jyxy-ly@xaut.edu.cn

J. Ma
e-mail: 307394845@qq.com

© Springer Nature Singapore Pte Ltd. 2019
B. K. Panigrahi et al. (eds.), *Smart Innovations in Communication and Computational Sciences*, Advances in Intelligent Systems and Computing 670,
https://doi.org/10.1007/978-981-10-8971-8_10

different weights for different objectives; multiple objectives were combined into a single objective. Shi et al. [5] took total workload, earliness/tardiness penalty, product qualification rate, and cost into MOFJSP. Wang et al. [6] proposed three-point satisfaction model for fuzzy job shop scheduling.

At present, most of researches are based on certain conditions and less on uncertain conditions. The manufacturing system itself has a variety of uncertainties [7], such as processing time, delivery time of parts. These uncertainties have great impact on the production [8] and cannot be described with accurate parameters [9].

Therefore, trapezoidal fuzzy number is used to represent the delivery time of the workpiece considering the uncertainty of the delivery time for IPPS problem, and the mathematical model of this problem is presented, which takes the maximizing workpiece delivery satisfaction weighted and minimizing the maximum completion time as the optimization objective. Meanwhile, the genetic algorithm is designed. Finally, an example is given to validate the model and algorithm.

2   Problem Modeling

2.1   Parameter Setting

In order to establish a mathematical model of process planning and job shop scheduling, the following parameters are introduced.

The symbol n represents workpiece number; J is workpiece set; J_i represents the workpiece i ($i = 1, 2, ..., n$); Q_i represents the number of possible process routes for the workpiece J_i; R_{ij} represents the feasible process route j for the workpiece J_i, ($j = 1, 2, ..., Q_i$); O_{ij} represents the number of processes in the process route R_{ij}; O_{ijp} is the process p of the j process route for the workpiece J_i; S_{ijpk} represents the starting time of process O_{ijp} in the machine k; T_{ijpk} represents the processing time of process O_{ijp} in the machine k; C_{ijpk} represents the earliest completion time of process O_{ijp} in the machine k; d_i represents the delivery term for the workpiece J_i; C_i is the completion time for the workpiece J_i; C_{imax} represents the makespan for the workpiece J_i; if the workpiece J_i chooses process route j, $X_{ij}= 1$, otherwise $X_{ij}= 0$; if the process Q_{ijq} chooses machine k to process, $Z_{ijpk}= 1$, otherwise $Z_{ijpk}= 0$.

2.2   Expression of Uncertain Due Date

Trapezoidal fuzzy number is used to represent due date of parts, and the fixed due date value is set in a certain range, which is conducive to improve the flexibility of scheduling. Figure 1 shows the linear membership function diagram of the trapezoidal fuzzy due date.

Fig. 1 Linear membership
function of fuzzy due date

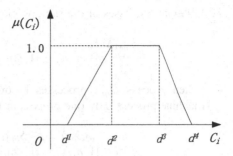

As shown in Fig. 1, if workpiece processing is completed during $[d^2, d^3]$, the satisfaction is 1 without penalty, otherwise the satisfaction will be less than 1 and linear variation. The earliness and tardiness penalty of the workpiece is determined according to the actual situation. The satisfaction degree of the workpiece is represented by linear membership function, and $\mu_i(C_i)$ represents the membership function of workpiece J_i as shown in Eq. (1).

$$\mu_i(C_i) = \begin{cases} 0 & C_i < d^1, C_i > d^4 \\ \frac{C_i - d^1}{d^2 - d^1} & d^1 \le C_i < d^2 \\ 1 & d^2 \le C_i \le d^3 \\ \frac{d^4 - C_i}{d^4 - d^3} & d^3 < C_i \le d^4 \end{cases} \tag{1}$$

2.3 Modeling of IPPS Problem with Fuzzy Due Date

The maximum sum of the weighted satisfaction of delivery of all jobs is expressed as follows.

$$F_{\max} = \sum_{j=1}^{n} w_j u\left(C_j\right) \tag{2}$$

where W_j is weight of workpiece J to indicate the importance of the workpiece.

The constraint conditions of the workpiece in the process are as follows.

1. Only one process route can be selected in each process

$$\sum_{j=1}^{Q_i} X_{ij} = 1 \quad \forall_i \in [1, n], \forall_j \in [1, Q_i] \tag{3}$$

2. Only one machine can be selected in each process

$$\sum_{k=1}^{m} Z_{ijpk} = 1 \quad \forall_i \in [1, n], \forall_j \in [1, Q_i] \forall_p \in [1, Q_{ij}], \forall_k \in [1, m] \tag{4}$$

3. Different processes of the same workpiece cannot be processed simultaneously

$$S_{ijpk1}X_{ij}Y_{ijpk1} - S_{ij(p-1)k2}X_{ij}Y_{ijp1k2} \geq T_{ij(p-1)k2}X_{ij}Y_{ij(p-1)k2}$$
$$\forall i \in [1,n], \forall j \in [1,Q_i], p \in [1,O_{ij}], k_1, k_2 \in [1,m] \tag{5}$$

4. When process O_{ipj1} processes before the process O_{ipj2} to make the same machine process only one process at the same time, it can be met:

$$S_{ijp2k}X_{ij}Y_{ijp2k} \quad S_{ijp1k}X_{ij}Y_{ijp1k} \geqslant T_{iijp1k}X_{ij}Y_{ijp1k}$$
$$\forall i \in [1,n], \forall j \in [1,Q_i], p_1, p_2 \in [1,O_{ij}], k \in [1,m] \tag{6}$$

5. Aiming at first process of the first process route for the workpiece i

$$C_{ij1k} \times X_{ij} \times Z_{ij1k} \geq T_{ij1k} \times X_{ij} \times Z_{ij1k}$$
$$\forall_i \in [1,n], \forall_j \in [1,Q_i], \forall_k \in [1,m] \tag{7}$$

6. Aiming at last process of the first process route for the workpiece i

$$C_{ijo\,ijk} \times X_{ij} \times Z_{ijo\,ijk} \leq Z$$
$$\forall_i \in [1,n], \forall_j \in [1,Q_i], \forall_k \in [1,m] \tag{8}$$

7. The processing time of all processes is greater than or equal to 0

$$T_{ijpk} \geq 0 \quad \forall_i \in [1,n], \forall_j \in [1,Q_i], \forall_p \in [1,O_{ij}], \forall_k \in [1,m] \tag{9}$$

3   Solving Algorithm

A genetic algorithm based on four-layer coding structure is presented, and the coding, selection, crossover, and mutation operations of the algorithm are described as follows.

Coding. ① Workpiece layer: coding of each piece of work; ② machine layer: coding of individual machines; ③ processing time layer: the processing time of different processes on each machine; ④ process route layer: process routes for different workpieces.

Fitness function. The fitness value of the chromosome is the reference objective function as shown in Eq. (2).

Selection operation. The selection mechanism of genetic algorithm which is based on the combination of League Selection and Elitist Selection is chosen.

Crossover operation. The priority crossover strategy is used to cross the individual in order to avoid the illegal solution.

Mutation operation. The mutation operation based on neighborhood search is selected.

4 Example

An enterprise will produce a batch of shaft parts, including a total of three kinds of workpiece. Figure 2 represents the optional processing route of workpiece 1, and Table 1 represents the optional machine and processing time for each process of workpiece 1. Similarly, relevant data of workpiece 2 and 3 is shown in Figs. 3, 4 and Tables 2, 3.

1. Workpiece parameters

Fig. 2 Optional processing route of workpiece 1

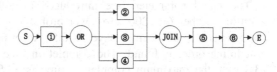

Table 1 Optional machine and processing time for each process of workpiece 1

Process	Optional machine (M)	Processing time
1	1, 2, 5	2, 1, 2
2	1, 3, 5	2, 3, 2
3	2, 5, 6	2, 3, 2
4	1, 4	3, 3.5
5	3, 5	2.5, 2
6	2, 3	4, 3

Fig. 3 Optional processing route of workpiece 2

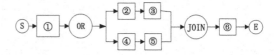

Table 2 Optional machine and processing time for each process of workpiece 2

Process	Optional machine (M)	Processing time
1	1, 2, 5	3, 2.5, 3
2	1, 4	1.5, 2
3	3, 5	2.5, 4
4	1, 4	2.5, 2
5	2, 5, 6	4, 3, 5
6	3, 5	6, 4

Fig. 4 Optional processing route of workpiece 3

Table 3 Optional machine and processing time for each process of workpiece 3

Process	Optional machine (M)	Processing time
1	1, 2, 5	4, 4.5, 3
2	1, 4	4.5, 5
3	2, 3	4, 6
4	2, 6	2, 4
5	6, 4	3, 3.5

2. Results

The initial parameters are introduced as follows. Population size $N = 100$, iteration number $G = 200$, crossover probability $Pc = 0.9$, $Pm = 0.05$, and mutation probability of machine layer is 0.5. According to the algorithm, the optimal scheduling order is found: The completion time of workpiece 1, 2, 3 is 11, 12, 13.5 min, the maximum completion time is 13.5 min, and the weighted sum of satisfaction is 0.75 (Figs. 5 and 6).

If delivery time is considered as a fixed value, the optimal scheduling order should be given when the delivery time of the workpiece 1, 2, 3 is 11, 10, 14 min, the completion time is 12, 13, 14 min, the maximum completion time is 14 min, and the weighted sum of satisfaction is 0.33 (Figs. 7 and 8).

If delivery time is a fixed value, the maximum completion time becomes greater, and customer satisfaction value is decreased compared with the fuzzy due date.

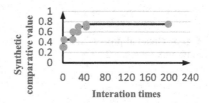

Fig. 5 Algorithm iteration chart under fuzzy due date

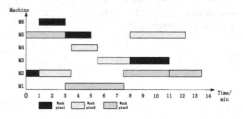

Fig. 6 Scheduling Gantt chart under fuzzy due date

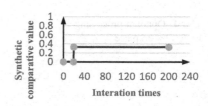

Fig. 7 Algorithm iteration chart under definite delivery date

Fig. 8 Scheduling Gantt chart under definite delivery date

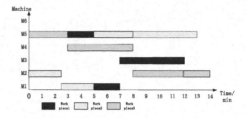

5 Conclusion

Aiming at uncertain delivery time in the process planning and scheduling, trapezoidal fuzzy numbers are introduced to present uncertain delivery time. At the same time, maximizing workpiece delivery satisfaction weighted and minimizing the maximum completion time are taken as the optimization objective. The optimal scheduling is made based on genetic algorithm for four-level coding structure and conforms the validity and feasibility of the algorithm. By using these methods, the problem can be solved better, and it has a great guiding role and practical reference value for the actual production scheduling

Acknowledgements This research is supported by the National Natural Science Foundation of China (Grant No: 61402361, 60903124); project supported by the scientific research project of Shaanxi Provincial Department of Education (Grant No: 14JK1521); Shaanxi Province Science and Technology Research and Development Project (Grant No: 2012KJXX-34).

References

1. Lv, S. P., Qiao, L. H.: Research Status and Development Trend of Process Planning and Job Shop Scheduling. J. Computer Integrated Manufacturing System. 02, 290–300 (2014).
2. Song, S. Q., Ye, C. M.: Using QPSO Algorithm to Solve Fuzzy Flow-Shop Scheduling Problem. J. Computer Engineering and Applications. 06, 246–248 (2009).
3. Palacios, J. J.; González, M. A.: Genetic Tabu Search for the Fuzzy Flexible Job Shop Problem. J. Computers & Operations Research. 54, 74–89 (2015).
4. Gutiérrez, C., García-Magariño, I.: Modular Design of a Hybrid Genetic Algorithm for a Flexible Job-shop Scheduling Problem. J. Knowledge-Based Systems, 24, 102–112 (2011).
5. Shi, J. F., Jiao, H. J.: Multi-objective Pareto Optimization on Flexible Job-shop Scheduling Problem About Due Punishment. J. Journal of Mechanical Engineering. 12, 184–191 (2012).
6. Wang, B., Li, Q.Y., Yang, X. F.: Three Point Satisfaction Model for Fuzzy Job Shop Scheduling. J. Control and Decision. 07, 1082–1086 (2012).
7. Wang, C., Liu, J. P., Chang, W. T.: Job shop scheduling problem under uncertainty. J: Equipment manufacturing technology. 154, 114–148 (2011).
8. Shahsavari-Pour, N., Ghasemishabankareh B.: A Novel Hybrid Meta-heuristic Algorithm for Solving Multi Objective Flexible Job Shop Scheduling. J. Journal of Manufacturing Systems. 32, 771–780 (2013).
9. Zhang, L. P., Wong, T. N.: Solving Integrated Process Planning and Scheduling Problem with Constructive Meta-heuristics. J. Information Sciences. 01, 340–341 (2016).

A Model for Computing User's Preference Based on EP Algorithm

Shan Jiang, Zongwei Luo, Zhiyun Huang and Jinqun Liu

Abstract In this paper, we address the problem of identifying target user through the model of computing user preference for a certain item or service. The model we present works for a specific domain through online behavior analysis which considers user's attentiveness of the entire area and the specific item combination style combining features of the specific industry. The model is evaluated by predicting users' behavior and advertising click-through rate in the real application environment. The results show that this model is successful in precision recommendation, especially for the dynamic data analysis.

Keywords Data mining · Accurate marketing · Target user identification

1 Related Works

Motivated by success application in traditional recommendation scenarios (e.g., basket analysis), collaborative filtering algorithm (CF) has been widely applied in precision recommendation. For instance, Chen et al. [1] proposed a method of using O2O neural network for precision recommendation. Chung et al. [2] proposed a knowledge-based recommendation model for agriculture precision. However, collaborative filtering also faces a few of issues, for instance how to screen out target user precisely, how to get the point-of-interest (POI) of users', and how to advertise more efficiency. These issues are also the focus of research in precision.

In order to screen out proper users, we should construct the users' social connective. Zhai et al. [3] presented a social graph-based collaborative filtering method. Multiage social network recommendation algorithms (MSN) and a new assessment

S. Jiang · Z. Luo (✉)
Department of Computer Science and Engineering, Southern University of Science and
Technology, Shenzhen 518055, China
e-mail: luozw@sustc.edu.in

Z. Huang · J. Liu
Shenzhen Aotain Technology Co., Ltd, Shenzhen 518055, China

B. K. Panigrahi et al. (eds.), *Smart Innovations in Communication and
Computational Sciences*, Advances in Intelligent Systems and Computing 670,
https://doi.org/10.1007/978-981-10-8971-8_11

method, called similarity network evaluation (SNE), were both proposed in [4]. Get users' point-of-interest (POI) is another popular direction in precision recommendation. Liu et al. [5] proposed a novel geographical probabilistic factor analysis framework which strategically takes various factors into consideration. Gao et al. [6] and Liu et al. [7] expanded the location-based social networks (LBSNs) on find users' POI. Gao et al. [6] associated the spatial, temporal, and social information in online LBSNs with human behavior. Liu et al. [7] predicted users' preference over location categories to improve location recommendation accuracy. In Internet ads recommendation field, users' online behavior history and location information are the key factors to decide whether a user is the right candidate for the certain advertisement. Gandhi et al. [8] combined recommendation for tourism application and data mining techniques. Liu and Yang [9] improved advertisement recommendation by enriching user browser cookie attributes. Tao [10] designed a recommendation in large-scale mobile advertising.

We developed a set of statistical methods for target users filtering model from existing domain ontology and a user's preferences calculation method considering operating behavior of the user. The proposed target user filtering model has been incorporated into a semantic analysis and Web crawling techniques which originally utilizes domain knowledge to establish the database of users' access records. We evaluated the performance of the filtering model using a real-world data set, the Shenzhen property data. The experimental results support the efficacy of the filtering model which application in real world. The main contribution of the paper is the novel model for searching target user in different field which takes into account:

- unique features and promotion requirements of the different industry;
- the user's attention to the certain industry;
- the user's preferences for a particular product from the certain industry.

The rest of this paper is organized as follow. Section 2 provides a formulation of the model. Section 3 introduces all modules of the model. In Section 4, we present how our model can be used for target user filtering and experiment in real word. Finally, we conclude and highlight future work in Sect. 5.

2   Introduction

How to identify target users and understand the needs of them, and finally reach the realization of precision marketing, are currently facing problems for many companies. In the past, simple statistics analysis methods were the mainstream methods, but the historical data statements, and cannot provide a basis for decision maker, only extensive marketing. In recent years, as the data mining technology matures, it has been widely applied to many areas, such as banking, transportation, and the Internet industry, etc., where data mining technology has become a key pillar of technology. Most of the current such as clustering or association rules require the very clean sample and cannot recognize the noise in the sample and dirty data. But in real world, the classical data mining approaches are not suited for this situation. However, using traditional mining methods on uncertain data will bias the answer set, even is wrong, and hence cannot satisfy the user's needs. In the case of established marketing costs, select the best client base, improve marketing efficiency, and maximize marketing effectiveness.

After information crawling and content analysis of the target sites, it is necessary to calculate the preferences for the selected specific field of the users who have visited the target Web sites, in order to enrich users' portrait in this field. The model of computing users' preference in specific area based on EP (entire-property) algorithm is proposed, in order to meet the needs of the target user screening.

Users-filtered for a particular domain, the degree of the user interested in a particular field is determined by attentiveness of the Entire area which the item is included, and the property set which define and distinguish the item from other items in the same area. The proposed model is based on the overall and partial mining of user preference, which involves data mining and expert's knowledge. Based on the EP algorithm, the users' preference calculation model is divided into three modules: the user access record analysis module, the computing module based on EP algorithm, and the target user selection module. The specific framework, module division, and collaboration between modules can be shown in the schematic diagram. It can be concluded from the flowchart that the module requires some background knowledge for the specific area for the product and user modeling, parameter adjustment (Fig. 1).

In many application domains like high-tech electronic products, which is important to make timely follow-up of user interests' change, and get the most possible consumers. User information can be updated promptly and regularly while time-sensitive industry through time-factors and experts knowledge for the industry. To meet different customers' promotion needs.

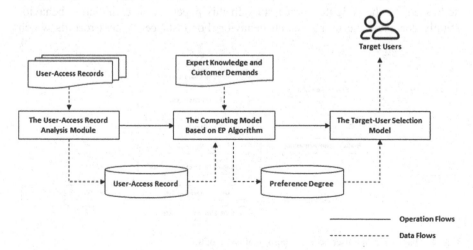

Fig. 1 Schematic diagram of computing user preference based on EP algorithm

3 The Model Based on EP Algorithm

3.1 The User Access Record Analysis Module

The user access record analysis module obtains users' surf data as input from the data platform, that is, users' access to the specific domain target site records, through this module to obtain access to users' accesses which are stored in the user's information table. Considering business characteristics and timeliness, user's behavior recording table as input to the next module is dynamically updated. The main work of this part is feature selection and feature value acquisition. And this module consists of three steps: feature selection, calculation of feature weight calculation, and record table adjustment (Fig. 2).

Feature selection: Any product or service j for the specific areas of A can be described in terms of product characteristics and customer needs with a set of features; that is, $j = \{j_1, \ldots, j_M\}$ where M is the amount of dimensions chosen by the system. Feature selection affects the accuracy of the following forecast. This part of work needs to combine expert knowledge and experience, and the selected features need to meet the requirements of a comprehensive description of the product and low coupling between features.

Feature weight calculation: The features of each dimension will be given a certain number of eigenvalue which is analyzed and calculated through users operating behaviors. This step mainly based on the user i interview with the specific domain of the Web pages, adjusted statistical frequency acquired and stored in freq(i_{j_k}), and all users' characteristic values are stored in the database D.

Record table adjustment: The main consideration is two factors, updating of users' behaviors and the specific business characteristics.

The main consideration of users' active behavior and the general behavior of the user needs to describe the importance of difference. In general, we believe that the behavior of the user is more responsive to the needs of the user needs to be addressed, so the weighted processing. In this paper, the user initiative behavior mainly considers the user's search behavior. For instance, Access records which

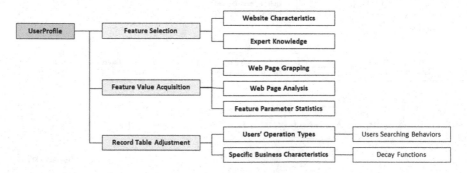

Fig. 2 Diagram of the user access record analysis module

generated by the active search of the user after analyzing web operating record, the eigenvalue value needs to be changed on the basis of the original value. The detail information is determined by industry characteristics and experience, and we temporarily set the parameter to 3.

Considering the characteristics of the industry and the timeliness of information, we have to adjust the user's access record table: The concept of the data stream has two characters: accuracy and timeliness. The information of the data stream will be continuous change, and the importance of the expert knowledge will be continuous decay. Therefore, it is necessary to distinguish between current and historical transactions, and weakening the impact of historical events on the results of data mining. In order to get the change of the users' hot-interest currently and provide more personal service, we set different attenuation functions for different industries. Attenuation functions' selection needs relevant expert knowledge; choose different value θ to control the information decay. Generally speaking, we think the users' browsing history of the traditional commodity has longer analyzation cycle compared to fast-moving commodity. The method is based on the time decay model, decay the supporters of this event depends on time t, and the attenuation function is $f(t) = 2^{-\theta t}, \theta > 0$. Using full attenuation support number to reduce the information, end to the time t_c,

$$D_a(P, t_c) = \sum_{T_j \in \text{Time}_w(P)} 2^{-\theta(t_c - T_j)}$$ (1)

Taking into consideration of the timeliness of information, in order to accurately captures the changes in users' interest in order to provide a more personalized service, we set different attenuation function for different industries. The selection of attenuation function needs the support of relevant industry experts. Upon completion of the above work, the real-time access record for each user will be updated according to the following formula (2).

$$\text{freq}_{t_c}(i_{jk}) = D_a(P, t_c) \cdot \text{freq}_{t_{c-1}}(i_{jk}).$$ (2)

3.2 The Computing Model Based on EP Algorithm

After this, module will be obtained for each user's preference weight of a certain item, and the usage of this preference weight can provide data support for businesses to achieve precision marketing. A user i's preference for the product or service j of area A, which is symbolized W_{ij}, is determined by two factors: E_{iA}, user I's interest of area A; P_{ij}, user I's interest of item j that is characterized by a method of attributes optimization. The preference can be calculated by formula (3).

$$W_{ij} = \alpha E_{iA} \times P_{ij} \tag{3}$$

where: α as adjustable parameter for threshold setting and W_{ij} adjustment, the value of which is the reciprocal of expectation, as shown in formula (4); N is the total number of users.

$$\alpha = \frac{1}{\left(\frac{1}{2}\right)^M \cdot \frac{1}{N}} = N \cdot 2^M \tag{4}$$

E_{iA} can be calculated by formula (5), which from the overall level of consideration for degree of concern for a particular field, that is, does not consider a certain attributes or characteristics of the product or service. A specific user i's browsing frequency for all products in a specific area is represented as Freq(i);

$$E_{iA} = \frac{\text{Freq}(i)}{\sum_{i=1}^{N} \text{Freq}(i)} \tag{5}$$

P_{ij} only analyzes the access data with attributes and features information of the certain user i. Through the analysis of the user's access to the attributes of the commodity, the quantitative analysis is made to determine the preference degree of user i to the specific attribute.

$$P_{ij} = \prod_{k=1}^{M} (P_{ij_k})^{\gamma_k} \tag{6}$$

where M represents the attribute dimensions for item j; γ_k which is given by the specific domain A and k's importance to describe j and is the weight of k attribute for product j; P_{ij_k} is determined by the times the user i view the Webs about item j. We set the parameter β to 0.8 temporarily. P_{ij_k} is calculated by changing the classical normalization, which avoids invalid calculation results.

$$P_{ij_k} = \frac{\log[\text{freq}(i_{j_k}) + 1] - \beta \cdot \log\{\min[\text{freq}(i_{j_k})] + 1\}}{\log\{\max[\text{freq}(i_{j_k})] + 1\} - \beta \cdot \log\{\min[\text{freq}(i_{j_k})] + 1\}}. \tag{7}$$

3.3   The Target User Selection

The last module of this model for marketing and product promotion is user selection based on their preference for the product. This module mainly solved two problems: the determination of the number of the target users and the selection of the target user.

According to the customer demand target, target users' screening is divided into two cases: (1) When the customer does not determine the number of target users, the number of filtered users is decided by the strength of preference of user i to item j, which means that all users meet the demand of $W_{ij} \geq 1$ which will be chosen by the EP method as target user; (2) When the number of target users is specified, users are sorted by the value of W_{ij}, according to the calculation module based on EP algorithm.

We use precision rate and recall rate to measure model's effect. The precision rate reflected relevant instances among the retrieved instances. The recall rate reflected relevant instances that have been retrieved over total relevant. Finally, use the *F-score* to evaluate this model.

4 Experiments

In order to demonstrate the effectiveness of the model within context-aware applications, a case study is presented which focuses on providing precision marketing services in the property sector. Target user identification is an integral part of precision marketing services which defines the object for advertising campaigns. A real-world document collection is used in our experiments, namely the Shenzhen property data. The details are given in the following sections.

Experiments data sets collected *10,000* records from users' Web explorer recordings from five popular real estate Web site in China (www.lianjia.com, www.anjuke.com, www.centanet.com, www.fang.com, www.qfang.com). The experimental trial has been conducted for 2 months. The records are divided into two groups, one using approach to construct user model and another using as test data sets.

Before the experiments, we give an example of a specific application environment that is the real estate industry to prove EP algorithm's advantageous in practical application. Expert knowledge for the real estate industry is collected by consulting works. We select three factors as SSS, location (L), price (P), and apartment type (T). Web crawler technology and Web mining are used to analyze the Web sites information.

According to the expert knowledge provided by the real estate industry, that is, the introduction of a real estate development model, assume that in this case any projects are available in three basic features in characterization, namely area (shorted as A), the price (shorted as P), and the room type (shorted as T), different characteristic values were marked for each feature, specific information as shown in the table. Users target site accesses record, real estate information, and customer needs, as input information. Target users were who may interest in the real estate developers launched, as output information (Table 1).

Table 1 Description of characteristic and characteristic value

Symbol	Character	Value	Implication
L	Location	R1	Nanshan
		R2	Baoan
		R3	Futian
		R4	Others
P	Price	P1	Below 4,000,000
		P2	4,000,000 ∼ 10,000,000
		P3	More than 10,000,000
T	Types	T1	One bedroom
		T2	Two bedrooms
		T3	Three bedrooms
		T4	Others

We used the offline data to get the recall–precision graph, which contains predict users' click situation. As we can conclude, the method achieves better performance compared to the statistical method especially during short period of time (Fig. 3).

In this paper, we use the online environment to test the performance of the proposed model from two aspects, the click-through rate and Web site staying time. During 30 days, a real estate promotion project, we got 10 thousands target users by EP model and other 10 thousands target users by statistical method of the real estate firm. As a contrast, 10 thousands users were randomly selected from the five Web site browsers within 2 months. Ads clicks and other online activities from the three groups were monitored and analyzed. From the online experiments, as shown in Fig. 4, the EP model achieves better performance.

Fig. 3 Diagram of recall–precision graph

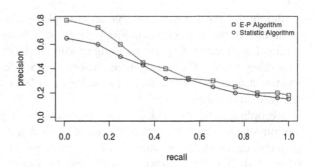

Fig. 4 Diagram of users' click behaviors for a real estate promotion within a month, which contain ten different content online advertisings

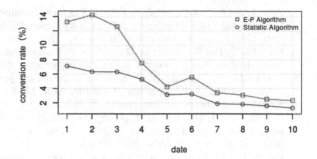

5   Conclusion

In this paper, we proposed a novel approach for target user selection for a certain industry based on surf behaviors.

Innovation point: Based on different application scenarios, by tracking user' behavior, depicted the level of users' preference degree and the consumer willingness. Combine the attention degree of the products and the characteristic of the products to assessment customers. Maintain the information of the user dynamically to accurately screen out the needs of target users. The method is integrated into the entire social environment, the users' attention (demand) of the certain industry, and the attention degree of the products which have specific attribute. Finally screen out the target user to prompt which meet the needs of enterprises.

Application prospect: In the new era of information explosion, it is of great significance for enterprises to excavate valuable information from massive data, for enterprises and institutions, when enterprises got potential customers' operation information. Take method to screen out target user, use these data efficiently and improve enterprise marketing hit rate, can reduce the cost and get higher income. For the ordinary users, personalized and targeted message push can make user get more useful information and reduce useless information disturb.

Acknowledgements This work was partially supported by GDNSF fund (2015A030313782), SUSTech Starup fund (Y01236215), SUSTech fund (05/Y01051814, 05/Y01051827, 05/Y01051830, and 05/Y01051839).

References

1. Chen Y C, Hsieh H C, Lin H C. Improved precision recommendation scheme by BPNN algorithm in O2O commerce [C]//e-Business Engineering (ICEBE), 2013 IEEE 10th International Conference on. IEEE, 2013: 324–328.
2. Chung N S, Kim C H, Oh T S, et al. Development of a Knowledge-Based Crop Recommendation Model for Precision Agriculture [J]. The Journal of the Korean Society of International Agriculture, 2013.

3. Zhai X, Jin F, Wang J, et al. A Kind of Precision Recommendation Method for Massive Public Digital Cultural Resources: A Preliminary Report [C]//Multimedia Big Data (BigMM), 2016 IEEE Second International Conference on. IEEE, 2016: 56–59.
4. Hu J, Gao Z, Pan W. Multiangle Social Network Recommendation Algorithms and Similarity Network Evaluation [J]. Journal of Applied Mathematics, 2013, 2013.
5. Liu B, Fu Y, Yao Z, et al. Learning geographical preferences for point-of-interest recommendation [C]//Proceedings of the 19th ACM SIGKDD international conference on Knowledge discovery and data mining. ACM, 2013: 1043–1051.
6. Gao H, Tang J, Hu X, et al. Exploring temporal effects for location recommendation on location-based social networks [C]//Proceedings of the 7th ACM conference on Recommender systems. ACM, 2013: 93–100.
7. Liu X, Liu Y, Aberer K, et al. Personalized point-of-interest recommendation by mining users' preference transition [C]//Proceedings of the 22nd ACM international conference on Information & Knowledge Management. ACM, 2013: 733–738.
8. Gandhi M, Mistry K, Patel M. A Modified Approach towards Tourism Recommendation System with Collaborative Filtering and Association Rule Mining [J]. International Journal of Computer Applications, 2014, 91(6).
9. Liu R S, Yang T C. Improving Recommendation Accuracy by Considering Electronic Word-of-Mouth and the Effects of Its Propagation Using Collective Matrix Factorization [C]// Dependable, Autonomic and Secure Computing, 14th Intl Conf on Pervasive Intelligence and Computing, 2nd Intl Conf on Big Data Intelligence and Computing and Cyber Science and Technology Congress (DASC/PiCom/DataCom/CyberSciTech), 2016 IEEE 14th Intl C. IEEE, 2016: 696–703.
10. Tao Y. Design of large scale mobile advertising recommendation system [C]//Computer Science and Network Technology (ICCSNT), 2015 4th International Conference on. IEEE, 2015, 1: 763–767.
11. Feng Y, Tang R, Zhai Y, et al. Personalized media recommendation algorithm based on smart home [C]//The Second International Conference on e-Technologies and Networks for Development. The Society of Digital Information and Wireless Communication, 2013: 63–67.
12. Li J, Xia F, Wang W, et al. Acrec: a co-authorship based random walk model for academic collaboration recommendation [C]//Proceedings of the 23rd International Conference on World Wide Web. ACM, 2014: 1209–1214.
13. Ju C, Xu C. A new collaborative recommendation approach based on users clustering using artificial bee colony algorithm [J]. The Scientific World Journal, 2013, 2013.
14. Bashir M A, Arshad S, Wilson C. Recommended For You: A First Look at Content Recommendation Networks [C]//Proceedings of the 2016 ACM on Internet Measurement Conference. ACM, 2016: 17–24.
15. Venkatraman V, Dimoka A, Pavlou P A, et al. Predicting advertising success beyond traditional measures: New insights from neurophysiological methods and market response modeling [J]. Journal of Marketing Research, 2015, 52(4): 436–452.
16. Wang Y, Feng D, Li D, et al. A mobile recommendation system based on logistic regression and Gradient Boosting Decision Trees [C]//Neural Networks (IJCNN), 2016 International Joint Conference on. IEEE, 2016: 1896–1902.

3D Face Recognition Method Based on Deep Convolutional Neural Network

Jianying Feng, Qian Guo, Yudong Guan, Mengdie Wu, Xingrui Zhang and Chunli Ti

Abstract In 2D face recognition, result may suffer from the impact of varying pose, expression, and illumination conditions. However, 3D face recognition utilizes depth information to enhance systematic robustness. Thus, an improved deep convolutional neural network (DCNN) combined with softmax classifier to identify face is trained. First, the preprocessing of color image and depth map is different in removing redundant information. Then, the feature extraction networks for 2D face image and depth map are, respectively, build with the principle of recognition rate maximization, and parameters about neural networks reset by a series of tests, in order to acquire higher recognition rate. At last, the fusion of two feature layers is the final input of artificial neural network (ANN) recognition system, which is followed by a 64-way softmax output. Experimental results demonstrate that it is effective in improving recognition rate.

Keywords 3D face recognition · Depth map · Deep convolutional neural network
Feature extraction · Feature fusion

J. Feng (✉) · Y. Guan · X. Zhang · C. Ti
School of Electronics and Information Engineering, Harbin Institute of Technology,
Xidazhi Street. 92, Harbin 150001, China
e-mail: fjy_fengjianying@163.com

Y. Guan
e-mail: 157673503@qq.com

X. Zhang
e-mail: 16S105149@stu.hit.edu.cn

C. Ti
e-mail: 823231451@qq.com

Q. Guo
Beijing Automation Control Equipment Research Institute, Beijing 100074, China
e-mail: 13810106152@139.com

M. Wu
School of Information Engineering Branch, Yangling Vocational & Technical College,
Weihuilu. 24, Yangling 712100, Shanxi, China
e-mail: 13131724330@163.com

© Springer Nature Singapore Pte Ltd. 2019
B. K. Panigrahi et al. (eds.), *Smart Innovations in Communication and
Computational Sciences*, Advances in Intelligent Systems and Computing 670,
https://doi.org/10.1007/978-981-10-8971-8_12

1 Introduction

Face recognition has been widely applied because of its characteristics of being easy to acquire, direct, interaction, and so on [1]. It is a valuable research direction for pattern recognition and machine learning. Two-dimensional face recognition is relatively complete at present, but it is not enough robust when it confronts various conditions. Three-dimensional face recognition utilizes depth information to avoid the inflation of environmental factors, which exist in 2D face recognition. More and more researchers focus on 3D face recognition, while the recognition rate of traditional superficial layer methods has trend saturation, especially when the training sets are enormous. Thus, some people try to utilize neural network to solve the above-mentioned problem [2, 3]. It is proposed that the cascade correlation neural network with multi-core programming model realizes 3D facial recognition system [4]. A method for automatically recognizing expression using DCNN is proposed. This method can reduce the time of extracting feature by general-purpose graphic processing unit [5]. Traditional face recognition methods are susceptible to various illuminations, different expressions, and other changeable conditions. Multilayer constructions, especially deep convolutional network, can have flexible reaction to these nonlinearity problems.

In this paper, DCNN is applied in 3D face recognition. Firstly, appropriate methods are taken to preprocess original data, including elimination of redundancy information and normalization. Secondly, an improved DCNN combined with softmax classifier is trained. As the principle of recognition rate maximization, the processions of extracting 2D and 3D features are independent. Moreover, the output of two feature layers is considered as the input of ANN recognition system. Finally, the recognition system is tested and compared with other methods. Excellent experiment performances on testing set show the great superiority of this method.

The rest of the paper is composed as follows: The second part elaborates the construction of DCNN in detail; the corresponding experiments are demonstrated in third part; and conclusion will be summarized in the last part.

2 3D Face Recognition Method Based on DCNN

2.1 Construction of DCNN

DCNN for color image is shown in Fig. 1, where the input image is 64×64, the first convolutional layer with 15 kernels is 5×5, the second convolutional layer includes 30 kernels of size 4×4, the size of the third convolutional layer with 45 kernels also is 4×4, and the kernel of two max-pooling layers is defined as $2 \times 2, 3 \times 3$, respectively. Figure 2 displays the DCNN of exacting depth feature. The kernels' size of two convolutional layers both is 5×5, and two max-pooling sampling layers is 2×2 Considering the speed and complexity computation, the activation function of two DCNNs is ReLU rather than Sigmoid [6].

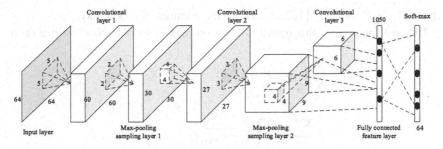

Fig. 1 Construction of DCNN for color image

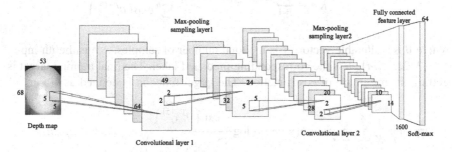

Fig. 2 Construction of DCNN for depth map

2.2 Softmax Regression Layer and Training Algorithm

Softmax layer outputs n parameters, which express the probability of n kinds of face samples. Given the predicted label $y \in \{1, 2, \ldots, n\}$, the probability of object belonging to the ith classify is

$$\phi_i = P(y = i | \phi) \tag{1}$$

where $\sum_{i=1}^{n} \phi_i = 1$

The predicted value of nth classify is

$$P(y = n | \phi) = \sum_{i=1}^{n1} \phi_i \tag{2}$$

Thus,

$$\phi_i = \frac{\exp\left(\theta_i^T x\right)}{\sum_{j=1}^{n} \exp\left(\theta_j^T x\right)} \quad i \in \{1, 2, \ldots, n\} \tag{3}$$

where $\theta_j = \{W_j, b_j\}, j \in \{1, 2, \ldots, n\}$ are the parameters of softmax.

The cost function of this network based on maximum likelihood estimation is

$$
\begin{aligned}
J(\theta) &= \frac{1}{K}\sum_{i=1}^{K} \log P(y_i \big| x_i^{(5)}\theta) \\
&= \frac{1}{K}\sum_{i=1}^{K} \log \prod_{j=1}^{n} \left(\frac{\exp\left(\theta_j^T x_i^{(5)}\right)}{\sum_j^n \exp\left(\theta_j^T x_i^{(5)}\right)} \right)^{\text{isture}(y_i=j)} \\
&= \frac{1}{K}\sum_{i=1}^{K}\sum_{j=1}^{n} \text{isture}(y_i = j) \cdot \log \frac{\exp\left(\theta_j^T x_i^{(5)}\right)}{\sum_j^n \exp\left(\theta_j^T x_i^{(5)}\right)}
\end{aligned}
\tag{4}
$$

where θ is a alterable vector, K is the total number of samples, $x_i^{(5)}$ is the ith input, $i \in \{1, 2, \ldots, m\}, y_i$ is the corresponding predicted label. The final training target is getting the minimum of cost function. The target function is defined as

$$
\arg \min_{\theta} \frac{1}{K}\sum_{i=1}^{K}\sum_{j=1}^{n} \text{isture}(y_i = j) \cdot \log \frac{\exp\left(\theta_j^T x_i^{(5)}\right)}{\sum_j^n \exp\left(\theta_j^T x_i^{(5)}\right)}, \quad i \in (1, 2, \ldots, n)
\tag{5}
$$

The partial derivation of Eq. (5) is

$$
\nabla_{\theta_i} J(\theta) = \frac{1}{K}\sum_{i=1}^{K} \left[x_i^{(5)} \cdot \left(\text{isture}(y_i = j) P\left(y_i = j \big| x_i^{(5)}\theta \right) \right) \right]
\tag{6}
$$

Softmax is n dimension, but its degree of freedom is $n - 1$, and thus, there is redundancy δ. Redundancy leads possibly to overlarge update. Usually, λ is added to cost function. Give an example as follows

$$
J(\theta) = \frac{1}{K}\sum_{i=1}^{K}\sum_{j=1}^{n} \left[\text{isture}(y_i = j) \cdot \log \frac{\exp\left(\theta_j^T x_i^{(5)}\right)}{\sum_j^n \exp\left(\theta_j^T x_i^{(5)}\right)} \right] \frac{\lambda}{2}\sum_{i=1}^{n}\sum_{j=0}^{120} \theta_{ij}^2
\tag{7}
$$

The partial derivation of Eq. (7) is

$$
\nabla_{\theta_i} J(\theta) = \frac{1}{K}\sum_{i=1}^{K} \left[x_i^{(5)} \cdot \left(\text{isture}(y_i = j) P\left(y_i = j \big| x_i^{(5)}\theta \right) \right) \right] \lambda \cdot \theta_j
\tag{8}
$$

It can be proved that $J(\theta)$ is a convex function. In order to reduce computational expense and speed up the training procession, stochastic gradient descent algorithm is taken. Training samples are randomly divided into specific groups to reconstitute the cost function with these groups.

In the real training, there is a coefficient of momentum to control iteration,

$$v^{(i1)} = 09v^{(i)} \alpha \nabla_{\theta_i} J(\theta)^{(i)} \tag{9}$$

where 0.9 is a constant, v is a momentum, α is studying rate.

2.3 Feature Fusion

The fusion of two feature layers is the final input of ANN recognition system. The structure is summarized in Fig. 3, where the input of 2D face feature is 1050 dimension, depth feature is 1600 dimension, partially connected layers is the half of corresponding input, and the final output layer is a 64-way softmax. Moreover, the number of fully connected layer is set as 450 by tests.

3 Experimental Results and Discussion

The experiment has been conducted using CASIA-3D FaceV1 collected by the Chinese Academy of Sciences' Institute of Automation (CASIA) [7]. The database contains 4624 scans of 123 people; each person has 37 or 38 scans, where the one who wears glasses has one additional scan. Furthermore, it considers various poses, illuminations, and expressions. Three-dimensional facial data is saved as wrl file,

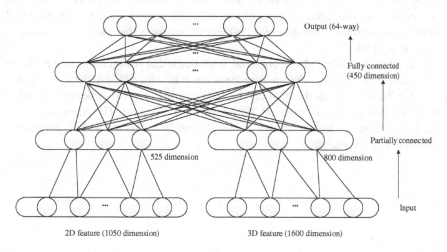

Fig. 3 Structure of ANN

Fig. 4 Some depth maps after preprocessing

which gives some information such as the number and space coordinates of point cloud, each point color (RGB) and normal vector, and the number of triangular patches. There are 64 persons selected as the final experimental data, and everyone includes 16 color images and 10 depth maps.

AdaBoost algorithm combined with Haar feature is used to remove redundancy information of 2D face images [8, 9]. Convert color images to gray images, then normalized 64×64. The 3D grid model is built according to the point cloud data to get depth map. Then, Otsu's method is applied to remove redundancy [10]. Figure 4 shows some depth maps which are normalized to 68×53.

Learning rate can be regarded as the range of parameters update. Data preprocessed is divided into training set, verifying set, and testing set. To find the most suitable learning rate, respectively, checkout learning rate $\alpha = 01, 007, 006 \cdot 005, 004, 001$ Finally, learning rate is set to 0.06 according to the maximization of recognition rate.

Different features' number between DCNN and softmax can result in different recognition rate. So based on a series of test, 2D and 3D feature numbers are, respectively, reset to 1050 and 1600, namely the dimensions of the input layer. In addition, Fig. 5 displays 3D feature map of every layer output for better comprehension. Two-dimensional feature map is not given to economize space, and it also makes 3D feature more and more abstract with the increasing of layers.

To illustrate the stability of the method, five different test sets were used for testing. The recognition rate is 96.87, 98.44, 100.00, 95.31, and 98.44%. There are two reasons for the different recognition rate. Objective reason is that pose and illumination conditions are different. Subjective reason is that the preprocessing is not enough ideal.

In the end, this method is compared with single 2D recognition, depth recognition, and some others. The final recognition rate with CASIA database is displayed in Table 1. From the data can be seen that this method is better than other methods when system is based on 2D feature or 3D feature. The recognition rate with fusion feature is higher about 5% than 2D feature in this method. Experimental results are commendable.

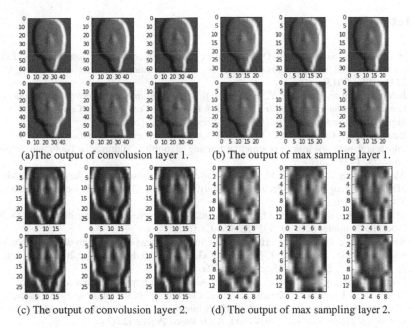

(a)The output of convolusion layer 1.　　(b) The output of max sampling layer 1.

(c) The output of convolusion layer 2.　　(d) The output of max sampling layer 2.

Fig. 5 Visualization output of depth feature

Table 1 Recognition rate of different methods

	Recognition rate (%) (2D)	Recognition rate (%) (3D)	Recognition rate (%) (2D and 3D)
PCA	80.100	70.430	–
ICA&SVM	87.000	78.940	–
Traditional CNN	83.570	78.280	–
Improved Fisher	91.800	80.430	–
Method in this paper	93.750	85.938	98.440

4 Conclusion

Face recognition has been widely used in many fields, but 2D recognition technology is confronted with the bottleneck of development and extension as it is instable of constantly changing factors. Facial depth map embodies 3D feature, which is beneficial for face recognition. In this paper, a 3D face recognition method with DCNN is designed. The fusion of 2D and 3D feature as the final input of softmax makes the best of all information. Experimental results demonstrate that it is effective in improving the recognition rate. With the development of 3D technology, 3D face recognition will be popularized in human–computer interaction, public security, entertainment, and so on.

References

1. Holland, C.D., Komogortsev, O.V.: Complex Eye Movement Pattern Biometrics: The Effects of Environment and Stimulus. Transactions on Information Forensics & Security. 8(12), 2115–2126 (2013).
2. Jadon, S., Kumar, M., Rathi, Y.: Face recognition using Som Neural Network with Ddct facial feature extraction techniques. In: 2nd IEEE International Conference on Communications and Signal Processing, pp. 1070–1074. IEEE Press, Chennai (2015).
3. Boughrara, H., Chtourou, M., Amar, C.B., et. al.: Facial expression recognition based on a mlp neural network using constructive training algorithm. Multimedia Tools and Applications. 75, 1–23 (2016).
4. Alqatawneh, S.M., Jaber, K.M.: Parallel Cascade Correlation Neural Network Methods for 3D Facial Recognition: A Preliminary Study. Journal of Computer & Communications. 3, 354–373 (2015).
5. Mayya, V., Pai, R.M., Pai, M.M.M.: Automatic Facial Expression Recognition Using DCNN. Procedia Computer Science. 93, 453–461 (2016).
6. Krizhevsky, A., Sutskever, I., Hinton, G.E.: ImageNet classification with deep convolutional neural networks. In: Huang, T.W., Zeng, Z.G., Li, C.D. (eds.) Euro-Par 2012. NIP, vol. 25, pp. 1097–1105. Springer, Heidelberg (2012).
7. Chinese Academy of Sciences' Institute of Automation, http://biometrics.idealtest.org.
8. Erdem C.E., Ulukaya S., Karaali A., et al. Combining Haar Feature and skin color based classifiers for face detection. In: 36th IEEE International Conference on Acoustics, Speech and Signal Processing, pp. 1497–1500. IEEE Press, Prague (2011).
9. Huang, C.C., Tsai, C.Y., Yang, H.C.: An Extended Set of Haar-like Features for Bird Detection Based on AdaBoost. In: Kim, T.H., Adeli, H., Ramos, C. (eds.) Euro-Par 2011. CIP, vol. 206, pp. 160–169. Springer, Incorporated (2011).
10. Yan, L.V., Gong, Q.: Application of weighting 3D-Otsu method in image segmentation. Application Research of Computers. 28, 1576–1579 (2011).

Color-Guided Restoration and Local Adjustment of Multi-resolution Depth Map

Xingrui Zhang, Qian Guo, Yudong Guan, Jianying Feng and Chunli Ti

Abstract The depth map obtained by Microsoft Kinect is often accompanied with a large area information loss called black hole. In this paper, an optimal algorithm for image restoration with multi-resolution anisotropic diffusion is proposed to fill the black areas. At the same times, the restriction of anisotropic diffusion algorithm can be broken through by using multi-resolution. Using the color map as a guidance to refine the details of the image. Then according to local adjustment, the error rate can be effectively reduced. At the end of this paper, the joint bilateral filter (JBF) is introduced as a contrast, and the PCL open source point cloud database is used for 3D reconstruction. Compared with competing methods, the proposed algorithm can fill the larger black holes significantly. The experiments show that it also has good performance on reducing the error rate and preserving the edge details.

Keywords Computer vision · Depth map enhancement · Anisotropic diffusion
3D reconstruction

PCL is a open project for point cloud processing in 2D/3D image.

X. Zhang (✉) · Y. Guan · J. Feng · C. Ti
School of Electronics and Information Engineering, Harbin Institute of Technology,
Xidazhi Street 92, Harbin 150000, China
e-mail: 16s105149@stu.hit.edu.cn

Y. Guan
e-mail: 157673503@qq.com

J. Feng
e-mail: 2360968637@qq.com

C. Ti
c-mail: 823231451@qq.com

Q. Guo
Beijing Automation Control Equipment Research Institute, Beijing 100074, China
e-mail: 13810106152@139.com

© Springer Nature Singapore Pte Ltd. 2019   131
B. K. Panigrahi et al. (eds.), *Smart Innovations in Communication and
Computational Sciences*, Advances in Intelligent Systems and Computing 670,
https://doi.org/10.1007/978-981-10-8971-8_13

1 Introduction

In the field of computer vision, for instance, target tracking, image recognition, and 3D reconstruction, a high-quality image is required. However, due to the limitations of the acquisition equipment and acquisition condition, the image acquired directly cannot meet the requirements of high quality. Therefore, it is necessary to restore and optimize depth map.

Over the past several years, there were many scholars having done a lot of exploration on the restoration algorithm of depth map. The methods can be split into two groups: one group is using filter; another one is based on filling holes. (1) Based on filter, Buades et al. [1] use a nonlocal mean filtering scheme to fill the missing areas, but it makes the edges blur. Min et al. [2] put forward a pattern filtering algorithm. He et al. [3] put forward a new bootstrap filter (linear shift filtering algorithm). Camplani et al. [4] proposed an iterative joint bilateral filter (JBF) to restore the missing areas. And the weight coefficient of the filter is affected by three elements: visual data, depth values, and time consistency. But, those methods cannot fix the larger black holes.

In contrast, the results of the depth map will not only have different degrees of defects after using filtering, but also it is easy to cause the depth data inconsistent. (2) Based on filling holes, in this field, the color image is usually used as a guide to repair the image [5–7], but this method does not take into account the depth of discontinuity at the edge of the object. The corresponding color image is used to refine the missing depth pixels, which can solve the problem of edge contour blur. There are also some scholars who repair the image based on geometric model, which belongs to partial differential equation (PDE) [8, 9]. This algorithm has good results on the smaller black holes. In this paper, a color-guide restoration and local adjustment of multi-resolution depth map for partial differential equation algorithm are proposed, which can repair larger black holes. The algorithm has two steps, forward repair and backward local adjustment.

The features of this algorithm are as below.

1. The multi-resolution anisotropic diffusion based on color image guidance is proposed to fill the image with larger black holes, which lost a lot of depth value;
2. Image pyramid is used to extract more detail information from the image. Multi-layer structure of mask provides the basis for local adjustment;
3. The key step of the proposed algorithm is local adjustment. The spatial distribution characteristics have been taken into account. Row priority and column priority pixel correction methods are both used in local adjustment. It effectively prevents the emergence of iterative error.

1.1 Outline

Other arrangements for this article are organized as follows. The color-guide restoration and local adjustment methods are described in Sect. 2. To confirm the algorithm's effectiveness, the experiment results for various repaired image are presented in Sect. 3. The last section is a summary.

2 Optimized Algorithm

The advantage of anisotropic diffusion algorithm is edge preservation. The results show that the edge of the restoration image is good, and the problem of edge blurring and unevenness is solved. As shown in Algorithm 1, in order to eliminate edge error caused by edge not aligned. Firstly, align and adjust two maps that acquired from the Kinect in the same moment. Then sub-samples the maps. Next, compute the average of the four neighborhoods to fill for the lowest resolution depth map. The color map is used as a guidance to do anisotropic diffusion [10]. Before the up-sampling, the removal of the error pixels is to be done under the guidance of mask, seeing in Algorithm 2. The spatial distribution characteristics of image are taken into consideration in local adjustment. After local adjustment, repeat the steps above until the image is recovered to original resolution.

Algorithm 1: Multi – resolution fillinng black holes

Input: Read color image C, depth map P, the number of layers L.
1. Down-sampling
   For $n=0$: L-1.
     Down-sampling C_n, P_n by a factor of 2^n and store in C_{n-1}, P_{n-1}.
     Get mask M_n from P_n by using adaptive threshold.
   End for
2. Hole filling
   Filling holes on P_{L-1} through adaptive neighborhood filling to get $\hat{P}_{L-1}$.

2.1 Optimized Multi-resolution Anisotropic Diffusion

There are many black holes in depth map, but if using filter on the original resolution image, it will cause the increase of errors rate and image blurry. So, the image of Pyramid is introduced. By down-sampling, the number of black holes is reduced in the depth map.

Process the lowest resolution image with mean filtering. In Eq. (1), N is a variable parameter, and it depends on whether there is a black pixel in neighbor areas. L is the number of layers.

$$P_{L-1}(i,j) = \frac{1}{N}(P_{L-1}(i-1,j) + P_{L-1}(i+1,j) + P_{L-1}(i,j-1) + P_{L-1}(i,j+1))$$

$$(1)$$

Anisotropic diffusion is widely applied in computer vision and edge-preserving algorithm.

$$\begin{cases} P_0 = P(x,y,0) \\ \frac{\partial P(x,y,t)}{\partial t} = \mathrm{div}(g(|\nabla P|) \bullet \nabla P) \end{cases} \tag{2}$$

In Eq. (2), P_0 is original image, $P(x,y,0)$ is diffusion image of original image at $t = 0$, div is the divergence operator, ∇ is a gradient operator, $g(r)$ is a const, if $g(r) = 1$, the equation is heat conduction equation. This can be discretized using four neighborhoods to calculate the center pixel.

$$P_D^{t+1}(i,j) = P_D^t(i,j) + \lambda[C_N\delta_N + C_S\delta_S + C_E\delta_E + C_W\delta_W] \tag{3}$$

In Eq. (3), N, S, E, W is shorthand of North, South, East, and West. The bigger λ is the more smooth the edge of image. t is the number of iterations, and C is thermal conductivity. δ is the gradient threshold, the bigger the value is, the smoother the image is.

Algorithm 2: Up-sampling and local adjustment

Input: mask M_n, color map C_n, lowest resolution depth map $\hat{P}_{L-1}$.

For $n=L-1$: 0.

Compute the coefficient of conductivity from the color image C_n.

Perform anisotropic diffusion of t iterations to compute $\hat{P}_n$.

Remove error pixels form $\hat{P}_n$ by using mask M_n to get updated $\hat{P}_n$.

Up-sampling $\hat{P}_n$ by a factor of 2^n and store in $\hat{P}_{n-1}$.

End for

2.2 Local Adjustment

The mask value is composed of 0 and 1. When the point $P_{i,j}$ is in the confidence region, the value of $M_{(i,j)}$ is 1, otherwise the value is 0, as shown in Eq. (4).

$$M_{(i,j)} = \begin{cases} 0, & P_{(i,j)} < \text{threshold} \\ 1 & P_{(i,j)} \geq \text{threshold} \end{cases} \tag{4}$$

In theory, the value of threshold should be 0, but the edge of the black holes should be labeled in mask too. So, the choice of threshold should be bigger than 0. The gray histogram distribution can be used to choose a threshold. In order to improve the computational efficiency, filling operation is carried out only in the black holes.

According to the obtained mask, the value of pixel is adjusted locally. If $M_{(i,j)} = 1$, use the original value of pixel replace the computed value. Detection mode depends on the distribution law of information in low frequency. In general, Fourier transform is used to obtain the law. If the low-frequency information is mainly distributed in the horizontal direction, the row priority adjustment algorithm is used, and vice versa.

3 Experiment Results and Analysis

The final results repaired on the depth map are acquired in this section. In Fig. 1b, in the original depth map, some pixels didn't been obtained. Optimized multi-resolution anisotropic diffusion algorithm aims to fix the larger black holes, which break through the limitation of original anisotropic diffusion algorithm. In order to highlight the superiority of this algorithm, we introduce joint bilateral filter method (JBF). Compared Fig. 1c, e, the JBF algorithm failed in the larger black holes, but in our algorithm, the depth data has been significantly restored. Compared Fig. 1d, e, although black holes have been filled, the pixels are not smooth with neighbors. According to the local adjustment, the result has been improved obviously.

As you can see above, the joint bilateral filter cannot repair the image with larger black holes, so in this group, we processed the depth map with smaller black holes. The four datasets are Books, Laundry, dolls, Art from the Middlebury datasets [11] with various synthetic degradations [12]. As shown in Fig. 2, from the view of local image, there are obvious advantages when processing the complex texture of image. Quantitative results of different methods are shown in Table 1.

As shown in Fig. 3, 3D reconstruction is also completed to verify the effectiveness.

Three types of representative images are used. The first column is the original depth map, from top to bottom the number of black holes from few to many, the size of black holes from small to large. The second column is depth map after restoration. The third column is the results using the original depth map for 3D reconstruction. Last column is the 3D module using the restored depth map. Directly compare the third column with the last column, the missing area has been reduced a lot. Since all the model generated by a frame image with point cloud

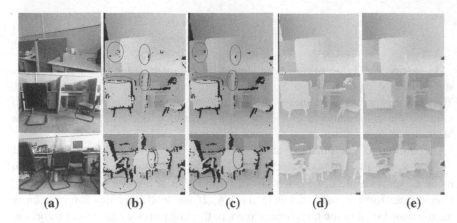

Fig. 1 Result of restoration, **a** color map, **b** depth map, **c** repaired depth map (JBF), **d** repaired depth map (before local adjustment), **e** repaired depth map (ours)

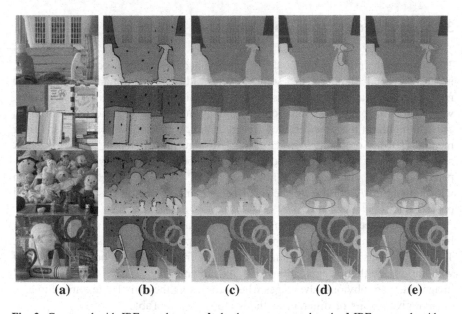

Fig. 2 Compared with JBF, **a** color map, **b** depth map, **c** ground truth, **d** JBF, **e** our algorithm

Table 1 Quantitative results of different methods

		Books	Laundry	Dolls	Art
JBF	PSNR	26.4246	26.5456	27.5674	24.1413
	SSIM	0.9355	0.9294	0.9486	0.9387
Our	PSNR	32.7932	30.3345	28.3654	26.3562
	SSIM	0.9367	0.9377	0.9507	0.9447

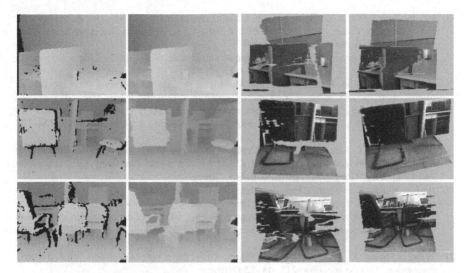

Fig. 3 Examples of three-dimensional reconstruction

data, and without surface triangulation, the model is discontinuous. From the perspective of three-dimensional reconstruction, the proposed methods will reduce the demand number of video frames for 3D reconstruction and reduce computation.

4 Conclusion

The application of depth map has been developed rapidly in recent years. As a cheap depth camera, Kinect is widely used in the bone tracking, face recognition, and other aspects. However, the depth map generated by Kinect often accompanies many black holes. The main research direction of this paper is to repair the depth map, which mainly analysis of the restoration problem of single frame image. The depth map restoration algorithm is evaluated in terms of 3D reconstruction and the value of SSIM, PSNR. In brief, algorithm proposed can efficiently restore the image with larger black holes, and break through the limitation of original anisotropic diffusion algorithm.

References

1. Buades, A, B., Coll, and J. M. Morel.: A non-local algorithm for image denoising. IEEE Computer Society Conference on Computer Vision and Pattern Recognition IEEE Computer Society, 60–65, (2005).
2. Min, D., Lu, J., Do, M.N.: Depth video enhancement based on weighted mode filtering. IEEE Trans. Image Process. 21, 1176–1190 (2012).

3. He, K., Sun, J., Tang, X.: Guided Image Filtering. In: Daniilidis K., Maragos P., Paragios N. (eds) Computer Vision – ECCV 2010. ECCV 2010. Lecture Notes in Computer Science, vol 6311. Springer, Berlin, Heidelberg (2010).

4. Camplani, M., Salgado, L.: Efficient spatio-temporal hole filling strategy for Kinect depth maps. Proceedings of SPIE, 8290, 13 (2012).

5. Jin-yi, Liu., Xiao-jin, Gong., Ji-lin, Liu.: Guided in painting and filtering for Kinect depth maps // international Conference on Pattern Recognition, Tsukuba, IEEE, 2055–2058, (2012).

6. S, Liu., Y, Wang., et al.: Kinect depth restoration via energy minimization with TV21 regularization // International Conference on Image Processing, Melbouene, IEEE, 724–727 (2013).

7. NE, Yang., YG, Kim., RH, Park.: Depth hole filling using the depth distribution of neighboring regions of depth holes in the Kinect sensor // Signal Processing, Communication and Computing, Hong Kong, IEEE, 658–661 (2010).

8. Jia-lun, Kang., Xiang-hong, Tang.: A direction image in painting algorithm based on fast marching method. Journal of Hangzhou Dianzi University, 32, 147–150 (2012).

9. Janoch, A., Karayev, S., Jia, Y., et al.: A category-level 3-D object dataset: putting the Kinect to work // international Conference on Computer Vision Workshops, Barcelona, IEEE, 1168–1174 (2011).

10. Wang, Z., Song, X., Wang, S, Z., et al.: Filling Kinect depth holes via position-guided matrix completion. Neurocomputing, 215, 48–52 (2016).

11. Scharstein, D., Pal. C.: Learning conditional random fields for stereo. In IEEE Computer Society Conference on Computer Vision and Pattern Recognition (CVPR 2007), Minneapolis, MN, June (2007).

12. Jingyu, Yang. et al.: Color-Guided Depth Recovery From RGB-D Data Using an Adaptive Autoregressive Model. IEEE Transactions on Image Processing 23, 3443–3458 (2014).

A Stacked Denoising Autoencoder Based on Supervised Pre-training

Xiumei Wang, Shaomin Mu, Aiju Shi and Zhongqi Lin

Abstract Deep learning has attracted much attention because of its ability to extract complex features automatically. Unsupervised pre-training plays an important role in the process of deep learning, but the monitoring information provided by the sample of labeling is still very important for feature extraction. When the regression forecasting problem with a small amount of data is processed, the advantage of unsupervised learning is not obvious. In this paper, the pre-training phase of the stacked denoising autoencoder was changed from unsupervised learning to supervised learning, which can improve the accuracy of the small sample prediction problem. Through experiments on UCI regression datasets, the results show that the improved stacked denoising autoencoder is better than the traditional stacked denoising autoencoder.

Keywords Deep learning · Stacked denoising autoencoder · Supervised learning · Regression forecast

X. Wang · S. Mu (✉) · Z. Lin
College of Information Science and Engineering, Shandong Agricultural University,
Taian 271018, China
e-mail: msm@sdau.edu.cn

X. Wang
e-mail: wxmsdau@163.com

Z. Lin
e-mail: 2207800939@qq.com

A. Shi
College of Chemistry and Material Science, Shandong Agricultural University,
Taian 271018, China
e-mail: saj31402@163.com

© Springer Nature Singapore Pte Ltd. 2019
B. K. Panigrahi et al. (eds.), *Smart Innovations in Communication and Computational Sciences*, Advances in Intelligent Systems and Computing 670,
https://doi.org/10.1007/978-981-10-8971-8_14

1   Introduction

The artificial neural network (ANN) constructed a model with self-learning by abstracting the neural network of human brain. Support vector machine (SVM) was proposed by Vapnik based on statistical learning theory in 1995. SVM has some advantages in solving small sample, nonlinear and high-dimensional pattern classification. However, both the traditional ANN and the SVM are shallow learning models, which have a strong dependence on the features. Therefore, their representation ability is limited [1]. In 2006, the work of Hinton made deep learning (DL) a new research hotspot in the field of machine learning [2]. Deep learning could extract feature automatically by constructing a neural network with multiple hidden layers, and it reduces the dependence of the prediction model on the feature extraction.

Since deep learning was put forward, it has been widely concerned by academia and industry. In academia, deep learning performed better than the traditional shallow model in image classification [3], speech recognition [4], and natural language processing [5]. In the industry, Google, Baidu, and other large companies are using deep learning as the main supporting algorithm of their intelligent products [6].

With the increase in the amount of data, unsupervised learning algorithm plays an important role in the pre-training stage. But the monitoring information provided by the sample label is still very important [7]. There are some areas in the real world, the data collection is difficult, and the amount of data is small. At this time, the supervised learning algorithm can make full use of the information provided by the label to extract features.

In this paper, the pre-training stage of stacked denoising autoencoder (SDAE) will be changed from unsupervised learning to supervised learning, and some regression experiments are made in UCI dataset. The results show that the improved SDAE is better than the traditional SDAE in regression prediction.

2   Methodology

2.1   Denoising Autoencoder

As one of the commonly used structural modules for deep learning, autoencoder (AE) is an unsupervised learning model, which was put forward by Rumelhart in 1986 [8]. The goal of AE is to learn new features from original input, and the new features could reconstruct the original input information. Therefore, the objective output information of the AE is the input information. The learning process of AE is structured by its encoder part and decoder part, as shown in Fig. 1.

For an AE in Fig. 1, assuming that the number of nodes is n in input layer and m in output layer, respectively, the nonlinear activation function is $f(x) = (1 + e^{-x})^{-1}$ in

Fig. 1 Learning process of
AE

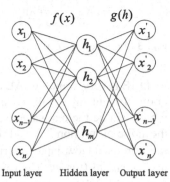

hidden layer and $g(h) = (1 + e^{-h})^{-1}$ in output layer, respectively. For the input sample $x = (x_1, x_2, \ldots, x_n)$, the values of the ith hidden node h_i and the jth output node x_j' are as follows:

$$h_i = f(w_i x + b_i) = f\left(\sum_{j=1}^{n} w_{ij} x_j + b_i\right) \tag{1}$$

$$x_j' = g(w_j' h + b_j') = g\left(\sum_{k=1}^{m} w_{jk}' h_k + b_j'\right) \tag{2}$$

where w_i is the weight between the input layer and the hidden layer, w_j' represents the weight between the hidden layer and the output layer, b_i, b_j' represent the bias of hidden layer and output layer, respectively. So we must find the optimal parameters $\theta = \{w, b, w', b'\}$ which could make the error of x and $x' = (x_1', x_2', \ldots x_n')$ as small as possible, namely minimize $L(x, x') = \sum_{i=1}^{n} (x_i - x_i')^2$.

Aiming at that there are many noises for the most information in the real world and AE cannot learn the real information from the noisy information, Vincent proposed the denoising autoencoder (DAE) through introduced the noisy information to AE in 2008 [9]. The structure of DAE is as shown in Fig. 2.

Fig. 2 Structure of DAE

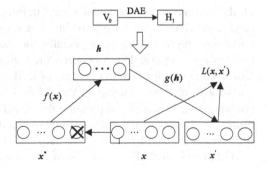

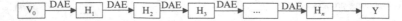

Fig. 3 Structure of SDAE

DAE is very similar to AE except the real used information of input layer. The node of input layer will be set to 0 according to a certain proportion randomly, which will be the real value of input layer in encoder part in DAE. Assuming that the noise proportion is k and the original input information is $x = (x_1, x_2, \ldots, x_n)$, the corrupted new input information is $x^* = (x_1, 0, \ldots, x_n)$ which the number of node equal to 0 is nk. Then x^*, instead of x, is fed to the input layer in AE. The target value of the output layer still is x'. Then, the training process will be made according to AE. A well-trained DAE can recover its corresponding clean version.

2.2 SDAE

In order to learn more complex features, it is necessary to stack the DAE to form a deep learning model, named stacked denoising autoencoder (SDAE). The training process of SDAE is divided into two parts: layer by layer pre-training and fine-tuning. Its structure is as shown in Fig. 3.

The layer by layer pre-training is an unsupervised learning progress. Firstly, the input layer and the first hidden layer would be seen as a DAE. After the first DAE trained, the first hidden layer, as the input layer in the second DAE, would construct the second DAE with the second hidden layer. And so on until the nth DAE trained.

The fine-tuning is a supervised learning progress. In this process, the multilayer DAE and the output layer in the top layer will be considered as a multilayer back propagation neural network (BPNN). The multilayer BPNN uses the stochastic gradient descent (SGD) to optimize the parameters.

3 Adaptive Deep Learning Model Based on SDAE

DAE is an unsupervised model whose training samples are unlabeled. These features from DAE could keep most information of the input data. However, for this supervised regression tasks, especially for the regression tasks which lack data, there is a strong correlation between the feature extraction and label information. Considering that the learning process of DAE is irrelevant to the label information, so we add label information to the DAE. The structure of improved DAE is as shown in Fig. 4 where y represents the actual label value of the sample x and y' represents the label value calculated by the model. The improved DAE could make the features learned by the new model better for regression and prediction.

The improved DAE is as shown in Fig. 4, and the training process is as follows:

Fig. 4 Structure of improved DAE

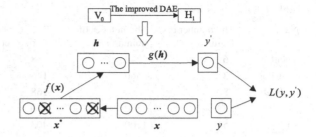

Step 1: Adding noise. Assuming that the noise proportion of the improved DAE is k, for the sample of input layer, we add noise according the noise proportion k. The new input sample is $x^* = (x_1, 0, \ldots, x_n)$, where the number of node equal to 0 is nk.

Step 2: The computation from input layer to hidden layer. Compute the feature of hidden layer $h = (h_1, h_2, \ldots, h_m)$ using Eq. (1).

Step 3: The computation from hidden layer to output layer. Compute the value of output layer y' using Eq. (2).

Step 4: Compute loss function and update parameters using SGD algorithm. Compute the loss function $L(y, y')$ and its derivatives, update the parameters until convergence, according to reverse direction of gradient.

The improved DAE model not only keeps the strong robustness of DAE, but also makes full use of the label information. For the dataset with small amount of data and labeled, the model could find more information which could contribute to regression.

Using the improved DAE in Fig. 4 instead of DAE in Fig. 3, we can get the improved SDAE. The training process of the improved SDAE is similar to the SDAE, which includes pre-training and fine-tuning, so there is not much detailed description.

4 Experimental Results

In order to verify the effectiveness of the improved algorithm, there are five groups comparison tests in five regression datasets from UCI standard library. The number of samples and attributes of each dataset are shown in Table 1.

In order to avoid numerical problems caused by large gaps of data magnitude and to improve the learning efficiency of the algorithm, we did normalization processing for the datasets and all the data were normalized to the interval [0,1].

Table 1 UCI regression datasets

Name	The number of features	The number of training sets	The number of testing sets	The total number of samples
Yacht	6	240	68	308
Housing	13	360	146	506
ENB2012	8	600	168	768
Concrete	8	800	230	1030
Airfoil	5	1100	403	1503

4.1 Selection of Hyper Parameters

There are many hyperparameters in SDAE and its improved algorithm, which plays an important role in the final results. In this paper, we first set the value of some hyperparameters such as epoch, batch size, learning rate, and so on depending on experience. Then for the momentum, we select the best results by doing some tests. We set ten different values ([0.1:0.1:1]) for the momentum in the pre-training and fine-tuning phase, respectively; so we did 100 tests for every dataset.

4.2 Results and Analysis

In this paper, the mean square error (MSE) is used as the index to evaluate the prediction ability of the model. In the process of model building and parameter optimization, the data we used are normalized, because the MSE calculated by normalized data is more representative for the predictive ability of the model. The anti-normalized MSE represents the predictive ability of the model for the specific dataset, which is relative to the numerical value range of predictive factor. The experimental results are shown in Tables 2 and 3, the MSE in Table 2 is calculated by normalized data, and the MSE in Table 3 is anti-normalized.

The data used in SDAE and improved SDAE are all normalized. The data of different attribute are in the same order of magnitude for the same dataset, and the data are in the same order of magnitude for different datasets, which avoids numerical problems caused by large gaps of data magnitude. We can see from

Table 2 Normalized MSE

Dataset	MSE	
	SDAE	The improved SDAE
Yacht	8.29E−04	5.49E−04
Housing	1.21E−02	1.14E−02
ENB2012	4.60E−03	4.35E−03
Concrete	1.52E−02	1.47E−02
Airfoil	2.90E−02	1.40E−02

Table 3 Anti-normalized MSE

Dataset	MSE	
	SDAE	The improved SDAE
Yacht	3.23	2.14
Housing	15.83	15.04
ENB2012	6.32	5.99
Concrete	98.04	94.90
Airfoil	40.99	19.75

Table 2 that the predictive ability of the prediction model is different for different datasets; the MSE of the yacht dataset is significantly smaller than the other four datasets, which shows that the prediction model has better prediction effect on the yacht dataset than other datasets.

The MSE in Table 3 represents the actual prediction error of the prediction factor, which is related to the difference between the maximum value and minimum value of the prediction factor. As can be seen from Table 3, for SDAE and improved SDAE, the MSE of concrete dataset is both the largest and significantly higher than the other four datasets MSE. In Table 2, the MSE of the airfoil dataset is the largest for the SDAE model, and the MSE of the concrete dataset is the largest for the improved SDAE. There are two main reasons for this difference: One is the error from the calculation process of prediction model, and the other is the difference of the actual value of the prediction factors of different datasets.

From Tables 2 and 3, we can see that the MSE of the improved SDAE is smaller than the MSE of SDAE for the five regression datasets in Table 1. Therefore, for the small sample dataset, the SDAE which the pre-training stage uses supervised learning can extract better features and further improve the prediction accuracy.

5 Conclusion

As one of the most popular research directions in the field of machine learning, deep learning provides powerful algorithm support for the realization of artificial intelligence. For unsupervised learning, the ability of its feature extraction is limited when the amount of data is limited, so the pre-training stage of SDAE was changed from unsupervised learning to supervised learning, which can make full use of the information provided by the label. Through experiments on five UCI regression datasets, the experimental results show that the improved SDAE has smaller prediction error and stronger generalization ability than the traditional SDAE. Considering that parameter selection consumes a lot of time, how to improve the efficiency of parameter selection is important for future research.

References

1. Bai Y., Chen Z., Xie .J, li C.: Daily reservoir inflow forecasting using multiscale deep feature learning with hybrid models. Journal of Hydrology. 532, 193–206 (2015).
2. Hinton G., Osindero S., Teh Y.: A fast learning algorithm for deep belief nets. Neural Computation. 18(7), 1527–1554 (2006).
3. Verbancsics P., Harguess J.: Image classification using generative neuron evolution for deep learning. Winter Conference on Applications of Computer Vision. 488–493 (2015).
4. Li D., Hinton G., Kingsbury B.: New types of deep neural network learning for speech recognition and related applications: an overview. International Conference on Acoustics, Speech and Signal Processing. IEEE 8599–8603 (2013).
5. Chen Y., Zheng D., Zhao T.: Chinese relation extraction based on deep belief nets. Journal of Software. 23(10), 2572–2585 (2012).
6. Yu k., Jia L., Chen Y., Xu W.: deep learning: Yesterday, Today, and Tomorrow. Journal of Computer Research and Development. 50(09), 1799–1804 (2013).
7. Jiang Z., Chen Y., Gao L.: A supervised dynamic topic model. Acta Scientiarum Naturalium Universitatis Pekinensis. 51(02), 367–376 (2015).
8. Hu Q., Zhang R., Zhou Y.: Transfer learning for short-term wind speed prediction with deep neural networks. Renewable Energy. 85, 83–95 (2016).
9. Vincent P., Larochelle H., Bengio Y., Manzagol P.: Extracting and composing robust features with denoising autoencoders. International Conference, Helsinki, Finland, June. Hu Q., Zhang R., Zhou Y.: Transfer learning for short-term wind speed prediction with deep neural networks. Renewable Energy. 85, 83–95 (2016).

Part II
Web and Informatics

When Things Become Friends: A Semantic Perspective on the Social Internet of Things

Nancy Gulati and Pankaj Deep Kaur

Abstract The Internet of Things (IoT) has become an integral component of modern day computing. It is evident that majority of applications will be associated with number of objects (things) in the coming future, whose sole purpose would be to provide required services to users. The current IoT enactment is retracting from quintessential smart objects (things) to smart as well as social objects (things). This unprecedented paradigm shift is certain to usher a remarkable furtherance in the epoch of modern day computing. This paper divulges the pristine phenomena of Social Internet of Things (SIoT) along with uncovering its various aspects and defining a generalized architecture for the same. It also contributes by providing a semantic description of its various components in terms of ontological structure. Finally, the paper concludes by proposing research directions in the context of SIoT accompanied by envisaging its future scope.

Keywords IoT · Social networking · SIoT · SIoT architecture
SIoT ontology · Social relationships

1 Introduction

The IoT has been trending in the field of IT for past few years. The idea was first highlighted when Kevin Ashton coined the term "Internet of Things" in 1999. Till date, no universal definition of IoT exists. In fact, different communities of people have defined the term differently. To generalize, IoT may be defined as an interconnection of numerous everyday physical and virtual objects (things) possessing unique identities as well as the propensity to communicate over a network without requiring any sort of human involvement or participation. Forecasts for the number

N. Gulati (✉) · P. D. Kaur
Department of CSE, GNDU Regional Campus, Jalandhar, India
e-mail: nancygulati91@gmail.com

P. D. Kaur
e-mail: pankajdeepkaur@gmail.com

© Springer Nature Singapore Pte Ltd. 2019
B. K. Panigrahi et al. (eds.), *Smart Innovations in Communication and Computational Sciences*, Advances in Intelligent Systems and Computing 670,
https://doi.org/10.1007/978-981-10-8971-8_15

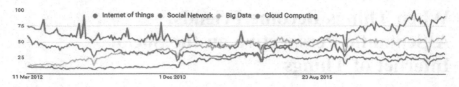

Fig. 1 Google search trends since 2012 for terms IoT, Social Networks, Big Data, and Cloud Computing [4]

of connected devices on the IoT for the year 2020 vary considerably, from 26 billion (Gartner), to 30 billion (ABI Research), to 50 billion (Cisco), to 75 billion (Morgan Stanley) [1–3]. Regardless of which prediction is closest to reality, it is generally believed that an IoT will become ubiquitous in the coming decade. Previous studies in IoT were mainly focused on human-to-object interaction where humans depend upon objects for services. Figure 1 portrays technology trends for past five years clearly unfolding the significance of IoT and social networks [4]. However, current research is being shifted from prevailing object-to-human interaction toward object-to-object interaction where objects rely on each other for services. To realize this paradigm shift, the notion of Social Internet of Things (SIoT) [5] was burgeoned.

Also, the small-world phenomenon depicted in [6] backs for the existence of SIoT.

The main contributions of this research article are:

The notion of SIoT, its architecture, characteristics, and applications are explored. A generalized architecture for SIoT based on various existing architectures is proposed. Further, ontological and semantic explanation of SIoT components is deliberated. The paper is arranged into different sections as outlined below:

- Section 2 puts forward an introduction to the novel paradigm of SIoT. Further, SIoT benefits and applications are discussed.
- Section 3 presents a generalized architecture of SIoT.
- Section 4 provides a semantic explanation of a SIoT network.
- Section 5 concludes the research work along with visualizing its future scope.

2 The SIoT Paradigm

The SIoT can be designated as an IoT where things or objects can build their own social association autonomously without the intervention of human. The objects in a SIoT network try to mimic human behavior while establishing relationships in order to build their own social network. Therefore, the prototypes built to analyze human social behavior and social networks can be implemented on SIoT as well. Usually, SIoT exploits four basic relational models (community, authority, equality, and market-pricing relationship) based on social relationship theory [7] to establish

Table 1 Types of relationships in SIoT

Relationship type	Description
Parental object relationship (POR)	Established if the maker of objects is same and they are being built during same period
Co-location object relationship (CLOR)	Established if the objects share their location
Co-work object relationship (CWOR)	Established when the nature of job for the objects is same
Ownership object relationship (OOR)	Established if the objects are owned by same individual
Social objects relationship (SOR)	Established when liaison between the objects is frequent or recurring

new relationships. Different kinds of relationships [8] that can possibly be developed among the objects in a SIoT network on the basis different relational models are depicted in Table 1.

2.1 SIoT Over IoT

In this subsection, we explore some characteristics of SIoT that explains the motivation behind selecting it over an IoT system. The characteristics and potential benefits of a SIoT network over IoT systems can be:

1. Enhanced Network Navigability: In [9], formal conditions for network navigability have been defined. In general, a network is said to be navigable if each node in the network is reachable from every other node which backs for the existence of a connected component in the network. Using the concepts of social networking and small-world phenomenon [6], it is possible to find a shorter path between the nodes as depicted in [8] without the global knowledge of network structure.
2. Boosted Object and Service Discovery: SIoT can potentially enhance object and service search in an IoT. Centralized networks have a global view of the network which is not scalable in case frequent changes in network are expected, whereas SIoT maintains a decentralized organization of network [8]. Service discovery is easier with distributed approach adopted in SIoT.
3. Improved communication: SIoT leads to improved communication and collaboration among objects. Different nodes can share different type of relationships and according to that an organized structure for communication among the nodes can be defined.
4. Added Flexibility: Adding the concepts of friendships in an IoT system makes it more flexible having additional advantage over the existing system. Existing centralized systems needs extensive information of the network in, whereas a SIoT supports a distributed organization of nodes.

5. Better Trustworthiness Evaluation: Various trust models [10] exist for trust evaluation of an object in SIoT. This helps in estimating the reliability of a node and in turn enhances the security of the network. Only trusted nodes will be selected as friends in a SIoT and malicious nodes will remain isolated.

2.2 Recent Advancements and Applications

The research work in the field of SIoT is gaining momentum. Various projects have been proposed and developed in past few years focusing on social perspective of IoT. In [11], a description of existing SIoT platforms such as Toyota Friend Network, Nike+, Xively, Paraimpu, Social Web of Things, Everything, and their implementation is elicited.

Recently, SIoT has been used in integration with other technologies to extend its existing capabilities. In [12], the integration of SIoT and Social Cloud has been proposed. And strategies for managing social relationships among objects have been presented. The idea has great potential and can help in improving resource sharing in SIoT. In [13], knowledge-based SIoT architecture is proposed, i.e., a system that exploits already existing communication patterns in order to generate new knowledge using the concepts of reasoning. The work in [13] is supported by COSMOS project [14]. In [15], the idea of social network of vehicles is presented, which is basically a vehicular instance of SIoT. In [16], the concept of human behavior analysis using data generated from a SIoT is introduced.

3 SIoT Architecture

In this section, some prevailing architectures for SIoT are illustrated. Also, a generalized architecture incorporating key characteristics from the prevailing ones is proposed. In [17], a three-layered architecture for IoT is defined consisting of sensing, communication, and application layer. In [5], three different architectures are defined for SIoT server, gateway, and the object. In [18], the social network platform architecture is tailored to define an architectural model for SIoT networks. Also, a three-layered client side and server side architecture for SIoT is proposed. In [19], distributed storage-based architecture for SIoT is defined. While in [20], a framework and architecture for SIoT client and server are defined. On the basis of services and components defined above, we propose a generalized four-layered architecture for SIoT (consisting of object, communication, management, and application layer) rendered in Fig. 2.

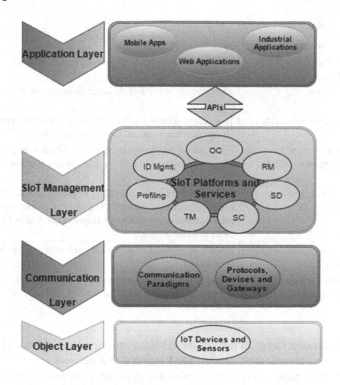

Fig. 2 SIoT reference architecture

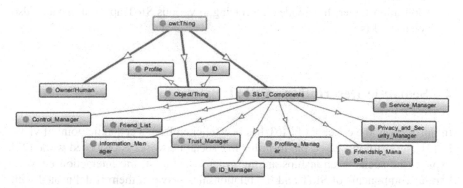

Fig. 3 Class diagram for SIoT ontology

The functioning of respective layers is depicted as:

1. Object Layer: The basc laycr where IoT objects and devices equipped with sensors are present. These devices can collaborate and communicate with each other via local sensor networks.

Table 2 Essential components of a SIoT network

S. No.	Component	Purpose
1	ID management (ID Mgmt)	To uniquely identify every object in the network
2	Profiling	To configure static and dynamic information about the objects
3	Owner control (OC)	To manage schemes defined by the owner for access control, objects activities and relationships building
4	Relationship management (RM)	Establishing and maintaining relationships between objects. Specific rules are defined for friendship selection among the objects
5	Service discovery (SD)	To search for objects providing required services from the social relationship network. The objects mimic human behavior while searching for friend objects able to provide them desired services
6	Service composition (SC)	Allows the objects to deal with each other using different approaches
7	Trustworthiness management (TM)	Responsible for trust evaluation of different objects for information sharing and relationship management

2. Communication Layer: It defines the communication technologies and protocols to enable communication among SIoT devices and networks.
3. SIoT Management Layer: It defines the required platform to manage SIoT services. The essential components comprising SIoT services are delineated in Table 2.
4. Application Layer: It provides interfacing to various SIoT applications and also includes APIs.

4 Semantic Description of SIoT

In this section, the concept of SIoT has been elicited from a semantic point of view. In [21], an ontology-based platform for IoT social networks is proposed while [22] exploits the social relationships among groups. In Fig. 4, the interaction between various components of SIoT and the relationships between them is delineated with the help of standard Web ontology language (OWL) and tool "Protégé" (Version 5.1.0) [23] while Fig. 3 displays the classes selected to build the ontology.

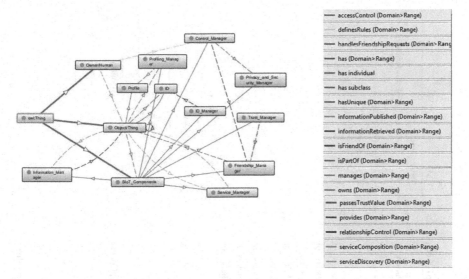

Legend (object properties):

- accessControl (Domain>Range)
- definesRules (Domain>Range)
- handlesFriendshipRequests (Domain>Rang
- has (Domain>Range)
- has individual
- has subclass
- hasUnique (Domain>Range)
- informationPublished (Domain>Range)
- informationRetrieved (Domain>Range)
- isFriendOf (Domain>Range)
- isPartOf (Domain>Range)
- manages (Domain>Range)
- owns (Domain>Range)
- passesTrustValue (Domain>Range)
- provides (Domain>Range)
- relationshipControl (Domain>Range)
- serviceComposition (Domain>Range)
- serviceDiscovery (Domain>Range)

Fig. 4 Schema of SIoT ontology

There are three major classes—Owner, Object, and SIoT_Components—while Object further consists of Profile and ID subclasses and SIoT_Components class includes the components of SIoT (Control_Manager (CM), Information_Manager (IM), ID_Manager (IDM), Profiling_Manager (PM), Service_Manager (SM), Friendship_Manager (FM), Trust_Manager (TM), Privacy_and_Security_Manager (PSM)) as subclasses. The relationship between the classes is described by object properties as enlisted in Table 3.

Semantic language queries enable the user ask questions from the semantic Web similar to the query-response phenomena with databases. Many languages are available for querying RDF/OWL data such as SPARQL or DL queries. The SIoT ontology we have built could be used for querying the semantic database. Query reasoning in our case is supported by the Reasoner "HermiT 1.3.8.413." For example, consider the scenario illustrated in Fig. 5. We define 11 different objects as participants of SIoT network (Fig. 5a). Data properties for the class Object/Thing are defined (Fig. 5b). Random values for data properties of these objects are set as enlisted in Table 4.

Suppose, a new device (say Samsung Mobile) is registered to the network and a query is generated to discover prospective friends for the given device according to some conditions. The conditions we have taken into account are trustworthiness and POR relationship. The snapshot of DL query results is pictured in Fig. 6.

Table 3 Semantic description of relationships in SIoT

	Owner	Object	ID	Profile	SIoT_Components	CM	IM	IDM	PM	SM	FM	TM	PSM
Owner		owns				definesRules	publishInformation						
Object		Owner	hasUnique	has	isPartOf								
ID		Subclass of								seviceDiscovery			
Profile		Subclass of											
SIoT_Components													
CM					Subclass of						relationshipControl		accessControl
IM		RetrievesInformation			Subclass of								
IDM			Provides		Subclass of								
PM				manages	Subclass of								
SM		serviceComposition			Subclass of								
FM		handlesFriendshipRequests			Subclass of								
TM					Subclass of							passesTrustValue	
PSM					Subclass of								

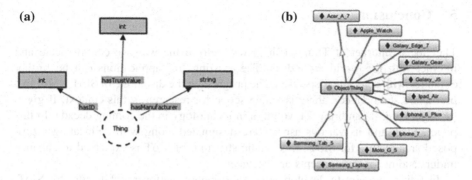

Fig. 5 Representation of **a** data properties of object/thing **b** object/thing instances

Table 4 Data property values for object/thing instances

Individual	hasID	hasManufacturer	hasTrustValue
Acer_A_7	0001	Acer	3
Apple_Watch	0002	Apple	2
Galaxy_Edge_7	0003	Samsung	2
Galaxy_Gear	0004	Samsung	4
Galaxy_J5	0005	Samsung	1
Ipad_Air	0006	Apple	1
Iphone_6_Plus	0007	Apple	2
Iphone_7	0008	Apple	1
Moto_G_5	0009	Motorola	0
Samsung_Laptop	0010	Samsung	1
Samsung_Tab_5	0011	Samsung	2

DL query

Query (class expression)

Object/Thing and hasManufacturer value "Samsung" and hasTrustValue **min** 1 xsd:int

Execute   Add to ontology

Query results

Instances (5 of 5)

- Galaxy_Edge_7
- Galaxy_Gear
- Galaxy_J5
- Samsung_Laptop
- Samsung_Tab_5

Query for

- ☐ Direct superclasses
- ☐ Superclasses
- ☐ Equivalent classes
- ☐ Direct subclasses
- ☐ Subclasses
- ☑ Instances

Fig. 6 Snapshot of an example DL query

5 Conclusion

The Social Internet of Things (SIoT) will reform the way we communicate and interact with the world around us. The existing IoT applications can be further extended with a social perspective. The progress in the discipline of SIoT is still in a premature state. Hence, there is ample scope for research in this context. If given the right path, it can bring a revolution in technology in the coming decade. In this paper, SIoT and its various aspects are designated along with delineating a proposed architecture for SIoT. A semantic structure of SIoT is explained to enhance understanding of the concepts and services.

In future, we aim to develop semantic-oriented platform architecture for SIoT. Also, we intend to study the integration of SIoT into different domains such as industrial manufacturing and agriculture along with developing theoretical framework for the same.

References

1. Gartner.: The Importance of 'Big Data': A Definition. https://www.gartner.com/doc/2057415/importance-big-data-definition/ (2014)
2. International Telecommunications Union.: ITU Internet Reports 2005: The Internet of Things. www.itu.int/internetofthings/ (2005)
3. Evans, D.: The Internet of Things: How the Next Evolution of the Internet is Changing Everything. Cisco White Paper. http://www.cisco.com/c/dam/en_us/about/ac79/docs/innov/IoT_IBSG_0411FINAL.pdf/ (2011)
4. https://www.google.co.in/trends/
5. Atzori, L., Iera, A., Morabito, G., Nitti, M.: The social internet of things (SIoT)–when social networks meet the internet of things: concept, architecture and network characterization. Comput. Netw. **56**(16), 3594–3608 (2012)
6. Kleinberg, J.: The small-world phenomenon: an algorithmic perspective. In: Proc. ACM Symposium on Theory and Computing (2000)
7. Fiske, A. P.: The four elementary forms of sociality: framework for a unified theory of social relations. Psychological review **99**, 689–723 (1992)
8. Nitti, M., Atzori, L., Cvijikj, I. P.: Friendship Selection in the Social Internet of Things: Challenges and Possible Strategies. IEEE Internet of Things Journal **2**(3), 240–247 (2015)
9. Boguna, M., Krioukov, D., Claffy, K.C.: Navigability of complex networks. Nat. Phys. **5**(1), 74–80 (2009)
10. Abdelghani, W., Zayani, C.A., Amous, I., Sèdes, F.: Social Media: The Good, the Bad, and the Ugly: Trust Management in Social Internet of Things: A Survey. Lecture Notes in Computer Science, Springer International Publishing, 430–441 (2016)
11. Atzori, L., Iera, A., Morabito, G.: From "smart objects" to "social objects": The next evolutionary step of the internet of things. IEEE Communications Magazine. **52**(1), 97–105 (2014)
12. Zhang, W., Jin, W., D., El Baz, D.: Enabling the Social Internet of Things and Social Cloud. IEEE Cloud Computing. **2**(6), 6–9 (2015)
13. Voutyras, O., Bourelos, P., Kyriazis, D., Varvarigou, T.: An Architecture supporting Knowledge flow in Social Internet of Things systems. In: 10th International Conference on Wireless and Mobile Computing Networking and Communications (WiMob), 45–50 (2014)

14. COSMOS project: http://iot-cosmos.eu/
15. Alam, K.M., Saini, M., Saddik, A.E.: Toward Social Internet of Vehicles: Concept, Architecture, and Applications. IEEE Access. **3**, 343–357 (2015)
16. Jara, A.J., Bocchi, Y., Genoud, D.: Social Internet of things: the potential of the Internet of Things for defining human behaviour. In: International Conference on Intelligent Networking and Collaborative Systems, 581–585 (2014)
17. Zheng L., et al. (Ed.): Technologies, applications and governance in the Internet of things. Internet of Things - Global Technological and Societal Trends, River Publisher (2011)
18. Atzori, L., Iera, A., Morabito, G.: SIoT: giving a social structure to the internet of things. IEEE Commun. Lett. **15**(11), 1193–1195 (2011)
19. Wu, J., Dong, M., Ota, K., et al.: Securing distributed storage for social internet of things using regenerating code and blom key agreement. Peer-to-Peer Netw Appl **8**(6), 1133–1142, (2015)
20. Tripathy, B.K., Dutta D., Tazivazvino, Chido.: (Ed.) Internet of Things (IoT) in 5G Mobile Technologies: On the Research and Development of Social Internet of Things. 153–173, Springer International Publishing (2016)
21. Byun, J., Kim, S. H., Kim, D.,: Lilliput: Ontology-Based Platform for IoT Social Networks. In: IEEE International Conference on Services Computing, Anchorage, AK, 139–146 (2014)
22. Kang, D. H., Choi H. S., Rhee, W. S.: Social Correlation Group generation mechanism in social IoT environment. In: Eighth International Conference on Ubiquitous and Future Networks (ICUFN), Vienna, 514–519 (2016)
23. http://protege.stanford.edu/

A Context-Aware Recommender Engine for Smart Kitchen

Pratibha and Pankaj Deep Kaur

Abstract Internet of Things (IoT) is the next wave of technological innovation that binds significant number of things (objects) together and produces considerable amount of services which people may use and subscribe for their convenience. As large numbers of objects are interconnected in IoT, collecting information related to the users and their preferences is challenging task. Thus, recommending services to the users based on the objects which are available with them is indispensable for the success of IoT. This paper proposes a context-aware recommender engine for suggesting possible recipes with available food items and time required to prepare the meal in the smart kitchen. It not only considers user's context but also other environmental context like weather, time, and energy consumption by appliances. The paper concludes with possible future research directions in the area under discussion.

Keywords Internet of Things (IoT) · Recommender systems (RSs)
Context-aware · Context-aware recommender system (CARS)

1 Introduction

Recent advancements in computing technologies have made it possible to identify, locate, and tract almost all the routine objects across the Internet. With numerous objects attached and communicating over the Internet, there is a compelling requirement to build an efficacious approach to search, recommend, and categorize among the collection of things in order to extract some patterns which are of interest to the user.

Pratibha (✉) · P. D. Kaur
GNDU Regional Campus, Jalandhar, India
e-mail: pratibha.cse05@gmail.com

P. D. Kaur
e-mail: pankajdeepkaur@gmail.com

© Springer Nature Singapore Pte Ltd. 2019   161
B. K. Panigrahi et al. (eds.), *Smart Innovations in Communication and Computational Sciences*, Advances in Intelligent Systems and Computing 670,
https://doi.org/10.1007/978-981-10-8971-8_16

The traditional (2D) RS focused on suggesting meaningful items to users by understanding their preferences without considering any additional contextual information, for instance location and time [1]. It is not adequate to consider only users and items for applications like tourism and e-learning which require additional information regarding user's context.

Therefore, it is essential to integrate contextual information into RS under certain situations [1]. As a consequence of adding contextual information to 2D recommender systems, the new system so devised is known as the multi-dimensional (MD) RS or CARS.

In past few years, context-aware computing has gained a substantial interest among researchers and several studies have been performed in this field. Context-aware computing makes use of data related to the current context (specified time, event, etc.) to run services. Contextual information was combined with RS for generating more relevant recommendations and enhancing user's experience.

Recommending things and services is the important measures taken to possess full privilege of IoT; it not only benefits individuals but also businesses and organizations. If things and service recommendations in IoT can be made context-aware, it can help provide more relevant things and services to users according to their interest, current context, and preferences. Furthermore, it can benefit user in searching information and reducing search time in an IoT network.

With growing amendments in the food section and lifestyle, making healthier and preferable food choices is becoming difficult for many people. Deciding a meal to prepare is a typical daily activity. In order to choose what to cook, people usually select food choices that fulfill their preferences. However, because of the substantial range of food varieties and working culture of both men and women, deciding food choices to cook by manually traversing a long recipe record can be unproductive and annoying. Many of the times, food items required to prepare the recipe which is selected by the user are not available in the kitchen. We believe that there is a need for recommender system that can suggest food options while taking into account the food items available in the kitchen, user's preferences, and eating history of last few days. In this paper, the use of CARS in making recommendations in IoT has been explored. Moreover, a context-aware recommender engine for making suggestions about possible recipes has been proposed.

The main contributions are presented as follows:

- We show the difference between traditional RS and CARS.
- We propose a recommender engine for suggesting recipes with the available food items in kitchen and various other contexts/situations of user like time to prepare meal, a different recipe from what user has consumed in last few days.
- We also design a flowchart for presenting the steps involved in the recommendation process.
- We present conclusion and suggestions for future research directions.

Rest of the paper is structured in the following sections:

Section 2 provides background history. Section 3 provides survey of related works. Section 4 describes the architecture and detail of proposed recommender engine along with the flow of the recommendation process. Section 5 discusses the motivational example to illustrate need for the proposed engine. Section 6 presents conclusion and suggestions for the future research.

2 Background

2.1 Internet of Things (IoT)

IoT has attained considerable acknowledgment in industry and academia during the last decade. The phrase 'Internet of Things' was primarily introduced by Ashton [2] in 1998. Further, a formal definition was given by International Telecommunication Union (ITU) [3].

Research into IoT is still in its early stage. Therefore, IoT has been defined differently by distinct researchers in [4–6]. In general, IoT can be defined as worldwide network (wired or wireless) of interrelated objects those are provided with distinctive identifiers over the Internet to share or exchange information with each other without human intervention.

2.2 Recommender System (RS)

RS generates meaningful suggestions to a user or collection of users for items or products that might interest them [7]. The concept of RS grows out of the idea of information reuse. It is a system where one would know something by following the footsteps of others, also known as social navigation. It enhances user's experience by assisting in finding information and reducing search and navigation time. It performs better if system has good knowledge of user's needs and preferences. It is also called filtering system as it discards or filters out irrelevant items [8].

The RS captures user preferences for different items to measure R (Rating function) [7]. The primary set of ratings can be collected from users' previous browsing history or can be extrapolated by the system. After primary ratings are gathered, an RS assesses the function R, ($R \rightarrow$ User $\times$ Item) for the (user, item) unrated sets. Here, Rating is a totally ordered set, and User and Item are the domains of users and items, respectively. The RS can suggest the top-ranked k items for each user after the function R is estimated for the whole (User $\times$ Item) space. Such systems are known as traditional or two-dimensional (2D) RS since they consider only two dimensions (User and Item) in the process of generating recommendations.

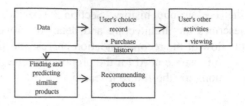

Fig. 1 Recommendation process in online shopping

If a customer has shown interest in a product by viewing or placing it in a shopping cart, he is likely to receive recommendations for some additional products as shown in Fig. 1. Suggestions of friends on Facebook, videos on YouTube, books on Amazon, or jobs on LinkedIn are some applications of recommender systems.

2.3 Context-Aware Recommender System (CARS)

While suggesting items to users if we analyze additional contextual information such as location, place, and time as additional categories of data, such type of systems is known as CARS [9]. Preferences of users are formed as the function of not only (user, item) sets, but context is included as well ($R \rightarrow$ User $\times$ Item $\times$ Context). The difference between traditional RS and CARS is presented in Table 1.

There are three different ways in which contextual information can be integrated into recommendation process as depicted in Fig. 2:

- *Pre-filtering*—Relevant contextual data is first selected with the help of current context, and rating function is predicted by the traditional RS (users × items) on selected data.
- *Post-filtering*—Anticipation of ratings on raw data is done using 2D RS (users × items), and then the resultant recommendation list is filtered by reviewing the context values.
- *Context modeling*—Context information is added in the 2D recommendation system. Thus, this algorithm makes use of multi-dimensional recommendation system.

Table 1 Difference between RS and CARS

Recommender system (RS)	Context-aware recommender system (CARS)
It considers only *User* and *Item* dimensions in recommendation process	It considers *User*, *Item*, and *Context* dimensions in recommendation process
Traditional RS does not incorporate details of the current situation of the user to make recommendations for the user	CARS deals with predicting user tastes by incorporating contextual information into recommendation process
RS needs to mine and manage the classification of items, similarities of users, and their interests, thus requiring big storage space	CARS takes user's contextual needs into consideration, but it does not improve the management technologies for items

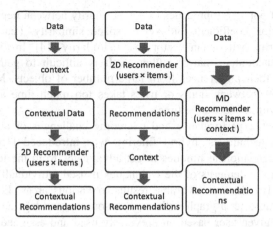

Fig. 2 Different ways to incorporate context in RS **a** pre-filtering, **b** post-filtering, **c** context modeling

3 Related Work

RS generates meaningful suggestions to a user or collection of users for items or products that might interest them [7].

RS can be classified into content-based filtering, collaborative filtering, hybrid filtering, and graph-based approaches [10].

Content-based filtering requires textual description of the items and preferences attained from browsing history of the user [11]. It recommends similar items to the user based on the items previously consumed or liked by them. It is based on characteristics of item defined in textual information and user's historical records. However, it requires a mechanism to associate content or description with many heterogeneous objects connected in a network. Moreover, in content-based filtering, items which are very similar to the previously consumed or liked items are recommended and therefore create an overspecialization problem [12]. In the IoT, analysis of objects and services content may not be a good idea since it will cost complex and expensive computation of similarity and take a long time.

In contradiction to content-based filtering, collaborative filtering does not depend on item description and content. This method takes into account opinions of other people who share similar interests. This technique first identifies users who share similar interest and then recommend the items they have liked. Collaborative filtering can be item-based and user-based [13]. Item-based collaborative filtering identifies similarity between items based on ratings given by users who have rated both the items. User-based collaborative filtering identifies similar users with similar ratings, and then predictions for the active user are calculated based on their ratings. Different popular approaches to find similarity between items and users are Pearson correlation coefficient and vector space similarity. Pearson correlation coefficient performs better than vector space similarity [14]. In the IoT, collaborative filtering encounters many challenges. It is difficult to gather information about users and their preferences from large number of objects. Moreover, computing similarities between items or users takes too much time and thus cannot make decision within acceptable time.

To achieve better performance and overcome challenges of above-mentioned recommendation techniques, hybrid filtering was introduced [15]. It combines multiple recommendation techniques to get advantages from the techniques combined and lessen the demerits of the techniques if used individually.

Graph-based RS showed favorable results in various areas. For example, tag recommenders construct a graph with users, resources, and tags and recommend a set of tags for a given user based on previously used and assigned tags [16, 17].

4 Recommender Engine

Figure 3 provides a graphical representation of the recommendation process. The first step is to determine the food items available in refrigerator and cupboard in the smart kitchen. The method then extracts only those recipes which can be prepared from the available items in the smart kitchen out of all the possible recipes. Finally, user and environmental contextual features can be used to extract the more diverse recipe options for the user.

The recommendation engine can be divided into four parts: food items taxonomy, recipe taxonomy, condition feature groups, and recommendation model as shown in Fig. 4.

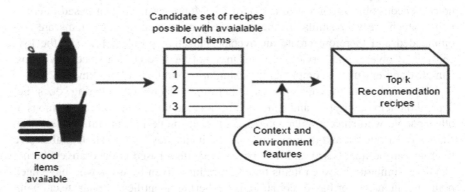

Fig. 3 Representation of the recommendation method

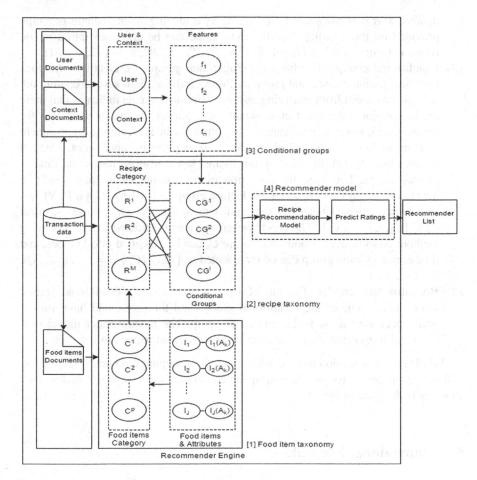

Fig. 4 Recommender engine

(1) **Food item taxonomy**: The food items in the refrigerator and cupboard can be represented as I_j, $j = 1, 2, ..., J$. Let $\alpha_{I_j} = [A_1, A_2, ... A_k, ..., A_K]I_j$ be an attribute vector of I_j, and then the set of food items in the database can be represented as $I = \{I_j(A_k)| j = 1, 2, ..., J\}$. Further, each food item can be classified into one of the mutually exclusive food item categories as $C^i = \{I_j^i A_k |$ $j^i = 1^i, 2^i, ..., J^i, i = 1, 2, ..., t\}$. The attributes of food item can be identified by referring item titles or labels on food item or existing categories already defined for a food item, and then attribute similarity with food item categories can be computed and that food item can be classified into one of the categories with highest similarity score.

(2) **Recipe taxonomy**: The recipes in the market represent as J', with each recipe denoted as $r_{j'}$, $J' = 1, 2, ..., J'$. Let $\beta_{r_{j'}} = [B_1, B_2, ... B_k, ..., B_K]r_{j'}$ be an ingredient vector of $r_{j'}$, and then the set of recipes in the database can be

denoted as $R = \{I_{j'}(A_{k'})|\ j' = 1, 2, ..., J'\}$. By analyzing the ingredients of recipe provided by the cooking experts, each recipe can be classified into a single recipe category as $R^m = \{I_{j'm}(A_{k'})|\ j'^m = 1^m, 2^m, ..., J'^m, m = 1, 2, ..., M\}$.

(3) **Conditional groups**: To obtain the conditional groups, we identify conditions requiring similar recipes and grouping those conditions into several conditional groups. The result from analyzing customer transaction data indicates that there are two major condition factors which may impact the customer choice for recipe: One is environment features, and other is context features. Environment features include weather, time, energy consumption by appliances. Context features include diet provided by nutritionist last consumed meals and time to prepare recipe. Let N be the condition of transactions, with each condition denoted as $c_{gq}, q \in N$. A condition feature group CG $= \{c_{gq}\ (C_s)|\ q \in N\}$ is a set of transactions labeled by the environment features and context features $C_s \in \{C_1, C_2, ..., C_s\}$. The condition groups are constructed by the following method: Assume every condition feature C_s can be classified into D types, and then each condition group can be represented as $\{CG^{qi}\ (C_s)|\ qi = 1i, 2i, ..., Qi, i = 1, 2, ... I\}$.

(4) **Recommender model**: The model assumes that a user's conditional feature must fall into one of the conditional groups and the user could have similar preferences with those within the same group. The recommender model uses historical transaction data to predict and recommend recipes to the user.

Whole recommendation process which includes grouping food items, identifying different conditional groups, and finding the conditional group to which current user belongs is depicted in Fig. 5.

5 Motivational Example

To illustrate the importance of the proposed recommender engine, the following example has been described. Consider a smart kitchen scenario where the user John wants to taste something new for dinner which is compliant to his own taste's preferences, diet provided by his nutritionist, and the food items which are available in the kitchen. Traditionally, John asks for suggestions from other family members and friends. In order to get the recipe options which are possible to prepare with the food items available in the kitchen, if John installs a number of sensors to identify different food items and to monitor the weather conditions (if it is cold John prefers something warm otherwise fresh food is desired). However, this is not enough to satisfy his needs. Without a powerful service that can interact with user to know what he has consumed in the last few days in order to recommend the diverse options to cook for dinner. How long it will take to prepare the meal before dinner is also one of the major concerns. John asks suggestions to the proposed recommender engine about possible recipes that can be prepared of the food items he has

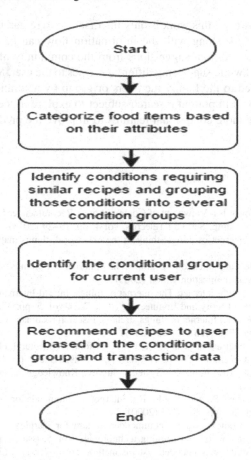

Fig. 5 Flowchart for suggesting recipes in smart kitchen

in his kitchen. Recommender engine recommends him a set of recipes which best meets his preferences. Proposed engine also enables John to monitor consumption of energy by household appliances.

6 Conclusion and Future Work

Recently, CARS has received significant attention in recommending items or services to users based on the contextual information. Contextual information was integrated with 2D RS to increase the effectiveness of traditional 2D system. In this article, it has been recognized that context awareness is a fundamental IoT research demand for generating more relevant recommendations to users by understanding the user's current context. This paper proposed a context-aware recommender engine for a smart kitchen. However, this work is only the first step to research about this issue. In future, a framework along with the information flow can be designed for the proposed system. In addition, suggestions from the community of cooking experts can also be included while suggesting different recipes to the user. Moreover, recipes can be recommended to the family members or group by neutralizing their preferences which is also an important research subject to explore in the future. It is also worth to investigate more conditional features to improve the prediction accuracy.

References

1. Asabere, N. Y.: Towards a Viewpoint of Context-Aware Recommender Systems (CARS) and Services. Int'l J. of Comp. Sci. and Telecom. vol 4, pp. 19–29 (2013)
2. Ashton, K.: That 'internet of things' thing in the real world, things matter more than ideas. J. RFID (2009)
3. International Telecommunication Union: The internet of things. Workshop Report, International Telecommunication Union (2005)
4. Lu, T., Neng, W.: Future internet: The internet of things. In: 3rd International Conference on Advanced Computer Theory and Engineering (ICACTE), vol. 5, pp. 376–380. (2010)
5. Guillemin, P., Friess, P.: Internet of things strategic research roadmap. Technical Report, The Cluster of European Research Projects (2009)
6. Gubbi, J. et al.: Internet of Things (IoT): A vision, architectural elements, and future directions. Future Generation Computer Systems. vol. 29, issue 7, pp. 1645–1660 (2013)
7. Bobadilla J., et al.: Recommender Systems Survey. Knowledge based systems. vol. 46, pp. 109–132 (2013)
8. Ansari, A., Essegaier, S. and Kohli, R.: Internet recommendation systems. Journal of Marketing research, 37, pp. 363–375 (2000)
9. Verbert K. et al.: Context-aware recommender system for learning: A survey and future challenges. IEEE transactions on learning technologies. vol. 5, issue 4, pp. 318–335 (2012)
10. Bouneffouf, D. Mobile recommender systems methods: An overview. (2013)
11. Lops P, de Gemmis M, Semeraro G: Content-based recommender systems: state of the art and trends. In: Ricci F, Rokach L, Shapira B, Kantor PB (eds) Recommender systems handbook. Springer, Berlin, pp 73–105 (2011)
12. Pazzani M and Billsus D: Content-based recommendation systems. In: Brusilovsky P, Kobsa, Nejdl W (eds) The Adaptive Web, vol 4321. Lecture Notes in Computer Science. Springer, Berlin, pp 325–341 (2007)
13. Lü L, Medo M, Yeung CH, Zhang YC, Zhang ZK, Zhou T: Recommender systems. Phys Reports 519, pp. 1–49 (2012)
14. Xue GR, Lin C, Yang Q, Xi W, Zeng HJ, Yu Y, Chen Z: Scalable collaborative filtering using cluster-based smoothing. In: 28th annual international ACM SIGIR conference on Research and development in information retrieval, Salvador (2005)
15. Burke R: Hybrid recommender systems: survey and experiments. User Model User Adap Inter 12, pp. 331–370 (2002)
16. Zhang ZK, Zhou T, Zhang YC: Tag-Aware recommender systems: a state-of-the-art survey. J Computer Science Technology, vol. 26, pp. 767–777 (2011)
17. Zhou X, Xu Y, Li Y, Josang A, Cox C. In: The state-of-the-art in personalized recommender systems for social networking. Artificial Intelligence Rev 37, pp. 119–132 (2012)

Analysis of Hypertext Transfer Protocol and Its Variants

Aakanksha, Bhawna Jain, Dinika Saxena, Disha Sahni and Pooja Sharma

Abstract With massive amounts of information being communicated and served over the Internet these days, it becomes crucial to provide fast, effective, and secure means to transport and save data. The previous versions of the Hyper Text Transfer Protocol (HTTP/1.0 and HTTP/1.1) possess some subtle as well as several conspicuous security and performance issues. They open doors for attackers to execute various malicious activities [1]. The final version of its successor, HTTP/2.0, was released in 2015 to improve upon these weaknesses of the previous versions of HTTP. This paper discusses the issues present in HTTP/1.1 by simulating attacks on the vulnerabilities of the protocol and tests the improvements provided by HTTPS and HTTP/2.0. A performance and security analysis of myriad of commonly used Websites has been done. Some of the measures that a Website must take to provide excellent performance and utmost security to its users have also been proposed in this paper.

Keywords HTTP/1.1 · HTTPS · HTTP/2.0 · Head-of-line blocking
Man in the middle attack · Sniffing attack

1 Introduction

Introduced in 1990, HTTP is an application-level protocol for distributed, collaborative, hypermedia information systems [1]. HTTPS is an extension of HTTP which is used to establish secure connections across the Internet.

In this paper, a detailed study of HTTP and its variants (HTTP/1.1, HTTPS, and HTTP/2.0) is done. Section 2 lays the groundwork by describing the HTTP/1.1

Aakanksha (✉) · B. Jain (✉) · D. Saxena · D. Sahni · P. Sharma
Shaheed Rajguru College of Applied Sciences for Women, University of Delhi,
Vasundhara Enclave, New Delhi 110096, India
e-mail: aakanksha_v@yahoo.co

B. Jain
e-mail: Bhawnanick14@gmail.com

© Springer Nature Singapore Pte Ltd. 2019   171
B. K. Panigrahi et al. (eds.), *Smart Innovations in Communication and Computational Sciences*, Advances in Intelligent Systems and Computing 670,
https://doi.org/10.1007/978-981-10-8971-8_17

protocol, discussing its vulnerabilities, and exploring the security features of HTTPS, respectively. The latest version HTTP/2.0 and its advantages over HTTP/1.1 are explored in Sect. 3. It further describes how certain design choices can allow an attacker to steal sensitive information from a Website and some steps that a Web developer can take to avoid these situations. Section 4 highlights the significant differences between the headers of HTTP/1.1, HTTPS, and HTTP/2.0. The results and discussions that were carried out to show the drawbacks of HTTP/1.1 and the advantages of using HTTP/2.0 over it are given in Sect. 5. A case study that was conducted to test security and performance parameters of 15 Websites is also tabulated in this section. Section 6 summarizes the tools that were used for the demonstrations. Section 7 concludes the paper by outlining the inferences that were obtained during the analysis. The references are listed in Sect. 8.

2   Background

Sir Tim Berners-Lee, a scientist at CERN, identified a major problem that the scientists there were facing regarding sharing their research work. To quote him, "in those days, there was different information on different computers, but you had to log on to different computers to get at it. Also, sometimes you had to learn a different program on each computer. Often it was just easier to go and ask people when they were having coffee..." [2]. To solve this problem, he began to tap into the potential of the fast-emerging technologies—the Internet and the Hypertext. In the October of 1990, he wrote the HTTP or the Hypertext Transfer Protocol (among two other technologies) that allowed the retrieval of documents from across the Web and thus the first version of HTTP was born. Over the years, it has been modified and updated to meet the security and performance concerns of the Web.

2.1   HTTP

HTTP is a stateless application layer protocol which means that it does not retain state between user requests. Every request is independent of the other. HTTP is called request–response protocol in which the client (browser) sends requests for the resource and the server gives back the responses.

In its first version, HTTP used only one method called GET which was used to request a page from a server [3]. Since then, HTTP has gone through a series of improvements. HTTP/1.0 came with the ability to transfer different types of documents including cascading style sheets, scripts, images, and videos. But this was not sufficient, and it was improved further to include the ability to reuse the already established connection with the server without the server terminating it for subsequent requests. The request that user makes from the browser or the response that

Message	=	<start-line> (<message-header>) CRLF (<message-body>)		
<start-line>	=	Request-line | Status-line		
<message-header>	=	FieldName':'Field-value		
Request-Line	=	Method	Resource	HTTP-verison
Stauts-Line	=	HTTP-version	Status-Code	description

Fig. 1 Format of HTTP request or response

comes from the server and the messages traveling over the network follow a particular format [4] as shown in Fig. 1.

2.2 Shortcomings of HTTP/1.1

Head-of-line Blocking. An average number of requests needed to properly display a Website rises at an alarming rate. HTTP/1.x does not perform well when there is large number of requests which arises due to head-of-line blocking. HTTP/2.0 comes to a rescue. Head-of-line blocking occurs when traffic waiting to be transmitted prevents or blocks traffic destined elsewhere from being transmitted [5]. Time to first byte, or TTFB, measures the duration from the client making an HTTP request to the first byte of the page being received by the client's browser. Till the time the server responds to the client's request, the client is idle. Thus, the client has to again wait for some time, which is not being used effectively. This problem is known as head-of-line blocking and reduces the performance up to a great extent. HTTP/1.x tries to solve the head-of-line blocking problem by opening up six parallel connections simultaneously. So if the first request is waiting to send its first byte, the second request can be served on the second connection. This is costly because each connection to be established requires TCP three-way handshake to be completed. Hence, head-of-line blocking serves as a bottleneck to Website's performance. After a request is served by the server, the connection is released and is made available for others to connect to the server. If the same client wants to send multiple requests to the server, it has to go through the three-way handshake process each time. This is a very time-consuming process. To solve this problem, HTTP introduces the concept of keepalive in the header. Keepalive maintains a persistent connection between the client and the server, thus preventing the server from closing the connection after successfully delivering the response. The client can reuse the already established connection for sending multiple requests. Also, the client can send request before the response of previous request is fully transferred.

Uncompressed Headers. To reduce the time to send the data, a lot of sites compress their assets or resources with compression algorithms that work on the Web like zip compression. Compression of data is good, but the major disadvantage is request and response headers which remain uncompressed. They repeat a lot across requests. The host headers and the cookies are always the same which makes them highly compressible.

Security. It is impossible to handle sensitive data like credit card details and contacts without TLS. Transport Layer Security (TLS) is a protocol that provides privacy and data integrity between two communicating applications. It is the most widely deployed security protocol used today and is used for Web browsers and other applications that require data to be securely exchanged over a network such as file transfers, instant messaging. There is an unencrypted version of HTTP/2.0, but browsers do not support it. Nothing changes while using TLS with HTTP/2.0.

3   HTTPS

HTTP/1.0 and HTTP/1.1 on their own are not secure protocols. Not only do they send the requests and responses in human readable format, but they also make communication vulnerable to attacks such as man in the middle or the phishing attacks. The threats and the probability of an attack increase several folds if the communication is taking place over a Wi-Fi network, which is widely used today. In order to secure the data being communicated, HTTPS was introduced, which intends to provide confidentiality, integrity, and authenticity of the data. To prevent the attacks, HTTPS uses encryption, to make the data being transmitted indecipherable, and authentication, to confirm that the user is connected to the right server.

HTTPS, also called HTTP over Transport Layer Security (TLS) or HTTP over Secure Sockets Layer (SSL), consists of communication over HTTP within a connection encrypted by TLS or SSL to provide secure communication over a computer network.

Transport Layer Security or TLS provides privacy and data integrity between two communicating applications [6]. TLS achieves this by using the following two processes:

3.1   Encryption

It is provided through symmetric key cryptography. The keys are generated uniquely for each connection based on a secret negotiated by another protocol (such as the TLS Handshake Protocol).

3.2 Hashing

To check the integrity of the message being transported, a keyed message authentication code (MAC) is used. Secure hash functions, such as SHA-1, SHA-256, are used for MAC computations (Table 1).

HTTPS combines the network protocol HTTP with the cryptographic protocol TLS. The TLS protocol updates the older public SSL protocol [7]. TLS provides a secured tunnel to a server, which is most commonly authenticated by an X.509 certificate [8]. HTTP contains some subtle security risks. The history and the backward compatibility of HTTP carry a lot of baggage from the time when security was not as important as it is now.

To understand some of these security concerns, it is important to first comprehend the concept of same origin policy.

Same Origin Policy. An origin is made up of three parts: data schema, host name, and port. If we change any of these parts, we are on a different origin and different rules will apply.

Cross-Origin Resource Sharing. Cross-origin resource sharing gives Web servers cross-domain access controls, which enable secure cross-domain data transfers. It works by adding new HTTP headers that allow servers to describe the origins that are permitted to read information using a Web browser. For HTTP request methods that can cause side effects on user data, the specification mandates that browsers "preflight" their requests with an HTTP OPTIONS method and then upon approval from the server, send the actual request with the actual HTTP request method. Servers can also notify clients whether cookies or HTTP authentication data should be sent with requests.

Cross-Site Request Forgery. Cross-site request forgery (CSRF) is an attack which forces an end user to execute unwanted actions on a Web application in which she is currently authenticated [9]. CSRF attacks mostly target Web-based applications. For CSRF attack to take place, two elements are required, a Web application which performs the actions and the authenticated user. The action can cause a transfer of money from bank account or purchase of items online. The attacker uses social engineering to send a link that can invoke malicious actions. On clicking this link, an HTTP request is sent to the target destination under the hood of which the victim is not aware.

Table 1 SSL/TLS layer between HTTP and TCP

HTTP		HTTP
TCP		SSL/TLS
IP		TCP
		IP

Cross-Site Scripting. The Open Web Application Security Projection (OWASP) defines cross-site scripting (XSS) as "a type of injection problem, in which malicious scripts are injected into the otherwise benign and trusted Web sites." According to the Web Application Security Consortium, XSS "is an attack technique that involves echoing attacker-supplied code into a user's browser instance. A browser instance can be a standard Web browser client, or a browser object embedded in a software product" [10].

Validating the user input at server side or encrypting the data or disabling the scripts in the browser can help to protect the Websites from XSS attacks.

4 HTTP/2.0

Between 2010 and 2015, the amount of data transfer for single Web page has tripled. The number of requests to get all the data is on a steady rise. Some current best practices like concatenation solely exist to work around the short comings of HTTP/1.1. This is where HTTP/2.0 comes in. It is backward compatible and solves the biggest issues that HTTP/1.0 and HTTP/1.1 have.

HTTP/2.0 was developed from the earlier experimental speedy (SPDY) protocol, originally developed by Google whose goal was to speed up the Web. Most major browsers like Chrome, Internet Explorer 11, Opera, Firefox, Safari added HTTP/2.0 support by the end of 2015 [11]. It decreases latency and improves page load speed in Web browsers by considering the following:

4.1 Compression of HTTP Headers

The human readability of request and response headers is gone. This is the first step toward improving performance. However, tools like Wireshark can help us to see headers even with HTTP/2.0.

4.2 Fixing the Head-of-Line Blocking Problem

The second major problem that HTTP/2.0 solves is head-of-line blocking. It is solved by multiplexing, i.e., combining multiple signals into a new single signal. It is a method by which multiple HTTP requests can be sent and responses can be received asynchronously via a single TCP connection. Multiplexing is the heart of HTTP/2.0 protocol. Every HTTP/2.0 request and response is given a unique ID called as Stream ID, and an HTTP request and response is divided into frames. Frames are binary pieces of data. A stream is a collection of frames with the same Stream ID. All the streams share that single connection. Stream ID is used to identify to which request or response a frame belongs to. To make a request, first, the client divides the request into binary frames and assigns the Stream ID of the request to the frames. Then, it initiates a TCP connection with the server. After that, the client starts sending the frames to the server. Once the server has the response ready, it divides the response into frames and gives the response and frames the same response ID. Stream ID is necessary as multiple requests to an origin are made using a single TCP connection. The request and response both happen simultaneously, i.e., while the client is sending frames to the server, the server is also sending frames back to the client.

4.3 Multiplexing Multiple Requests Over a Single TCP Connection

In HTTP/1.x, header fields are not compressed. As Web pages have grown to require hundreds of requests, the redundant header fields in these requests unnecessarily consume bandwidth, thus increasing latency. To reduce the overhead and improve performance, HTTP/2.0 compresses request and response headers metadata using the HPACK compression format. The primary motivation for the transition into HPACK compression is to reduce bandwidth, while the other components are designed to reduce round trips and accelerate the loading time of complex Web pages [12]. HPACK compression format uses two techniques:

Allowing the transmitted header fields to be encoded by compression algorithm i.e. Huffman Code which reduces individual transfer size.

Requiring that both client and server maintain and update an indexed list of previously seen header fields which is then used as a reference to previously encoded transmitted values. For example, cookies have really long headers and can be compressed. Huffman coding allows the individual values to be compressed, while transferred and the indexed list of previously transferred values allow us to encode duplicate values by transferring index values that can be used to look up and reconstruct the full header. The HPACK compression consists of static and dynamic tables [13]. The static table provides a list of common HTTP header fields that all connections are likely to use. The dynamic table is initially empty and is updated based on exchanged values within a particular connection. As a result, the size of each request is reduced.

5 Header Comparison

On comparing the size of request and response headers, HTTP/2.0 has significantly smaller header sizes as HTTP/2 uses HPACK algorithm that was specifically designed to compress headers. It uses predefined tokens, dynamic tables, and

Huffman compression. HTTPS piggybacks HTTP entirely on top of TLS due to which the entirety of the underlying HTTP protocol can be encrypted. It includes headers, request/response loads, while HTTP/1.1 headers are neither encrypted nor compressed.

6 Results and Discussions

Using tools like Wireshark and DevTools, various attacks on Websites were assessed. The following subsection discusses the simulation of these attacks and their outcomes.

6.1 Man in the Middle Attack

This demonstration aims to show how the vulnerabilities in HTTP/1.1 (and its predecessors) make the Websites served over this protocol susceptible to man in the middle attack. In this form of attack, a malicious user (or a computer bot) intercepts the communication channel between the server and the client, extracts the data being exchanged, and corrupts it. In order to mimic this attack, an executable binary was used that could revert images present on a Web page. Several Websites were then visited to check if they would fall prey to this attack. It was observed that the Websites using HTTP/1.1 were attacked, while those served over HTTPS were immune to it. Figure 2 demonstrates this attack.

6.2 Sniffing Attack

Since HTTP/1.1 does not encrypt its data, it is possible to use network analyzer tools and sniffers to listen to the communication between client and the server and find out the data being transferred between the two parties.

Fig. 2 a Web page before MITM attack. **b** Web page before MITM attack. *Source* http://www. rajgurucollege.com/

In this demonstration, Wireshark tool is used to analyze the network and listen to the packets being transferred. Figure 3 shows the user entering his username and password and trying to login to her account on the CSSS (Center Sector Scholarship Scheme, awarded by CBSE) Website. Since the login page is served over HTTP/1.1, the attacker can use Wireshark (or any such tool, for that matter) to find out the login credentials of the user. Figures 4, 5, 6, and 7 illustrate the process.

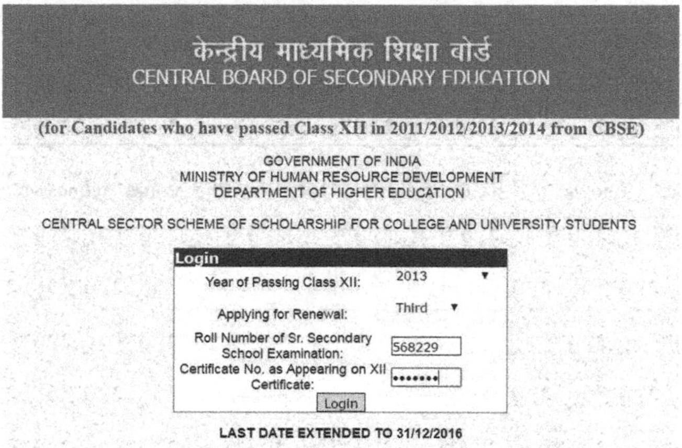

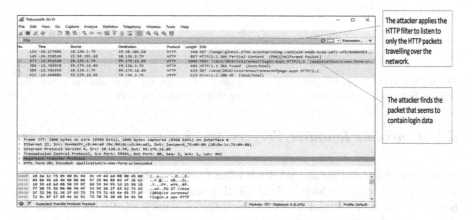

Fig. 3 User entered her login credentials on the Web page. *Source* http://59.179.16.89/2016/csssrenew/rlogin.aspx

Fig. 4 Attacker uses Wireshark to listen to all the (HTTP) packets traveling on the network

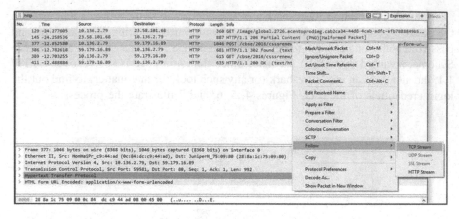

Fig. 5 Attacker follows the TCP stream to listen to the entire request–response cycle corresponding to the selected packet

Fig. 6 Wireshark shows the communication occurring between the client and the server. The attacker finds the password of the user (client) by simply reading this data

6.3 Cross-Site Resource Forgery Attack

In this demonstration, the attacker sends a link to the user that takes her to malicious site (in our case, game.html) a Website that was created to perform actions on behalf of the user without his/her knowledge. Figures 7, 8, 9, 10, 11, and 12 demonstrate this attack.

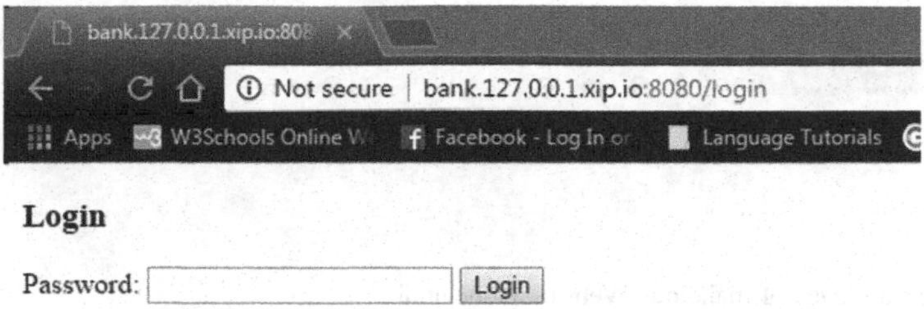

Fig. 7 Login page of the fake bank Website that was created

Fig. 8 Web page showing transaction of the logged-in user

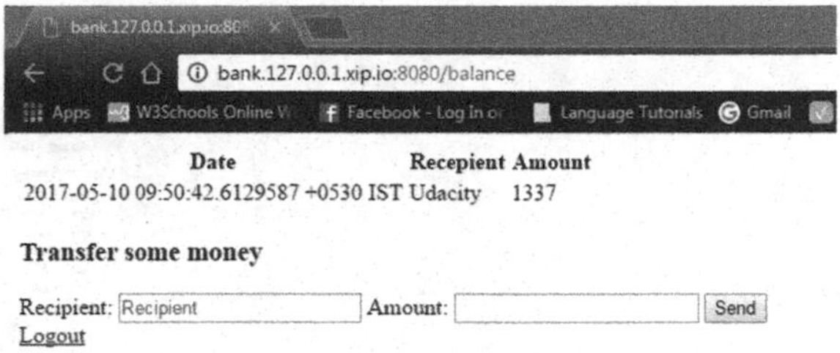

Fig. 9 Session ID of the session for which the user is logged into her bank account

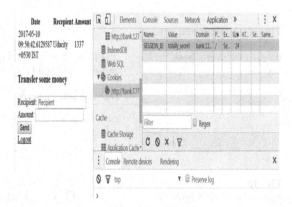

6.4 XSS Cross-Site Scripting Attack

In this demonstration, the XSS injection problem was simulated, which takes advantage of the vulnerabilities of the sites that do not validate user input. Figures 13, 14, 15, 16, and 17 demonstrate this attack.

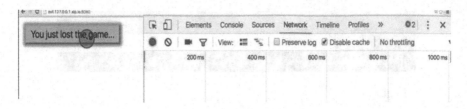

Fig. 10 Home page of malicious Website: game.html

Fig. 11 Attacker has gained access to the user's session ID for her bank account

Fig. 12 A transaction has been made by the attacker through the user's bank account

Date	Recipient	Amount
2016-05-16 17:05:02.412528328 -0700 PDT	Udacity	1337
2016-05-16 17:05:15.157422748 -0700 PDT	Umbrella Corp	666

Transfer some money

Recipient: Recipient   Amount: Send

Logout

Fig. 13 A Website vulnerable to XSS problem

← → C ⌂ ⓘ badwebsite.127.0.0.1.xip.io:8080

⚏ Apps ☐ ASP.NET ☐ Language Tutorials 🟩 W3Schools Online We

Hello Anonymous!
Want to tell us your real name?
☐   Send

```
⌜⌐ ⌐⌐  |  Elements   Console   Sources   Network

⊘  ▽  top    ▼  ⬜ Preserve log

>  document.cookie
<  "SESSION_ID=DEADBEEF"
>  document.cookie.split('=')
<  ["SESSION_ID", "DEADBEEF"]
>  document.cookie.split('=')[1]
<  "DEADBEEF"
>  document.cookie.slice(
        document.cookie.indexOf('SESSION_ID')
   ).split('=')[2]
<  "DEADBEEF"
>
```

Fig. 14 Extracting the session id of the browser

```
File  Edit  Selection  Find  View  Goto  Tools  Project  Preferences  Help
◀ ▶   xss.js        ✕
1  <script>
2  fetch(http://decoder.127.0.0.xip.io:8088/?key='
3      +document.cookie.slice
4      (document.cookie.indexof("cookieTest")).
5      split('=')[1])
6  </script>
```

Fig. 15 Code entered into the textbox instead of normal text

← → C ⌂ | ⓘ badwebsite.127.0.0.1.xip.io:8080

⦂⦂⦂ Apps   ☐ ASP.NET   ☐ Language Tutorials   ⬛ W3Schools Onl

Hello Anonymous!
Want to tell us your real name?
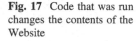 Send

Fig. 16 Code pasted into the textbox the Website does not validate the text, and the code is run

Fig. 17 Code that was run changes the contents of the Website

← → C ⌂ | ⓘ badwebsite.127.0.0.1.xip.io:8080/?

⦂⦂⦂ Apps   ☐ ASP.NET   ☐ Language Tutorials   ⬛ W3School

Hello !

6.5   Analysis of Security and Performance Parameters of Websites

In this activity, the security and performance parameters of 15 commonly used Websites were tested. The results have been recorded in Table 2. Column 3 shows the number of requests sent to fetch and render the home page of the Website, and column 4 shows the bytes transferred in the process. Each of the Websites shown below was accessed 5 times in private/incognito mode in the browser window. An average of the time taken to load the page each time was taken. This average load time is shown in column 5. Column 6 and 7 show the version of HTTP that the Website is served over at the time of writing this paper. Since the users submit their most crucial information on the login page of a Website, this page was particularly checked for security during the case study. The results are recorded in column 8. To test whether or not the Websites are susceptible to man in the middle attack, they were accessed over a proxy network which was made to mimic a network that was compromised by an attacker (in this case, a program) that could revert/flip the images. If the images of the Websites flipped, it meant that those pages on the Websites were not fully secure. The results of this security test are shown in column 9. All the errors shown in the console window when the Website was accessed are summarized in the last column.

7   Tools Used

Following tools were used to conduct the demonstrations in the paper.

Wireshark. It is a free and open source software analyzer used for network troubleshooting.

DevTools. The developer tools are a set of Web authoring and debugging tools built into Google Chrome. They provide Web developers deep access into the internals of the browser and their Web applications. Using DevTools, we can track down layout issues, set JavaScript break points, among other things.

BadSSL.com. It is a great site to see how browser behaves when the connection has some issues. It has its own valid certificates, but it also has intentionally invalid certificates and invalid setups so that we can see what the browser does in different situations.

Table 2 Results of the case study on 15 Websites served over different versions of HTTP

S. No.	Websites	No of requests	Bytes transferred	Average load time (s)	Served over HTTP/1.1/ HTTP/2.0	Served over HTTPS	Is the login page secure?	Susceptible to man in the middle attack?	Error
1	Shaheed Rajguru College of Applied Sciences for Women http://www.rajgurucollege.com/	70	777 KB	5.15	HTTP/1.1	No	Not applicable[a]	Yes	Several resources not avail/not found jquery not defined
2	University of Delhi http://www.du.ac.in/	27	1.6 MB	1.98	HTTP/1.1	No	No	Yes	None
3	Hansraj College http://www.hansrajcollege.co.in/	26	1.1 MB	3.17	HTTP/1.1	No	No	Yes	Resources not found
4	CBSE cbse.nic.in	53	9.3 MB	6.17	HTTP/1.1	No	No	Yes	Uncaught type errors, resource loading fields
5	MIT Open Courseware ocw.mit.edu	113	1.7 MB	10	HTTP/1.1	Yes	Not applicable[a]	No	None
6	IIT Bombay http://www.iitb.ac.in/	82	2 MB	1.83	HTTP/1.1	No	Not applicable[a]	Yes	Several resources not avail/found
7	IIT Delhi http://www.iitd.ac.in/	67	466 KB	1.56	HTTP/1.1	No	Not applicable[a]	Yes	None
8	NASA https://www.nasa.gov/	136	7.2 MB	7.56	HTTP/1.1	Yes	Not applicable[a]	No	Uncaught reference error
9	ISRO http://www.isro.gov.in/	49	5.7 MB	22.05	HTTP/1.1	No	Not applicable[a]	Yes	Resource not found
10	LPI https://www.lpi.usra.edu/ lpiintern/	40	3.8 MB	7.8	HTTP/1.1	Yes	Yes	Yes	Uncaught reference error
11	Digital India http://digitalindia.gov.in/	82	913 KB	3.05	HTTP/1.1	No	No	Yes	None

(continued)

Table 2 (continued)

S. No.	Websites	No of requests	Bytes transferred	Average load time (s)	Served over HTTP/1.1/ HTTP/2.0	Served over HTTPS	Is the login page secure?	Susceptible to man in the middle attack?	Error
12	DMRC http://www.delhimetrorail.com/	65	6.6 MB	6.83	HTTP/1.1	No	Not applicable[a]	Yes	Uncaught error, resource not found
13	NTPC http://www.ntpc.co.in/	65	1.3 MB	4.09	HTTP/1.1	No	Yes	No	None
14	GAIL http://www.gailonline.com/final_site/index.html	50	2.3 MB	7.05	HTTP/1.1	No	Not applicable[a]	Yes	None
15	IRCTC https://www.irctc.co.in/eticketing/loginHome.jsf	80	1.4 MB	5.97	HTTP/1.1	Yes	Yes	No	Uncaught reference error

[a]Login page on these websites is not available

8 Conclusion

The experiments and simulation tests done so far in this paper show why HTTP/2.0 needs to become a Web standard. It was observed that several Websites including those of banks, commonly used e-shopping portals, universities, hospitals, and research institutes are still using HTTP/1.1. Some of these Websites use HTTPS only for pages where the user logs in data or for when she enters sensitive information. However, from the results, it can be said that all the resources and pages of a site should be encrypted (i.e., they must use HTTPS) and be served over HTTP/2.0 to protect the security of their users and data. Failure to do so makes a Website vulnerable to several hacks/attacks such as man in the middle, phishing, sniffing, spoofing. In addition to that, HTTP/2.0 decreases the load time of a site and improves its performance parameters.

Several browsers, including Google Chrome, have already announced that they shall begin to give a "red lock of shame" to Web pages that do not run upon HTTPS to create a secure channel between the client and server; in future, such unsecured pages will also be blocked.

While it is usual for Web developers to serve a site that does not require user input over HTTP/1.1, it is important to realize that even in such situations an attacker can execute attacks that deface the sites or post ad links to malicious sites on their home page.

In addition to this, Web developers must follow the usual good practices such as validating user input to provide more security to their visitors and to the data that the site uses.

References

1. Fielding, R., Berners-Lee, T.: RFC 2616—Hypertext Transfer Protocol–HTTP/1.1, https://tools.ietf.org/html/rfc2616#page-7.
2. History of the Web. (2017). World Wide Web Foundation. Retrieved 10 September 2016, from http://webfoundation.org/about/vision/history-of-the-web/.
3. Berners Lee, T.: Hyper Text Transfer Protocol, https://www.w3.org/History/19921103 hypertext/hypertext/WWW/Protocols/HTTP.html.
4. Podila, P.: HTTP: The Protocol Every Web Developer Must Know—Part 1, https://code.tutsplus.com/tutorials/http-the-protocol-every-web-developer-must-know-part-1–net-31177.
5. Jon C. R. Bennett; Craig Partridge; Nicholas Shectman (December 1999). "Packet reordering is not pathological network behavior". IEEE/ACM Transactions on Networking. 7 (6): 789–798. https://doi.org/10.1109/90.811445.
6. Rouse, M.: Transport Layer Security (TLS), http://searchsecurity.techtarget.com/definition/Transport-Layer-Security-TLS.
7. Prusty, N.: What is Multiplexing in HTTP/2?, http://qnimate.com/what-is-multiplexing-in-http2/.
8. Clark, J.: SoK: SSL and HTTPS: Revisiting past challenges and evaluating certificate trust model enhancements. IEEE Symposium on Security and Privacy (2013).

9. Wireman, M.: CSRF and XSS: A Lethal Combination—Part I, http://resources. infosecinstitute.com/csrf-xss-lethal-combination/#gref.
10. Chauhan, S.: Cross-Site Scripting (XSS), http://resources.infosecinstitute.com/cross-site-scripting-xss/#gref.
11. Usage Statistics of HTTP/2 for Websites, March 2017, https://w3techs.com/technologies/details/ce-http2/all/all.
12. HTTP/2: In-depth analysis of the top four flaws of the next generation web protocol. Imperva (2017).
13. Gmarkham: Same Origin Policy—Web Security, https://www.w3.org/Security/wiki/index. php?title=Same_Origin_Policy&oldid=2.

Spam Detection Using Rating and Review Processing Method

Ridhima Ghai, Sakshum Kumar and Avinash Chandra Pandey

Abstract In recent times, e-commerce sites have become an essential part of people lifestyle. Viewers give feedback and firsthand account of the online products, and these reviews thus play an important role in decision making of the other buyers. So, in order to increase or decrease sales of products, spam reviews are generated by the companies. Hence, there is a need to detect and filter the spam reviews to provide customers genuine reviews of the product. In this paper, a review processing method is proposed. Some parameters have been suggested to find the usefulness of reviews. These parameters show the variation of a particular review from other, thus increasing the probability of it being spam. This method introduced classifies the review as helpful or non-helpful depending on the score assigned to the review.

Keywords Rating deviation · Caps count · Reviewer's count · Data scrapping

1 Introduction

In today's era, reviews play a crucial role in one's decision to purchase any product or not. There is abundance of different varieties of product and higher competition, thus misleading customers from other competitive products by spamming reviews about them on e-commerce sites. Spam reviews are designed to give a deviated opinion about a product for increasing or decreasing its sales in market. Thus, it has become important to detect those spams and to give the customer a genuine review about the product and also maintain the company's stature.

Lin et al. [1] introduced a method which uses six different features to detect spam based on review content and reviewer's behavior. They have used supervised and

R. Ghai (✉) · S. Kumar (✉) · A. C. Pandey (✉)
Jaypee Institute of Information Technology, Noida, India
e-mail: ridhimaghai02@gmail.com

S. Kumar
e-mail: sakshumkumar@gmail.com

A. C. Pandey
e-mail: avish.nsit@gmail.com

© Springer Nature Singapore Pte Ltd. 2019   189
B. K. Panigrahi et al. (eds.), *Smart Innovations in Communication and
Computational Sciences*, Advances in Intelligent Systems and Computing 670,
https://doi.org/10.1007/978-981-10-8971-8_18

unsupervised methods to detect spam. Heydari et al. [2] presented a survey on detection of review spam through four approaches such as performing sentiment analysis of the reviews, characteristic-based perspective extraction that determines features of the reviewed product with the main purpose of seeking the attention, and perspective of the reviewer about that particular feature. Xie et al. surveyed and devised algorithm [3] for observing cryptic reviews with the help of interrelated patterns by indentifying multidimensional time series based on statistics. The algorithm is systematic and takes care of the time windows where spam attacks were encountered. Lin et al. proposed a method [4] for detecting spam review via unusually correlated temporal patterns by indentifying and creating multidimensional time series based on aggregate statistics in order to depict and mine such correlations.

In the study, an approach is devised that assigns a score to each review based on rating variation, caps score, and reviewer count. According to score, reviews are ranked in descending order of scores [5]. The higher ranked reviews are considered to be helpful and genuine, and the lower ranked are considered as potential spam.

The proposed method is applied to scrapped reviews from online Web site Amazon which contains customer reviews on various products. The evaluation of proposed approach shows a high accuracy, thus proving its efficiency.

The remaining paper is divided into the following sections: Sect. 2 discusses the preliminaries; Sect. 3, proposed work; Sect. 4, results and discussion; Sect. 5, conclusion of the paper.

2 Preliminaries

The following studies are carried out in the area of review spammer detection, and the findings from the same are illustrated:

1. Towards Online Review Spam Detection
   The main idea behind this study was to devise a strategy to attain spam reviews. Different features based on review content and reviewer's behaviour which includes personal content similarity, number of reviews on the product was developed [6].
2. Opinion Spam Detection Using Anomalous Rating Deviation
   In this previous work by researchers, an efficacious technique for recognizing spams based on manipulation of average rating of products has been proposed [9, 10]. A regression method [11] is considered to identify reviewers containing unusual proportions of rating that variate from opinion of the masses. The advantage of this method is its simplicity and its low computational cost.
3. Opinion Spam Recognition Using Ontological Features
   In this paper, the author analyzed review spam based on the review content which used ontology model [12] as the main model to identify them. The faulty reviews are categorized into four types: non-review—a review that does not contain any opinion; brand only review—review that assesses the company or service but not the actual product; off-topic review—a review which is not about the product or

service; and untruthful reviews—fake reviews which give positive or negative reviews to deceive users.

4. Detecting Spam Reviews with the Help of Temporal Patterns

It is observed that reviewer's reviewing about a product writes review in a manner which depicts a pattern. Spam attacks are busty and relatres to reviewing in a positive or negative manner. An algorithm is developed for the same. The singleton review spam detection problem [13] is joined with a pattern detection problem. Experimental results show that the approach is efficacious in detecting singleton review attacks. It is discovered that singleton reviews are a potential source of spam reviews and largely affect the ratings of online stores.

5. Hybrid spam detection based on unstructured datasets

Shao et al. [14] devised strategies to detect hybrid spams based on unstructured datasets. The study is based on detecting textual as well as image features of a potential spam based on sentiment analysis and local image properties. The research was able to distinguish between meta and real spam. The only limitation was the number of unstructured datasets taken that were limited as it involved intensive study and understanding of semantics.

6. Deceptive review detection using labeled and unlabeled data

Rout et al. [15] have implemented supervised as well as unsupervised techniques to identify online product review spam. Linguistic features, POS features, and the sentiment score model are collaborated together in order to make it efficient for real-life models. To get the best performance, some well-known classifiers were applied on labeled dataset. Due to privacy concerns, the user information on the mentioned Web sites cannot be obtained.

7. Detection of opinion spam with character n-grams Fusilier et al. [16] identified that the potential spam reviews are similar to normal reviews in terms of its content but vary in the terms of opinion (style) it expresses. The character n-grams feature is one which focuses on lexical as well as the stylistic information of a review. The result of this study depicts these character n-grams to be superior to word n-grams in detection opinion-based spams. Small training dataset is required for this research. The limitation is its inability to merge the character n-grams and word n-grams features.

8. Review Spam Detection using Active Learning

Ahsan et al. [17] introduced an active learning approach to detect review spam using the TF-IDF features of the review content and three classifiers—Linear SVM, SGD, and Perceptron. During the process, the model is trained from the best samples in multiple iterations.

9. Suspicious Behavior Detection: Current Trends and Future Directions

Jiang et al. [7] surveyed more than 100 advanced techniques for detecting suspicious behaviors. They bifurcated them broadly into four categories: traditional spam, fake reviews, social spam, and link farming. After analysis, tremendous progress in link-farming detection systems and trends in information integration are seen.

10. Trust-Aware Review Spam Detection

Xue et al. [18] researched a unique feature for spam detection social networking between users and potential spam reviewers. The method proposed focuses on how social relationship can be used as a factor to deviate users from their opinions. Two factors were considered as indicators for spamicity–proximity of user and reviewer for developing trust-based rating prediction and trustworthiness score.

3 Proposed Work

The challenge in spam detection is rhetorical and thus leads to new framework to solve the problem. In this paper, a Rating and Review Processing Method has been introduced to find the overall score of reviews for spam detection. The proposed method uses some parameters for spam detection, and these parameters show the variation of a particular review from other, thus increasing the probability of it being spam. This approach has been proposed which classifies a review as helpful or non-helpful depending on the score assigned to the review. The parameters to compute this score are discussed below. The detailed flowchart of the proposed method has been shown in Fig. 1. The detailed steps of the proposed method are given in Algorithm 1.

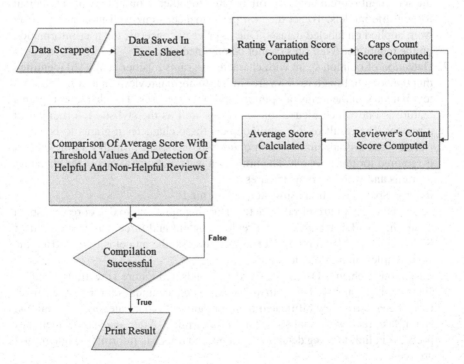

Fig. 1 Flowchart of the proposed method

Algorithm 1 Proposed Method

$-n$ (number of reviews))
$-\theta$ (threshold value
counter $= 0$
while *counter* $<= n$ or *stopping* $-$ *criterion* **do**
    Compute RV(i)
    if $RV(i) >= |2|)$ **then**
        RVW(i)=0.25
    else if $RV(i) >= |1|)$ **then**
        RVW(i)=0.5
    else
        RVW(i)=1
    end if
    Compute CC(i)
    Compute RC(i)
    if $RC(i) = 1$ **then**
        RCW(i)=0.5
    else
        $RCW(i) = 0.25$
    end if
    $OverallScore(i)=(RVW(i) + CC(i) + RCW(i))/3$
    if $OverallScore(i) >= \theta$ **then**
        i^{th} review is HELPFUL
    else
        i^{th} review is NON-HELPFUL
    end if
    counter =counter+1
end while
Rank the reviews using *OverallScore*

---

1. **Rating Variation**—It has been found that the rating of suspicious reviewer variates from the rating of other honest reviewer. This variation can be found by computing the average rating R_{avg} and then computing the variation of ith review RV(i) as:

$$RV(i) = |R_i - R_{avg}|. \tag{1}$$

Rating Variation Weight (RVW) is calculated as follows:
RVW $= 1$ if RV(i) $= 0$
RVW $= 0.25$ if $RV(i) \geq 2$
RVW $= 0.5$ if $RV(i) > 1$ and $RV(i) < 2$.

2. **Caps Count**—The use of capital letters in a review is considered suspicious. Thus, we count the number of words in capital letters A to compute caps count CC(i) as follows:

$$Y = A/N. \tag{2}$$

Y is normalized value between 0 and 1

$$CC(i) = 1 - Y \tag{3}$$

Caps count is computed by complementing the value of Y. This shows that larger the use of capital letter words, less is the caps count of a review.

3. **Reviewer Count**—The recursive review of same reviewer for the same product is considered to be a potential spam. Thus, the reviewer's name corresponding to the review and then the reviewer count are scrapped, and RC(i) is calculated for ith review. Reviewer Count Weight (RCW) is computed as follows:

RCW = 0.5 if RC(i) = 1
RCW = 0.25 if $RC(i) > 1$

The proposed algorithm provides a recursive approach to calculate overall score of each review. The inputs for the algorithm are product name, reviewer name, review, and rating. Based on this score, the algorithm categorizes each review as helpful, genuine reviews, or non helpful, potential spam reviews.

In this algorithm, the values assigned to RVW are 0.25, 0.5, or 1. The maximum variation found from average rating was 2. If variation lies between 1 and 2, the score assigned to RVW is 0.25, and if it lies between 0 and 1, it assigned a score as 0.5. The higher the deviation, the lower the score is. Similarly, the values assigned to RVW are 0.25 or 0.5. The higher the value, the lower the score is. Then, the caps count is normalized between 0 and 1 by taking the ratio of number of words in capital letters to the overall words in the review.

The threshold value θ is taken as 0.50, 0.55, 0.60, 0.65, 0.70, 0.75, and 0.80. On testing with these values, optimal result was found at 0.75. The reviews having score greater than threshold are categorized as HELPFUL and others as NON-HELPFUL. Then, a sample of 100 reviews is taken and compared theoretically and experimentally. Table 1 shows the number of non-helpful reviews detected by the proposed system. The reviews are arranged in ascending order of the overall score. Computed accuracy shows the percentage of non-helpful reviews programmatically, and theoretical accuracy is the percentage calculated manually. The accuracy is calculated using Eq. (4).

$$Accuracy = \frac{Experimentally\ Computed\ Number\ of\ Spam\ Reviews}{Theoretically\ Computed\ Number\ of\ Spam\ Reviews} \qquad (4)$$

The complexity calculated for the above recursive approach is O(n), where n is number of reviews.

4 Results and Discussion

To evaluate our approach, product review data is scrapped from online Web site Amazon.com using scrapping tool in Python. Each product has a profile page that links to set of reviews contributed by various reviewers. The attributes such as rating of a product, review data, and reviewer name are scrapped from the profile page of the product. Around 4000 and 2000 reviews of product, Lenovo K5 Note and Oppo F1S, respectively, are scrapped and stored in an excel file. On the fetched data,

Table 1 Detection of non-helpful reviews

Sr. No.	Dataset	Threshold value	Total reviews	Number of non-helpful reviews	Actual non-helpful reviews	Accuracy (%)
1.	Lenovo K5 note	0.50	100	62	94	65.95
2.		0.55	100	65	94	69.14
3.		0.60	100	69	94	73.40
4.		0.65	100	75	94	79.78
5.		0.70	100	82	94	87.23
6.		0.75	100	86	94	91.48
7.		0.80	100	89	94	94.60
1.	Oppo F1S	0.50	100	60	90	66.66
2.		0.55	100	63	90	70
3.		0.60	100	68	90	75.55
4.		0.65	100	72	90	80
5.		0.70	100	77	90	85.55
6.		0.75	100	81	90	90.32
7.		0.80	100	95	90	91.15

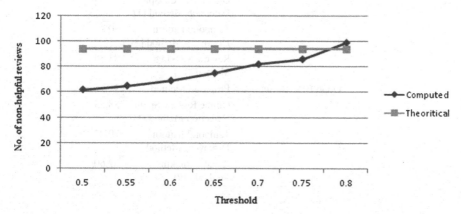

Fig. 2 Theoretical versus computed number of non-helpful reviews with respect to threshold value for Lenovo K5 Note dataset

our algorithm is applied and a score is given to each review. Each review is ranked according to the assigned score. It was found that the top ranked reviews described the aspects of a product in detail. They are genuine and focused on positive and negative features of the product. These reviews prove more relevant or helpful for users.

Moreover, the proposed method is also compared with three existing methods. From Table 2, effectiveness of proposed method can be easily observed. Further, the

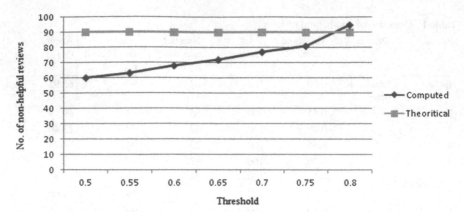

Fig. 3 Theoretical versus computed number of non-helpful reviews with respect to threshold value for Oppo F1S dataset

Table 2 Detection of non-helpful reviews

Sr.No.	Dataset	Methods	Accuracy (%)
1.	Lenovo K5 note dataset	Proposed Method	94.60
2.		Online Review Spam Detection Method [1]	90.08
3.		Temporal pattern Discovery Method [3]	80
4.		Review Ranking Method [19]	83.30
1.	Oppo F1S dataset	Proposed Method	91.15
2.		Online Review Spam Detection Method [1]	88.23
3.		Temporal pattern Discovery Method [3]	79.07
4.		Review Ranking Method [19]	82.60

line graph has also been plotted in Figs. 2 and 3 which validate the performance of the proposed method.

5 Conclusion

The factors considered in this algorithm are derived after studying many research papers. All the necessary factors are taken care of through which spam reviews can be detected, and as shown in the table, these factors gave an accurate account of helpful and non-helpful reviews at the threshold value of 0.75. Different studies are being carried out by researchers which have not been included till now, but are certain to be implemented in the future.

References

1. Y. Lin, T. Zhu, X. Wang, J. Zhang, and A. Zhou, "Towards online review spam detection," in *Proceedings of the 23rd International Conference on World Wide Web*. ACM, 2014, pp. 341–342.
2. A. Heydari, M. ali Tavakoli, N. Salim, and Z. Heydari, "Detection of review spam: A survey," *Expert Systems with Applications*, vol. 42, no. 7, pp. 3634–3642, 2015.
3. S. Xie, G. Wang, S. Lin, and P. S. Yu, "Review spam detection via temporal pattern discovery," in *Proceedings of the 18th ACM SIGKDD international conference on Knowledge discovery and data mining*. ACM, 2012, pp. 823–831.
4. Y. Lin, T. Zhu, H. Wu, J. Zhang, X. Wang, and A. Zhou, "Towards online anti-opinion spam: Spotting fake reviews from the review sequence," in *Advances in Social Networks Analysis and Mining (ASONAM), 2014 IEEE/ACM International Conference on*. IEEE, 2014, pp. 261–264.
5. E.-P. Lim, V.-A. Nguyen, N. Jindal, B. Liu, and H. W. Lauw, "Detecting product review spammers using rating behaviors," in *Proceedings of the 19th ACM international conference on Information and knowledge management*. ACM, 2010, pp. 939–948.
6. G. Wang, S. Xie, B. Liu, and P. S. Yu, "Identify online store review spammers via social review graph," *ACM Transactions on Intelligent Systems and Technology (TIST)*, vol. 3, no. 4, p. 61, 2012.
7. M. Jiang, P. Cui, and C. Faloutsos, "Suspicious behavior detection: Current trends and future directions," *IEEE Intelligent Systems*, vol. 31, pp. 31–39, 2016.
8. M. Sasaki and H. Shinnou, "Spam detection using text clustering," in *Cyberworlds, 2005. International Conference on*. IEEE, 2005.
9. R. Patel and P. Thakkar, "Opinion spam detection using feature selection," in *Computational Intelligence and Communication Networks (CICN), 2014 International Conference on*. IEEE, 2014, pp. 560–564.
10. X. Li and X. Yan, "A novel chinese text mining method for e-commerce review spam detection," in *International Conference on Web-Age Information Management*. Springer, 2016, pp. 95–106.
11. L. Wu, X. Hu, F. Morstatter, and H. Liu, "Adaptive spammer detection with sparse group modeling," 2017.
12. J. G. Thanikkal and M. Danish, "A novel approach to improve spam detection using sds algorithm," *International Journal for Innovative Research in Science and Technology*, vol. 1, no. 12, pp. 306–310, 2015.
13. S. Rayana and L. Akoglu, "Collective opinion spam detection: Bridging review networks and metadata," in *Proceedings of the 21th ACM SIGKDD International Conference on Knowledge Discovery and Data Mining*. ACM, 2015, pp. 985–994.
14. Y. Shao, M. Trovati, Q. Shi, O. Angelopoulou, E. Asimakopoulou, and N. Bessis, "A hybrid spam detection method based on unstructured datasets," *Soft Computing*, vol. 21, pp. 233–243, 2017.
15. J. K. Rout, S. Singh, S. K. Jena, and S. Bakshi, "Deceptive review detection using labeled and unlabeled data," *Multimedia Tools and Applications*, pp. 1–25, 2016.
16. D. H. Fusilier, M. Montes-y Gómez, P. Rosso, and R. G. Cabrera, "Detection of opinion spam with character n-grams," in *International Conference on Intelligent Text Processing and Computational Linguistics*. Springer, 2015, pp. 285–294.

17. M. I. Ahsan, T. Nahian, A. A. Kafi, M. I. Hossain, and F. M. Shah, "Review spam detection using active learning," in *Information Technology, Electronics and Mobile Communication Conference (IEMCON), 2016 IEEE 7th Annual*. IEEE, 2016, pp. 1–7.
18. H. Xue, F. Li, H. Seo, and R. Pluretti, "Trust-aware review spam detection," in *Trustcom/BigDataSE/ISPA, 2015 IEEE*, vol. 1. IEEE, 2015, pp. 726–733.
19. G. Ansari, T. Ahmad, and M. Doja, "Review ranking method for spam recognition," in *Contemporary Computing (IC3), 2016 Ninth International Conference on*. IEEE, 2016, pp. 1–5.

DDITA: A Naive Security Model for IoT Resource Security

Priya Matta, Bhaskar Pant and Umesh Kumar Tiwari

Abstract Information security has its own importance in information era. It forms the third pillar of information world after the performance upsurge and power issues. Security, as the term suggests, is the state of being free from threats. Resultantly Internet of Things receives almost all of the existing security threats from the world of Internet, along with some newly generated threats. In this paper, we are essentially and largely focussing on the security of data as well as resources involved in an Internet of Things system. In this paper, we propose a naive security model, namely DDITA (Definition, Design, Implementation, Testing and Amendment) that emphasizes on security policies, their implementation, their testing under various strategies and finally the amendments if required. In this paper, we have focussed on data involved in Internet of Things. We have classified data as *private data* and *public data*. We have also extended our studies toward the further classification of private data into *Stored Data* and *Data in Transit*. The security of *Stored Data* is proposed keeping encryption, authorization, authentication, attestation, and encryption using TPM under its umbrella.

Keywords Security · Internet of Things · Threats · DDITA model
Stored Data

P. Matta (✉) · B. Pant · U. K. Tiwari
Department of Computer Science and Engineering,
Graphic Era University, Dehradun, India
e-mail: mattapriya21@gmail.com

B. Pant
e-mail: pantbhaskar2@gmail.com

U. K. Tiwari
e-mail: umeshtiwari22@gmail.com

© Springer Nature Singapore Pte Ltd. 2019   199
B. K. Panigrahi et al. (eds.), *Smart Innovations in Communication and
Computational Sciences*, Advances in Intelligent Systems and Computing 670,
https://doi.org/10.1007/978-981-10-8971-8_19

1   Introduction

An organization, an industry, or a paradigm having secured data handling and protected resources would be preferred over the same which do not implement them. Hence, the security of data and resources is one of the most important aspects of the organization, industry, or paradigm. Improving the security of the data can have a positive impact on the business of the organization, growth of an industry and efficiency of a paradigm. More specifically, the competence and capabilities of a paradigm are directly proportional to its level of security.

Some common terminologies used in this context are:

Security: Information security can be outlined as a concept of maintaining data integrity, preserving confidentiality, and continuing the data availability. Information security is the practice of defending information from unauthorized access, use, disclosure, disruption, modification, perusal, inspection, recording, or destruction [1]. Security may be associated with any resource; it may be the physical security of some hardware, security of software, or the security of data and information. Security when associated with the data and information encompasses the security of digital or electronic data as well as the physical data.

Internet of Things: Internet of Things (IoT) is one of the important paradigms out of all already evolved and emerging paradigms like Distributed computing, Mobile computing, Cloud computing, Ubiquitous computing, Big data, Internet of Things, and even Web of Things. The Internet of Things (IoT) is the concept of interconnecting the physical entities, to make them able to communicate and transfer data. Entities are embedded with electronics, computing capabilities, and internetworking capacities.

According to Alex [2], the information world conceived the term IoT in 1999. The term was coined by Kevin Ashton, while he was working with Auto-ID Laboratories. ITU Global standard initiative [3] has outlined IoT as "the network of physical objects—devices, vehicles, buildings, and other items—embedded with electronics, software, sensors, and network connectivity that enables these objects to collect and exchange data."

IoT Security: The notion of IoT is making use of electronics, computing abilities, and interconnectivity in almost all physical devices. The idea behind this paradigm of IoT is to bind the entire physical and information world together. The IoT systems are skilled to congregate, correlate, and therefore contribute to a large amount of data, out of which sensitive and critical data raises thoughtful concerns. In many cases, security is not integrated to the IoT systems. The data, as well as resources, may easily be compromised with little efforts. And therefore one of the finest evolved paradigms will shape itself into a worst one.

This paper is divided into five sections. In the first section, we have discussed the introductory part of our domain of research, comprising of security, Internet of Things, and IoT Security. The second section involves motivation behind this work. Third section describes the focus of our work. The fourth section contains our proposed model for security of data and resources. This section also elaborates all the phases of our model. Finally, the fifth section concludes the paper, followed by references.

2 Motivation

In each organization, industry, business, or evolving technology, information security is an essential functional requirement. It is a fundamental factor that helps to avoid the intentional or unintentional leakage, loss, deletion, integrity compromise, or theft of the data and information. Many researchers have focussed on the security concerns regarding data as well as resources. Machara et al. [4] recognized two fundamental interests over the paradigm of IoT, namely privacy and Quality of Context (QOC). Renu et al. [5] were majorly concerned over RFID systems and their related security issues with respect to IoT paradigm. According to Bhattasali et al. [6], "the two most sensitive security concerns in cloud-based IoT are the transmission of critical data and storage of critical data." Therefore, these two aspects should be considered more attentively to improve the efficiency and efficacy of an IoT system.

According to Zorzi et al., advances in wireless technology permit real-time acquisition, transmission, and processing of critical data [7]. Therefore, the adversary can also have the same benefits using the same advances. It has also been observed that the number of interconnected devices is increasing rapidly. This pace of increasing interconnected devices and technology advances would definitely make an IoT system more prone to adversary attack. Thus, the threat to the IoT security becomes an issue to pay an appropriate attention. Markward [8] also identified some threats related to IoT systems. His key focus is on the security issues that are also related to IoT data acquisition. He recognized three threats, namely Data transfer between IoT devices, Data transfer between devices and other parties, such as users and Clouds, and Security issues related to the configuration made by users on their devices.

After observing his predictions, we can easily predict that data security should be handled with utmost importance during the implementation of an IoT system. Kawamoto et al. [9] have shown their intensive concerns over the authentication in an IoT system. They discussed authentication in terms of location-based authentication. According to them in location-based authentication systems, "ambient information" is collected from a significant number of diverse devices, which are deployed in an IoT society. In their work, they proposed a novel data collection method for authentication systems. Tsai et al. [10] claimed that open issues on privacy and security forms next major issue in IoT research. According to them, even though some of the camera-based detection and recognition technologies are mature enough, the privacy concern of users will make most people uncomfortable. According to Whitmore et al. [11], only securing the data is not the critical issue, but ownership of data must also be handled essentially. This will definitely give assurance to the end user about the notion of privacy, and therefore, end user feels comfortable to contribute and participate in IoT.

An example is applications related to the health care in a smart home or hospital. In this case, the sensitive information, such as the behavior of patients, needs to be protected from others. The confidential data must not go into wrong hands. Technical issues and challenges on how to handle these data and how to dig out the useful information have emerged in recent years. As we have already mentioned, IoT is fabricated over an existing layer of the Internet; therefore, all the security threats existing over there turn up here along with some new complications. Data protection emerges as one of the vital issues in IoT.

Various research issues/questions which cannot be left unattended are as follows:

(i) Whenever an IoT system is active and responding to some signals, it is necessary to check whether those signals are generated by the authorized entity or some malicious software.

(ii) Whenever an appliance being used in a bank, a hospital, an institute, an office, a parking place, a shopping mall or any other IoT destination, it is required to check whether the information generated is trustworthy or not.

(iii) When checking up the IoT device controls and backing up the data, it is required to check the liveliness of data. How long is the data effective? How long must the data be kept stored?

(iv) Whenever the data or signals are generated by an IoT device and sensed by a sensor, the integrity of the data is in question.

(v) Which policies and safety mechanisms are required and sufficient during the initial configuration of an IoT system?

(vi) Considering the privacy issues of the end user, another question to be addressed is that how much data and what type of information can be disclosed?

Analyzing the above concerns and realizing their effects on the efficacy of an IoT system, it can easily be estimated that security implementations with privacy are an important area to be explored.

3　Focus of This Work

Alike all the network-oriented applications, IoT also revolves around the data. All this data can be classified in the descending order of their relative sensitivity. These are *public data* and *private data*. Public data is the collection of all those facts and figures, which can be disclosed to every entity involved in the system, while private data is the confidential data with high intrinsic value, which cannot be disclosed to everyone, except the authorized one.

The private data can be further classified as *Stored Data* and *Data in Transit*.

Stored Data is the data residing in storage devices, data centers, and databases. Stored Data may be further classified as *Assembled Data* and *Active data*. *Assembled Data* is all those facts and figures, and processed information which is static and stored in the storage devices. *Active data* comprises all those data values which are being processed currently, and therefore continuously being a swap in and swap out from/to the computing device to/from the storage device. It is Stored Data but being used by the user or some application. *Data in Transit* is the data values which are being transferred on the network. It may also be termed as *data in motion*. It is the data that actually resides on the transmission medium, whether it is a wired one or wireless.

In this research work, our focus is basically on Stored Data. Our sole target in this paper is toward the "Stored Data," does not mean that "Data in Transit" is not important, or does not need security. Undoubtedly, "Data in Transit" has its own security vulnerabilities.

4　Proposed Security Model for IoT

There is a wide range of security models proposed by various researchers and academicians to provide security to different types of networks and network-oriented applications. Here, we propose a generalized and multidimensional security model that basically focuses on secure data handling in an IoT system. There are two pre-assumptions defined for the proposed model. First, there is no predefined application domain of the IoT system where this model is applicable. Second, the underlying network is a reliable one. The proposed model comprehends five phases, namely:

(i) Definition Phase, (ii) Design Phase, (iii) Implementation Phase, (iv) Testing Phase, and (v) Amendment Phase (Fig. 1).

Fig. 1 Outline of DDITA security model

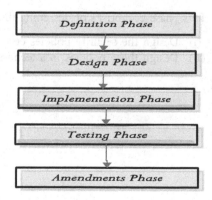

4.1 Definition Phase

This phase comprises of the different conducts to define the Security Policies. This phase is as essential as the successful deployment of any IoT system. Noticeably, one of the principal risks in any technology deployment is avoidance of security principles by the users. Failure to act according to the security principles, either intentionally or unintentionally, generates adverse effects; therefore, it is very important to define the security policies prior to the deployment of an IoT system. Here, we perform the following steps:

(i)   *Define Access Policies*
(ii) *Resource Recognition*: Identifying and listing all the resources, specifically the data, with high as well as low intrinsic value.
(iii) *Resource Valuation*: Evaluating the relative value of resources present in an IoT system. It can be represented numerically (1, 2, 3...) or in a subjective way (high, medium, low).
(iv) *Threat Recognition*: Identifying and listing different types of threats with respect to different resource types.
(v)   *Access Declaration*: Proclaiming the access guidelines for different users for all resource types.
(vi) *Privilege Definitions*: Proclaiming the privilege guidelines for different users for all resource types.

These policy definitions become a criterion for all the following steps. After analyzing and following these policies, apposite and counter steps can be taken as remedial measures, against adversaries. These policies can be correlated to the Policy management at IoT gateways as mentioned by Kim [12]. According to them, a secure gateway is an essential characteristic for successful deployment of IoT to detect mis-configurations and any policy conflict.

4.2 Design Phase

This phase comprises of designing the above-defined policies.

(i) Design the security policies considering all the above-mentioned definitions.
(ii) Design these policies in a technologically feasible manner (Fig. 2).

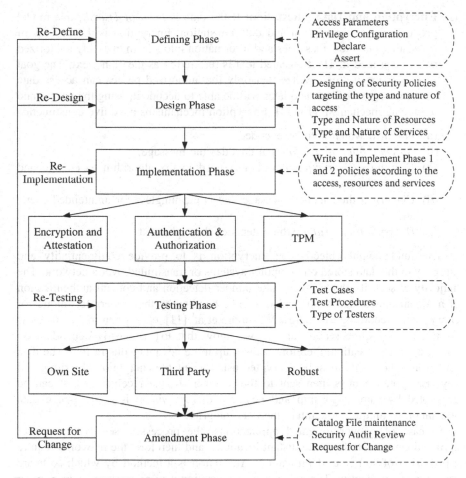

Fig. 2 Outline of DDITA security model

4.3 Implementation Phase

This phase comprises of implementing the above-defined and -designed security policies. It consists of: (i) write these policies in the IoT system, (ii) link them accordingly, and (iii) implement these policies.

Data, when stored or transmitted in a digital format in hard drive, seeks security. Some of the data stored in the hard disk or being transmitted on the medium of high intrinsic value and therefore demands high security. The implementation phase can be further broken down into three subsections, on the basis of data being involved. These three subsections are:

(i) **Encryption and Data Attestation**: If the data is *Data in Transit*, this model proposes to use encryption and data attestation. Encryption is the process of translating plain text messages and information into a form that only authorized parties can read it. The translated text is referred to as the ciphertext. The goal of encryption is to guarantee that only the authorized parties can access data (text messages or files), and later will be able to decode it, using the proper and legitimate mechanism [13, 14]. Encryption mechanisms have five constituents:

(a) *Plaintext* which is to be encoded;
(b) An *encryption algorithm* that encodes the message;
(c) *Encryption/Decryption key* that is used in the algorithm to encode and decode the message;
(d) *Ciphertext* the encoded message that is meaningless for unintended users; and
(e) *Decryption algorithm* which decodes the ciphertext.

The fundamental objective of encryption is to provide confidentiality and integrity to the data stored on computer systems or transmitted over a network. The majority of security threat issues, like tamper detection and content authentication of a digital image, audio, and video, can be handled by the modern techniques of encryption. According to Andrew Whitmore et al. [11], encryption makes wireless communication quite secure, and as we know that IoT is mostly dependent on wireless communication, therefore it turns up more advantageous in IoT. The data sent from sender side must be encrypted using a suitable encryption algorithm and key. Encrypted data is then sent to the receiver. At the receiver end, it can be decrypted by same algorithm and key. Another provision is to use some data attestation technique to maintain the data integrity.

In case of network-oriented applications, the processes, services, and user applications execute from the distant locations, and therefore, the networks require a provision of software identification. Attestation is a method by which software can prove their identity. The major aim of attestation is to prove the remote end that the software is complete and reliable. Attestation allows a program to authenticate itself, and remote attestation is a means for one system to make reliable statements about the software it is running on another system [15]. Berger et al. [16] have also proposed the remote attestation for cloud computing, and similarly, these provisions can be implemented in IoT system.

(ii) **Authentication and Authorization**: If the data is *active data* (i.e., Stored Data but under processing), this model proposes to use authentication and authorization. Authentication can be defined as the verification process of the truthfulness of properties related to an entity. It is an act to validate the claimed identification of the corresponding entity [17]. Authentication process involves a variety of measures and metrics including software constructs like passwords, and embedded hardware like biometrics.

(iii) **Encryption using Trusted Platform Module (TPM)**. If the data is *Assembled Data* (another category of *Stored Data*), this model proposes to use encryption using TPM. Prior to elaborate the encryption using TPM, we discuss some other countermeasures to keep the Stored Data safe and secure. It can be formerly achieved by keeping the storage device safe. Some former approaches may include: (a) Keep your device always password protected and be crafty and smart during password definition, (b) keep the limited data on your device, or keep them in a scattered manner, (c) backup your data frequently depending on its intrinsic value, and (d) use some encryption technique to keep the data safe.

If the Stored Data is taken or stolen, some counteractive steps can be performed, it may include practicing a tracking app to remove your data from the lost device. Here in this model, we propose the encryption technique involving the end user himself. As we say, encryption is a decent technique for Stored Data security, and it will become more effective if it is user-controlled. If an attacker gets access to your personal computer, laptop, or storage device, he/she can easily gain access to the data even without knowing your password. He/she can accomplish this task, by booting into another (their own) operating system by using the special bootable device or bootable disc. Another way he/she can follow is that he/she can remove the disc and put that in another machine running with some operating system. Therefore to avoid such problems, we suggest using the technique known as encryption using Trusted Platform Modules (TPM).

Trusted Platform Module (TPM) is a tamper-resistant piece of cryptographic hardware built onto the system board that implements primitive cryptographic functions on which more complex features can be configured [15]. Due to the latest developments in hardware technologies, the information world has achieved improved levels of trust via TPM [18]. Some of the tools supporting TPM are TrueCrypt, VeraCrypt, BitLocker, and in near future, CipherShed will also be in the scene.

4.4 Testing Phase

This phase comprises of testing the above-defined, -designed, and -implemented security policies with different test cases (including both own testers or third-party testers). The testing phase is again categorized into three categories on the basis of their testing procedures, test cases, and the individuals performing the tests.

Own Site Testing (Subjective Testing): In this testing, the different test cases are used to test the security, but tests are performed by the developer or the developing team. In other words, testing these policies via own data set and own users is called own Site Testing.

Third-Party Testing: In this testing, the different test cases are used to test the security, but tests are performed by the third-party people. The third party may be some potential end users or the potential customers who are going to purchase your security implement. In other words, testing these policies using third-party users is called third-party testing.

Robust Testing: In this testing, some particular test cases are used to test the security, and tests are performed by the developer or the developing team as in the case of Own Site Testing. Here, the major difference is testing is done by the test cases which are highly prone to security threats. In other words, explicitly trying to violate the defined policies is called Robust Testing.

4.5 Amendment Phase

This phase comprises of making the necessary amendments after testing the above-defined and -designed security policies with different test cases.

  (i) Make necessary amendments, if found and required.
 (ii) Make necessary changes in the system.
(iii) Reconfigure the security policies.
(iv) Make necessary amendments.

5 Conclusion

As the IoT tries not only to assure the improvement in living standards of people but also improve the capabilities already offered by the Internet. IoT promises to enhance the decision making and that too in lesser time limits. But the matter to be resolved is whether IoT will emerge and persist as a continuing paradigm, or it will be a temporary fascination for the user. The full potential of IoT can be exploited if it provides a secure feel to the end user.

Here in this paper, we have discussed the need for security in IoT implementation. A naïve model is proposed to achieve and guarantee the security in IoT system. Our major focus is on the security of data that has been categorized into Stored Data and Data in Transit. Two major concerns are over privacy and ownership of the data being involved in an IoT system. These two concepts reflect that the data cannot be accessed by any object or entity without the approval of the owner. Encryption, data attestation, authorization, authentication, and encryption using TPM are majorly emphasized. To facilitate all the above measures in an IoT system, the special effort should be put into designing the algorithms, which are more efficient and less power consuming.

References

1. http://en.wikipedia.org/wiki/Information_security.
2. Alex W., The Guardian, The internet of things is revolutionizing our lives, but standards are a must, http://www.theguardian.com/media-network/2015/mar/31/, 2015 (accessed 15.08. 2016).
3. ITU, Internet of Things Global Standards Initiative, http://www.itu.int/en/ITU-T/gsi/iot/ Pages/default.aspx, 2015 (accessed 05.06.2016).
4. Machara, S., Chabridon, S., and Taconet, C., Trust-based context contract models for the internet of things, Ubiquitous Intelligence and Computing, *Proceedings of IEEE 10th International Conference on Autonomic and Trusted Computing (UIC/ATC)*, pp. 557–562, 2013.
5. Renu, A. and Manik L.D.. RFID security in the context of internet of things, *Proceedings of the First International Conference on Security of Internet of Things*, ACM, 2012.
6. Bhattasali, Tapalina, Rituparna Chaki, and Nabendu Chaki., Secure and trusted cloud of things', *Annual IEEE India Conference (INDICON)*, IEEE, 2013.
7. Zorzi, M., Gluhak, A., Lange, S. and Bassi, A., From today's INTRAnet of things to a future INTERnet of things: A wireless- and mobility-related view, *IEEE Wireless Communication*, Vol. 17, No. 6, pp. 44–51, 2013.
8. Mark, W., Smart devices to get security tune-up', *BBC News, 23 September 2014*, Available Online: http://www.bbc.com/news/technology-34324247, Retrieved on 25 Jan, 2016.
9. Yuichi Kawamoto, Hiroki Nishiyama, NeiKato, Yoshitaka Shimizu, Atsushi Takahara, and Tingting Jiang., Effectively Collecting Data for the Location-Based Authentication in Internet of Things', *IEEE Systems Journal*, Vol. 99, pp. 1–9, 2015; https://doi.org/10.1109/jsyst.2015. 2456878.
10. Tsai, C.W., Lai, C.F., Chiang, M.C. and Yang, L. T., Data mining for internet of things: a survey, Communications Surveys and Tutorials', *IEEE.* Vol. 16, 2014, pp. 77–97.
11. Whitmore, A., Anurag A., and Li Da Xu., The Internet of Things—A survey of topics and trends', *Information Systems Frontiers*, Vol. 17, No. 2, 2015, pp. 261–274.
12. Jun, Y.K., Secure and Efficient Management Architecture for the IoT, ACM, 2015.
13. William Stallings, Cryptography and Network Security; Principles and Practices, *Pearson Education, Inc*, 4th edition, 2009.
14. Behrouz A. Forouzan and Debdeep Mukhopadhyay, Cryptography and Network Security, *McGraw Hill Education*, 3rd edition, 2016.
15. Bare, J. Christopher, Attestation and Trusted Computing', *CSEP 590: Practical Aspects of Modern Cryptography*, University of Washington, Washington, 2016.
16. Berger, S., C´aceres, R., Goldman, K. A., Perez, R., Sailer, R. and van Doorn, L., vTPM: Virtualizing the Trusted Platform Module, *Security Symposium.* USENIX, 2006, pp. 305–320.
17. https://en.wikipedia.org/wiki/Authentication.
18. Babar, S., Mahalle, P., Stango, A., Prasad, N., and Prasad, R., Proposed security model and threat taxonomy for the Internet of Things (IoT)', In N. Meghanathan et al. (Eds.), *Recent trends in network security and applications, communications in computer and information science*, Berlin: Springer, Vol. 89, 2010, pp. 420–429.

IO-UM: An Improved Ontology-Based User Model for the Internet Finance

Xinchen Shi, Zongwei Luo, Bin Li and Yu Yang

Abstract Building user model which can accurately reflect the user's preferences is necessary and important in personalized service systems. For financial platforms, user's focuses are regulated not only by his interests but also the environmental conditions. Considering the user's investment state and operation behaviors, we build a decay function for user modeling in the Internet financial area. In order to meet the requirements of the timeliness and accuracy, this paper presents an improved ontology-based approach to build user model (IO-UM) considering decay function, which constructs the domain ontology by text mining, and update user model to capture recently focuses by ontology learning. Experiments were taken to illustrate the usefulness of the IO-UM to provide personalization services. To prove the influence of decay function, we took different values for comparison in the experiments.

Keywords User model · Ontology · Decay function · Personalized system

1 Introduction

The Internet finance (ITFIN) is a new type of financial services, which is the traditional financial institutions and Internet companies to use Internet technology and information communication technology to achieve financial intermediation, payment, investment, and information intermediary services. In China, ITFIN develops rapidly in the scale of transactions [1], which takes an example in Internet insurance area [2]. Constantly, innovation and enrichment of Internet financial models provide plentiful services and selections, but also lead to information trek to

X. Shi · Z. Luo (✉)
Department of Computer Science and Engineering, Southern University of Science
and Technology, Shenzhen 518055, China
e-mail: luozw@sustc.edu.in

B. Li · Y. Yang
Shenzhen Aotain Technology Co., Ltd., Shenzhen 518055, China

© Springer Nature Singapore Pte Ltd. 2019
B. K. Panigrahi et al. (eds.), *Smart Innovations in Communication and
Computational Sciences*, Advances in Intelligent Systems and Computing 670,
https://doi.org/10.1007/978-981-10-8971-8_20

users. To provide useful information and products for users and improve the competitiveness of platforms, personalization service is necessary.

Personalized services aim to provide information or products that match a user's personal interests or needs from plentiful data in the Internet. Before personalized systems provide individual services, user model must be constructed firstly. Ontology represents knowledge within a domain as a set of concepts, and the relationships between those concepts [3]. Owing to abundant background information provided by domain ontology, ontology-based user model overcomes the shortage of lacking semantic information of traditional user model. Compared to keyword vector space model [4], ontology-based user model has been concerned by more and more researchers.

Because of the powerful knowledge representation formalism and associated inference mechanisms, user modeling method based on ontology [5] is widely used in personalized service systems. According to different ontology types, purposes, domains, and ways, Duong and Uddin [6], Zhou and Wu [7] ontology-based user modeling methods are varied. Mylonas and Vallet [8] combine contextualization and personalization methods to achieve personalized information retrieval. Cuan and Zhou [9] construct the domain ontology by text mining, and update user model by ontology learning. Skillen and Chen [10] present an ontology-based user model considering context-aware application personalization. Jiang and Tan [11] present a set of statistical methods to learn a user ontology from a given domain ontology and a spreading activation procedure for inferencing in the user ontology.

In this paper, we present an improved ontology-based user model (IO-UM) to meet the demands of financial platforms, which takes decay function into account to capture users' up-to-date focus. In this section, we discussed our chosen research domain and general approaches to construct the ontology-based user model. The rest of this paper is organized as follows: Section 2 illustrates an overview of the research efforts on IO-US method. In Sect. 3, the updating method of user ontology (UserO) will be described in detail. We put forward experiments to show IO-UM can improve user modeling and recommendation accuracy in Sect. 4. Conclusions and future works are presented in Sect. 5.

2 The IO-UM Method

In this section, IO-UM will be described in general, including its compositions and main process. DO construction will be depicted in detail for UserO building.

2.1 The Design of IO-UM

Accurate and sensitive user model especially in the finance field is necessary for developing other works for personalization services. To capture user's interests or

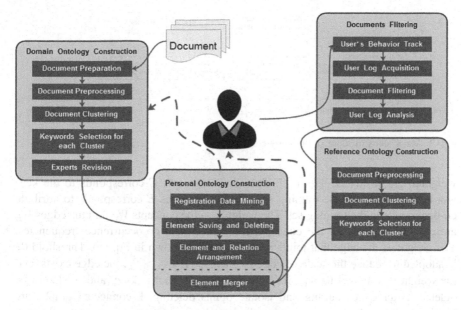

Fig. 1 Workflow process of IO-UM construction

needs timely and accurately, we choose to develop a user model with ontology integrating decay function, which is named as IO-UM. The development of IO-UM mainly contains four modules: domain ontology (DO) construction, documents filtering, reference ontology (RefO) construction, and personal ontology (UserO) construction. The main steps and correlations between them are shown in Fig. 1.

2.2 DO Construction

DO which provides abundant background knowledge for a specific domain contains basic terms and relations between those terms. In consideration of the limitations of artificial construction, text mining techniques were adopted to construct DO for the financial area. The process of DO construction, which includes document predisposition, clustering, and cluster description, is shown in Fig. 2.

After document predisposition [12], including denoising, segmentation, removal of stop words and frequent terms, a document i can be represented as a vector, as $d_i = (t_1, t_2, \ldots, t_m)$; all the documents from the training set can be denoted as a Boolean matrix, $D = \{d_1, d_2, \ldots, d_n\}$; m is the number of keywords in a document; n is the number of documents in the training set.

We structure keywords into clusters, on the base of cluster mining method [13]. Cluster mining discovers relationships between keywords, by looking for a

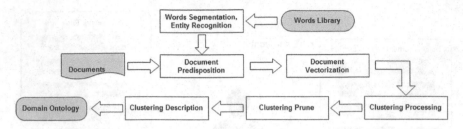

Fig. 2 Major steps for domain ontology construction

weighted graph $G(C, E, W_T, W_E)$. The set of concepts C corresponds to the keywords obtained from the training set. The set of edges E corresponds to attribute co-occurrence in the training set. The weights on the concepts W_C and the edges W_E are computed based on the concepts frequencies and co-occurrence frequencies. The weight of the edge that joints c_x and c_y, W_{xy} is shown in Eq. (1). Threshold θ_E is adopted to reduce the scale of edge in graph G: If $W_{xy} > \theta_E$, the edge exists, and the weight is assigned to W_{xy}; if $W_{xy} < \theta_E$, the edge that links c_x and c_y should be deleted. After edges cutting and isolate points deleting, k connected graphs are obtained. For each concept c_x, we calculate $W_x(i)$, its weight in graph i, through Eq. (2).

$$W_{xy} = \frac{N_{xy}}{\max\{N_x, N_y\}} \tag{1}$$

$$W_x(i) = \frac{N_x(i)}{N_x} \tag{2}$$

After cluster mining, the next step is selecting proper descriptive feature words for each cluster. The selection of feature words depends on this fact: The selected words' frequencies in this cluster are the highest of all clusters. Through preprocessing and clustering, basic concepts in particular area and relationship between them were obtained. After the revision of experts, *DO* which provides abundant background knowledge for *UserO* will be established.

3 UserO Constructions

User model generally as the input of intelligent engines is the foundation of realizing the function of personalization and recommendation. Different user modeling approaches were applied according to diverse requirements and objectives in different personalized platforms. For our personalized financial service platform, user model U is defined as a triples-oriented schema, $U = \{UserI, UserO, AO\}$, consisting of:

- $UserI = \{Name, Sex, Age, Job, Hobby\}$ contains a user's basic information and is collected when the user is registered. It is used to solve the cold start problems and obtain original $UserO$ from DO;
- $UserO$ defines a user's personal ontology. $UserO = \{C, R, \sigma, \theta_c, \theta_r\}$, where two sets C and R, whose elements c_x and c_{xy} are the concepts and semantic relations in the domain ontology, respectively; a function $\sigma: C \times C \to R$ defines the relationships between two different concepts; a function θ_c and a function θ_r assign weights to concepts and relationships in the DO, respectively;
- AO contains ontology a user might be interested in by association rules, which will be discussed in future works. For this paper, we will focus on $UserO$ construction and updating.

According to different needs and operation behaviors, $UserO$ was generated individually from DO. That is, $UserO \in DO$, $UserO$ is a subset of DO. User's needs are considered as planes of projection, while $UserO$ generation is a projection of DO as shown in Fig. 3. $UserO$ is considered as long-term interest in IO-UM method. $RefO$ was obtained from documents that the user has read recently and has been filtered. $UserO$ is updated according to $RefO$ which represents the user's short-term interest.

3.1 RefO Construction

In this part, we will discuss the construction of $RefO$, especially in the aspect of documents filtering. In order to better capture and describe the user's short-term interests, documents that used to construct $RefO$ should be filtered reasonably. In this part, documents filtering method will be discussed. As shown in Table 1, reading time and operations to the document are critical considerations when choosing documents to construct $RefO$.

$$w_{id} = \frac{T_{id}}{\max\{T_{in}\}} \tag{3}$$

User i performance for the document d can be calculated through Eq. (3); n is the number of documents the user read during a certain time period. Documents with top N w_{id} will be used to construct $RefO$. After confirming that the user pays more attention to filtered documents, $RefO$ which used to update $UserO$ can be constructed. $RefO$ construction method is the same as DO, while documents are from a user's reading data for a certain period of time instead of the training set. Ontology merging method will be introduced in the next part.

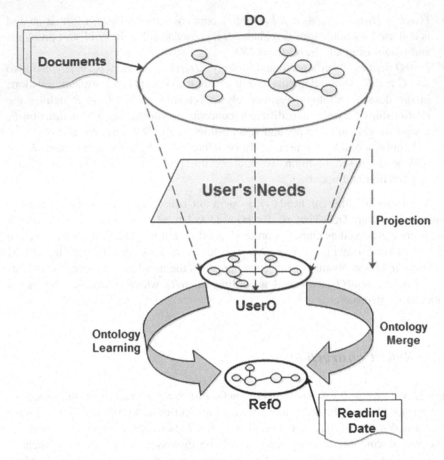

Fig. 3 Projection of ontology and UserO updating

Table 1 Operations according user behavior for document filtering

Behavior	Processing method
Reading	Mainly by computing time span the user i spends on the document d and recording it in a variable T_{id}, to evaluate the user's performance to a document
Collection	Accumulating the time span T_{id} that the user i reading and operating to the document d, to evaluate the user's performance to document
Comment	
Share	

3.2   UserO Updating

User's focus is kind of highly dynamic information, which was affected both by user's interests and external environment, especially in financial sector. External policies tend to affect customers' investment behaviors. In order to accurately reflect user's personal interests and predict user's next behaviors, user model should be updated timely. There are three situations where *UserO* from user model should be updated:

- A new interest appeared

   Ontology merges: $UserO = UserO \cup RefO'$

   RefO' means interesting ontology set that contains new user ontology, which is *RefO'* $\in$ *RefO*.

- User is no longer interested in the concept c_x that belongs to *UserO*

   When $w_x < \theta_C$, remove the concept c_x which w_x corresponds to, and delete the concepts and their relationship whose father node is c_x. Concepts eliminate method: $UserO = UserO - RefO''$, *RefO''* means ontology set that the user is no longer interested in, which *RefO''* $\in$ *RefO*.

- The interested level of a concept c_x has changed

$$I_t(t_{i+1}) = O_x(t_i) \times m_{xy} \times (1 - \alpha), \quad \alpha \in [0, 1] \tag{4}$$

$$w_x(t_i) = w_x(t_{i-1}) \times f(\tau, \varepsilon, N) + P_x(t_i) \tag{5}$$

$w_x(t_i)$ represents the preference weight of concept c_x for the user i at the time point of t_i, where $P_x(t_i)$ corresponds to the occurrence probability of concept c_x within the time period from t_{i-1} to t_i, which can be calculated via Eq. (6):

$$P_x(t_i) = \frac{\text{freq}(x)}{m * n} \tag{6}$$

θ_C, a certain threshold, is set by the system according to different systems and demands. While $w_x(t_i) < \theta_C$, it is considered that the concept c_x is still attractive to the user; while $w_x(t_i) > \theta_C$, which represents the concept without operations for an extended period of time, we consider that the concept is losing attraction for the user.

$$w_{xy}(t_i) = \frac{\beta \times w_{xy}(t_{i-1}) + \mathrm{freq}(r_{xy})}{\beta + \sum_y \mathrm{freq}(r_{xy})} \tag{7}$$

θ_R, a certain threshold, is set by the system according to different systems and demands. While $w_{xy}(t_i) < \theta_R$, it is considered that the relationship between concept c_x and c_y has a relatively strong correlation for the user i; otherwise, we will consider the relationships as unattractive to the user.

3.3 Decay Function

To meet the need of timeliness and flexibility for financial customers, IO-UM combines the concept of decay function for user model. In Sect. 3.2, we introduced the concept of decay function $f(\tau, \varepsilon, N)$ to adjust the ontology component of UserO. Different variables' assignment leads to varied decay function values which influence the ratios of long-term interests and short-term interest in the construction of UserO. For our works, three factors were taken into consideration: the time interval between real time and a user's recent transaction time; species number of products that the user maintains; and time interval the user logins. Those three parameters determine the decay function, using the expression method as given in Eq. (8).

$$f(\tau, \varepsilon, N) = \left(\log^{a+\tau(t)}\right)^N \varepsilon^{-b} \tag{8}$$

where $\tau(t)$ represents a time-dependent function, valued [0,1]. N represents the number of products types the user maintains at the time t_i. ε represents the time interval between t_{i-1} and t_i. a and b are constants, as $b \in [0, 1]$.

From Eq. (8), users with fixed investment objects will be assigned more attention to long-term interests; the one who does transactions with high frequency will be assigned comparatively decentralized points. In addition, decay function also prevents the information overflow of UserO.

3.4 An Illustration

A simple scenario for UserO construction and updating is given as follows: Suppose a user, Jackson's original UserO from searching his records and DO in financial area, as in Fig. 4a. In this case, we set $\theta = 0.2$, $\theta_r = 0.25$. Concepts weighted lower than 0.2 and relationships weighted lower than 0.25 were considered as unattractive information for Jackson, and should be deleted. After dimension descending, Jackson's UserO can be depicted in Fig. 4b.

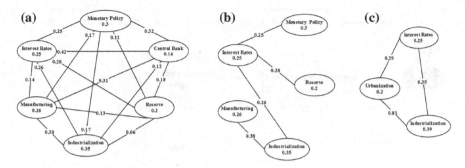

Fig. 4 An illustration of the user ontology updating

To illuminate the updating method of IO-UM, supposing during a certain period, Jackson's reading list which contains five documents, the frequencies of terms and relation between those terms were displayed in Table 1. We set decay function $f(\tau, \varepsilon, N) = 0.4$, $\beta = 0.2$. Assuming "industrialization" is marked as "1", and "urbanization" is marked as "2", then during $[t_{i-1}, t_i]$, Jackson's short-term attention degree for concept "industrialization" is calculated via Eq. (9), and the relation degree between "industrialization" and "urbanization" can be calculated via Eq. (10) (Table 2).

$$v_1(t_i) = v_1(t_{i-1}) \times f(\tau, \varepsilon, N) + P_1(t_i) = 0.35 \times 0.4 + \frac{5}{4 * 5} = 0.39 \qquad (9)$$

$$m_{12}(t_i) = \frac{\beta \times m_{12}(t_{i-1}) + \mathrm{freq}(r_{12})}{\beta + \sum_2 \mathrm{freq}(r_{12})} = \frac{0.2 \times 1 + 4}{0.2 + 5} = 0.81 \qquad (10)$$

After updating and pruning, the user Jackson's *UserO* has been altered according to his most recent document and long-term interests. According to Jackson's latest user model by IO-UM as shown in Fig. 6c, financial platforms can make recommendations for Jackson, including news and financial products that he might be interested in recently.

Table 2 Concepts and relations between them obtained from documents Jackson read recently

Concept	Freq	Relation	Freq
Industrialization	5	Industrialization, urbanization	4
Urbanization	4	Industrialization, interest rate	2
Interest rate	3	Urbanization, interest rate	1

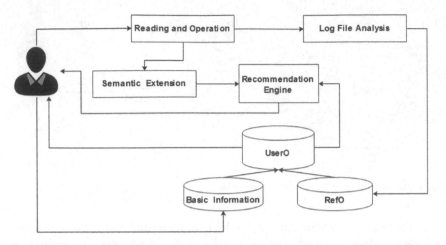

Fig. 5 Framework of the personalization system using IO-UM

4 Experiments

In the experiment, text mining method is used to construct the *DO* of finance. The initial personal ontology is constructed by ontology projecting based on the information submitted by experimenters. We employed five users in the financial circle to evaluate the personalized services which adopt IO-UM to describe user's interests. The experiment process is shown in Fig. 5.

The document collections from Huarun Bank (www.crbank.com.cn) were divided into the training set and the testing set randomly. Users can make investments and read information on that platform. To capture users' interests, users have to browse the top 20 documents returned and provide feedback on the documents that are relevant. After training, the user's interests can be included in *UserO*. We provide a recommendation list for the user which contains top ten documents according *UserO* from the testing set. To conduct a convincing evaluation, we adopted two other modeling methods: ontology UM [14] and keyword-based vector space model [15], to measure the performance of the IO-UM.

After training, the learnt user *UserO* is applied to provide recommendation for the test lists. Figure 6 depicts the average precision for the top five and top ten documents in the recommendation lists, respectively. We can see IO-UM consistently outperforms, or at least produces equal performances, compared with the other two user modeling methods. Especially for certain users, through IO-UM, the recommendation engine can provide better services.

In order to evaluate the impact of decay function $f(\tau, \varepsilon, N)$, three parameters $\gamma = f(\tau, \varepsilon, N)$ were selected to conduct simulation experiments using our IO-UM method. Statistics on five users' average precisions of recommendation lists which on the basis of IO-UM on the top ten documents retrieved from the financial Web

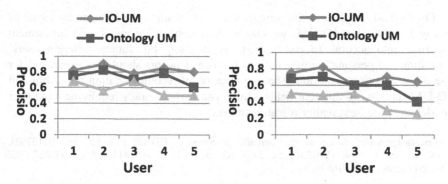

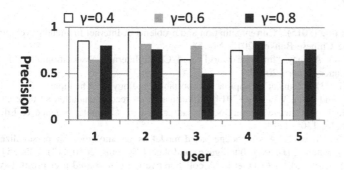

Fig. 6 Average precision of recommendation lists using three user model methods on the top five and top ten documents

Fig. 7 Average precision of recommendation lists of IO-UM method using three different decay function values on top five documents

site. From the statistical information in Fig. 7, we can draw the conclusions that individual decay function is needed to meet different users' information demands.

5 Conclusions and Future Works

This article presents an improved ontology-based user model (IO-UM) method, aiming to provide individualized services to meet the user's needs in financial field of personalization and sensitivity. Using text mining technology to construct DO, $UserO$ is selected from domain ontology considering the user's original submission. On the basis of traditional ontology-based modeling method, decay function is used to update and maintain users' ontology to fit users' focus changing. The decay function is mainly influenced by the user's investment condition and operation behavior. Experiments show that the model is easy to construct and does better in updating user's interest timely and accurately. The improved ontology-based user modeling method improves the precision in the personalized services.

On the basis of the existing researches, we will study the influencing factor of decay function. For IO-UM, we take operation behavior and user's investment positions into account to construct decay function. For future, different users' condition and personality, external environmental factors should be researched for the decay function evolution. The combination of recommendation algorithm and IO-UM applying to different fields which with high accuracy requirements in user modeling is the next reality research subject.

Acknowledgements This work was partially supported by GDNSF fund (2015A030313782), SUSTech Starup fund (Y01236215), SUSTech fund (05/Y01051814, 05/Y01051827, 05/Y01051830, and 05/Y01051839).

References

1. Wang Guoming (2015). "Current situation and problems of Internet financial development in China". The Chinese Banker (2015. No. 5).
2. Xinhua News Agency "financial world" and China Internet Association (2014) "China Internet Finance Report, 2014".
3. Gruber T (1993) A translation approach to portable ontology specifications.
4. Jiang Shan, and Hong Wenxing (2014). A vertical news recommendation system: CCNS—An example from Chinese campus news reading system. Computer Science & Education (ICCSE), 2014, 1105–1114.
5. Tao X, Li Y, Zhong N. A knowledge-based model using ontologies for personalized web information gathering [J]. Web Intelligence and Agent Systems, 2010, 8(3): 235–254.
6. Duong T H, Uddin M N, Li D, et al. A collaborative ontology-based user profiles system// Computational Collective Intelligence. Semantic Web, Social Networks and Multiagent Systems. Springer Berlin Heidelberg, 2009: 540–552.
7. Zhou X, Wu S T, Li Y, et al. Utilizing search intent in topic ontology-based user profile for web mining//Web Intelligence, 2006. WI 2006. IEEE/WIC/ACM International Conference on. IEEE, 2006: 558–564.
8. Mylonas P, Vallet D, Castells P, et al. Personalized information retrieval based on context and ontological knowledge. *The Knowledge Engineering Review*, 2008, 23(01): 73–100.
9. Cuan Q Z, and, Zhou Zh R (2007) "Research of Ontology-based user model" *Computer Applications*, (2007) 10.
10. Skillen K L, Chen L, Nugent C D, et al. Ontological user profile modeling for context-aware application personalization//Ubiquitous Computing and Ambient Intelligence. Springer Berlin Heidelberg, 2012: 261–268.
11. Jiang X, Tan A H. Learning and inferencing in user ontology for personalized Semantic Web search [J]. Information sciences, 2009, 179(16): 2794–2808.
12. Chen Yi-feng, Zhao Heng-kai, YU Xiao-qing, and, Wan Wang-gen (2010) "Research on Ontology-based User Interest Model Constriction". *Computer Engineering*, Vol 36, No. 21.
13. Paliouras G, Papatheodorou C, Karkaletsis V, et al. Clustering the Users of Large Web Sites into Communities[C]//ICML. 2000: 719–726.
14. Trajkova J, Gauch S. Improving Ontology-Based User Profiles[C]//RIAO. 2004, 2004: 380–390.
15. Li Zheng, Lei Li, Wenxing Hong, Tao Li. "PENETRATE: Personalized News Recommendation Using Ensemble Hierarchical Clustering". *Expert Systems with Applications*. 2012 40(6): 2127–2136.

Part III
Smart Hardware and Software Design

A Low-Voltage Distinctive Source-Based Sense Amplifier for Memory Circuits Using FinFETs

Arti Ahir, Jitendra Kumar Saini and Avireni Srinivasulu

Abstract SRAM is the key element used in digital circuits. It is also used as a cache memory in computers, automatic modern equipment like mobile phones, modern appliances, digital calculators, digital cameras, so that the requirements of high-speed advanced memory or embedded memory lead to the development of low-voltage SRAMs. The paper has introduced the design and simulation of the distinctive source-based sense amplifier which is a peripheral circuit for static random access memory (SRAM) that has to amplify the data which is present on the bit lines during the read operation. Simulation of the proposed design has been implemented using 20 nm FinFET technology on the Cadence Virtuoso Tool with a supply voltage of +0.4 V. The main advantages from the proposed design circuit are less power consumption and display of minimum sense delay in sensing the data from SRAM when compared to the existing design circuit. A comparison is drawn between the proposed circuit and the existing design circuit.

Keywords FinFET · Delay · Low voltage · Sense amplifier · SRAM

1 Introduction

Static random access memory which is also considered as CPU memory is used to enhance processor speed and memory cell interconnection accompanied by the progression improvement in VLSI. Nowadays, high clock rates are used which improves the delay characteristics. High clock rates require instruction and data

A. Ahir · J. K. Saini
Department of E.C.E, Birla Institute of Technology, Mesra, Jaipur Campus 302017, India

A. Srinivasulu (✉)
Department of Electronics and Communication Engineering, School of Engineering,
JECRC University, Jaipur 303905, Rajasthan, India
e-mail: avireni_s@yahoo.com; avireni@ieee.org

© Springer Nature Singapore Pte Ltd. 2019
B. K. Panigrahi et al. (eds.), *Smart Innovations in Communication and Computational Sciences*, Advances in Intelligent Systems and Computing 670,
https://doi.org/10.1007/978-981-10-8971-8_21

provided to processor with minimum delay or no delay in ideal situations. With the advancement in VLSI technology, logic gates speed has increased appreciably and also restricted layouts areas of memory circuit, whereas memory access times have not improved proportionally. For the reason stated above, SRAM and its peripheral circuitry play a vital role in realizing high-speed computers. One such peripheral circuit like sense amplifier helps in understanding and analyzing the speed and performance of memory [1–4].

Semiconductor memory chip stores the data in the small memory circuit called a memory cell. Volatile memory cell like static RAM or dynamic RAM cells is arranged on the chip in rows and columns structure [5, 6]. The whole cell in a row is connected to each line. These lines which sweep across the row are known as word lines, and the line that sweeps across the columns is known as bit line [7, 8]. By providing voltage on word lines helps in its activation. Two complementary bit lines which are called bit line and bit-line bar are linked to sense amplifier at the extremity of the memory cells array [9, 10]. They are countable number of sense amplifiers required for the memory chip circuit. For addressing purpose, at the crossroads of a selective word line and bit line, a memory cell exists. For reading and writing the data, these bit lines sweep across the head of the rows where columns are used [11, 12].

FinFET devices are also called multi-gate devices which have two gates in one device, and these are very fine alternates of scaled CMOS invention. Dual gates of FinFET possess various advantages like subdue value of sub-threshold and ejaculation of current from the gate dielectric simultaneously, through their only one gate comprised of MOSFET. It suppresses the short channel effects problem, well-organized control on the gate, and decreased leakage from the gate. Due to these supercilious properties of FinFET, the device shows dominance in terms of speed, transistors size, and better attainment in sub-micron or threshold regions.

Dual-gate FinFET configuration [1, 13, 14] and FinFET cipher are shown in Fig. 1.

Fig. 1 **a** Dual-gate FinFET configuration and **b** FinFET cipher

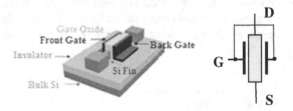

2 Conventional Circuit

Sense amplifier is an agile circuit which is used to sense or detect stored data from read selected memory. It abridges the signal propagation delay between logic circuitry and at the edge which transforms inconsistent logic level, which arises on the bit line, into binary logic level [2, 3].

One of the most vital parts of memory circuit is sense amplifier which is mainly used to reveal stored data or to sense data from the selected memory. Access time of memory and net dissipation of power are severely strained by the implementation of sense amplifiers. If the memory capacity is increased, then it results in increased bit-line capacitance which has negative side effects that makes memory slower. For making a high-speed memory circuit and to produce signals with the requirement of driving sense amplifier within the memory, understanding and evaluating the sensitive or peripheral circuit are necessary [15].

The major factors contributing to improve the performance of the circuit are time delay, power consumption, and power delay product. The delay depends on the number of transistors in the circuit as well as on the intra-cell wiring capacitances. Similarly, the density of the circuit depends on the number of transistors, their size, and design complexity.

Existing differential source-based sense amplifier shown in Fig. 2 is made up of two units comprised of reference generator and distinctive latched-based current sense amplifier. SA is used to synchronize the selection of column array for achieving additional maximum efficiency of the area. This sense amplifier has the prospects of differential amplification of the data by comparing the source voltage with bit-line voltage. The proposed circuit used in this paper is similar to the existing one, but designed with less number of transistors [3].

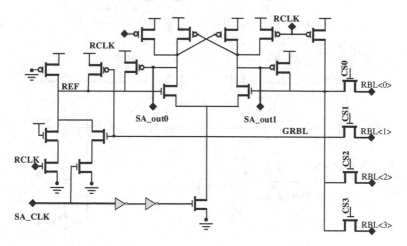

Fig. 2 Conventional differential source-based sense amplifier circuit

The contents of the paper are arranged as follows: Sect. 2 deals with conventional circuit; Sect. 3 deals with the proposed circuit structure and its operation; in Sect. 4, simulation results are coming up; and Sect. 5 has the conclusion of the work followed by references taken to proceed with the work presented.

3 Proposed Circuit and Operation

The proposed circuit shown in Fig. 3 is made up of two units comprised of source voltage producer and a distinctive latched-based current sense amplifier. The source voltage producer circuit has an ON p-FinFET (MP_0), pull-down networks of n-FinFETs (MN_0 and MN_1). The source voltage producer initially produces a voltage V_{REF} which is the result of the contention or conflict between the MP_0 and pull-down way (MN_0 and MN_1). Further, according to the global read bit-line (GRBL) voltage, the reference voltage V_{REF} can self-regulate. Due to this, a large amount of voltage difference amid the global read bit line and source voltage is produced. At the end, when the latched-based current SA is validated, the distinctive of voltage is strengthened for read purpose. The advantages of proposed design as compared to the existing design are use of small channel length of 20 nm, and hence, small transistor size as compared to channel length of 65-nm MOSFETs could be used in the existing design. Due to providing less number of transistors and their small size, the hardware implementation of sense amplifier requires less surface area using proposed design as compared to existing design. The sense delay is also less while performing the read action by the proposed sense amplifier.

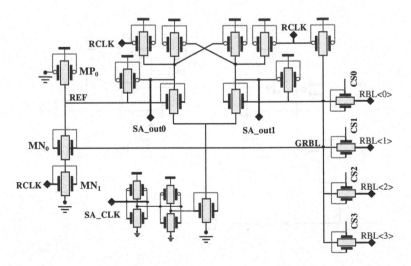

Fig. 3 Proposed distinctive source-based sense amplifier circuit

4 Simulation Results

The proposed low-voltage modified distinctive source-based sense amplifier for memory circuits was simulated on Cadence Virtuoso Tool. The n-FINFETs and p-FINFETs used to design the proposed circuit have the design parameter of a 20-nm FinFET technology model files as shown in Table 1. The proposed design is simulated with a supply voltage of $V_{DD} = +0.4$ V and frequency of $f = 70$ kHz.

Figure 4 demonstrates the timing illustration of read action performed by proposed circuit of SA. Before performing read action, first, V_{DD} is given to read bit lines (RBLs). Commencement of read operation is enabled by read clock (RCLK). For movement of data, the column selection signal is responsible for selecting particular RBL. Concurrently, the ON p-FinFET MP_0 is opposed to or at variance with the ON pull-down network. Due to this, a potential is established on junction REF. The voltage V_{REF} is produced at that junction REF which can be resolved by a ratio of operating current between the pull-down network and MP_0. When the sense amplifier signal clock (SA_CLK) is enabled, the SA launches into observing or sensing level.

While referring Figs. 3, 4 and Table 2, when SRAM is performing or executing a read '0' action, a low voltage is established on the global RBL (V_{GRBL}), this will enable MN_0 to OFF, and MP_0 to ON, and the outcome in the V_{REF} goes to high level. Eventually, a considerable voltage distinction is entrenched amid V_{REF} and V_{GRBL}, and '0' is interpreted out to SA_OUT0. On the other hand, a read '1' action effects in a 'V_{DD}' value on V_{GRBL}. When the SA_CLK is asserted, again because of outcomes of contention on junction REF, a voltage drop is observed on junction REF. Consequently, a considerable voltage distinction is entrenched between V_{REF} and V_{GRBL}, and '1' is read out to the SA_OUT0. A tabular form helps to quickly understand the operation of a distinctive source-based sense amplifier.

The uniqueness of proposed design lies in the fact that the proposed distinctive source-based SA has 21 FinFETs and it is composed of source voltage producer and a distinctive latched-based current SA. The source voltage producer is constituted by an ON p-FinFET (MP_0) and a pull-down network (MN_0 and MN_1). The propagation delay of the proposed circuit is 16.066 ps. The advantages are confined to less delay and the use of less number of transistors as compared to conventional circuit. In the existing design, the reference generator [3] is made up of six MOSFETs, while the proposed design comprised of three transistors only which work similarly as that of reference generator and named as source voltage producer in the proposed design (Table 3).

Table 1 Design parameters of n-FINFETs and p-FINFETs	Channel length	0.02 μm
	Channel thickness	0.01 μm
	Oxide thickness	0.0025 μm

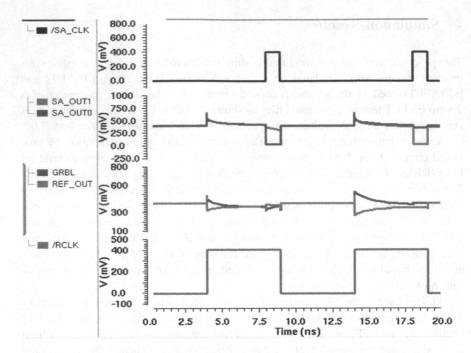

Fig. 4 Timing diagram of the read operation

Table 2 An operation performed by a distinctive source-based sense amplifier

Read	V_{GRBL}	V_{REF}	SA_OUT0	SA_OUT1
0	LOW	HIGH	0	1
1	HIGH	LOW	1	0

Table 3 Comparative analysis of candidates designs with existing design

S. No.	Parameters	Existing design with MOSFETs	Proposed design with FinFETs
1.	Technology used	65 nm	20 nm
2.	V_{DD}	0.36 V	0.4 V
3.	Average power	–	72.268 mW
4.	Sense delay against bit line	–	4.49 ps
5.	Number of transistors used	24	21
6.	Power delay product	–	324.483 fJ

5 Conclusion

The designed sense amplifier has the objective of minimal sense delay against bit lines, better unbeatable amplified signal, less power consumption, required less area due to small transistor size, highly authentic and better tolerance. From the proposed circuit, a new reference generator is designed which is a subpart of sense amplifier and provide better amplifying property as compared to existing design. It can be inferred that proposed design has better high-rated properties and minimal distorted output waveforms, when compared to existing circuit in terms of sense delay and power.

References

1. Jaydeep, P. K., John, K., Kyung-Hoae, K., Satyanand, N., Zheng, Guo., Eric, K., Kevin, Z.: 5.6 Mb/m^{m2} 1R1 W 8T SRAM Arrays Operating Down to 560 mV Utilizing Small-Signal Sensing With Charge Shared Bitline and Asymmetric Sense Amplifier in 14 nm FinFET CMOS Technology. IEEE Journal of Solid-State Circuits. vol. 52, issue. 1, pp. 229–239, (2017). https://doi.org/10.1109/jssc.2016.2607219.
2. Innocent Agbo., Mottaqiallah Taouil., Daniël Kraak., Said Hamdioui., Halil Kükner., Pieter Weckx., Praveen Raghavan., Francky Catthoor.: Integral Impact of BTI, PVT Variation, and Workload on SRAM Sense Amplifier. IEEE Transactions on Very Large Scale Integration (VLSI) Systems. vol. PP, issue: 99, pp. 1–11, (2017).
3. Liang Wen, Xu Cheng, Keji Zhou, Shudong Tian, and Xiaoyang Zeng.: Bit-Interleaving-Enabled 8T SRAM With Shared Data-Aware Write and Reference-Based Sense Amplifier. IEEE Transactions on circuits and systems-II, Express Briefs, vol. 63, no. 7, pp. 643–647, (2016).
4. Balakrishna, K., Srinivasulu, A., Sarada, M.: 7-T single end and 8-T differential dual-port SRAM memory cells. IEEE Conference on Information and Communication Technologies (IEEE ICT-2013), Kumaracoil, India, Apr 11-12, pp. 1243–1246, (2013). https://doi.org/10.1109/cict.2013.6558291.
5. M. R. Garg, Anu Tonk, "A study of different types of voltage & current sense amplifiers used in SRAM", International Journal of Advanced Research in Computer and Communication Engineering, vol. 4, Issue. 5, pp. 30–35, (2015).
6. Ya-Chun Lai., Shi-Yu Huang.: A Resilient and Power Efficient Automatic Power down Sense Amplifier for SRAM Design. IEEE Trans. on Circuits & Systems, vol. 55, no. 10, pp. 1031–1035, (2008).
7. Tiffany Moy., Liechao Huang., Warren Rieutort-Louis., Can Wu., Paul Cuff., Sigurd Wagner., James C. Sturm., Naveen, V:. An EEG Acquisition and Biomarker-Extraction System Using Low-Noise-Amplifier and Compressive-Sensing Circuits Based on Flexible, Thin-Film Electronics. IEEE Journal of Solid-State Circuits, vol. 52, Issue. 1, pp. 309–321, (2017). https://doi.org/10.1109/jssc.2016.2598295.
8. Yiping Zhang., Ziou Wang.,. Canyan Zhu., Lijun Zhang.: 28-nm Latch-Type Sense Amplifier Modification for Coupling Suppression. IEEE Transactions on Very Large Scale Integration (VLSI) Systems. vol. PP, issue. 99, pp. 1–7, (2017).
9. Sivakumari, K., Srinivasulu, A., Reddy, V.V.: A High Slew Rate, Low Voltage CMOS Class-AB Amplifier. IEEE Applied Electronics International Conference (IEEE AEIC-14), Pilsen, Czech Republic, Sep 9-10, 2014, pp. 267–270, (2014). https://doi.org/10.1109/ae.2014.7011717.

10. Chong, K. S., Ho, W-G., Lin, T., Gwee, B-H., Joseph, S. Ch.: Sense Amplifier Half-Buffer (SAHB) A Low-Power High-Performance Asynchronous Logic QDI Cell Template. IEEE Transactions on Very Large Scale Integration (VLSI) Systems. Vol. 25, issue. 25, pp. 402–415, (2017). https://doi.org/10.1109/tvlsi.2016.2583118.
11. Taehui Na., S. Woo., J. Kim., H. Jeong., S. Jung.: Comparative Study of Various Latch-Type Sense Amplifiers. IEEE Trans. On VLSI Systems, vol. 22, no. 2, pp. 425–429, (2014).
12. Wicht, B., Nirschl, T., Schmitt-Landsiede, D.: Yield and speed optimization of a latch-type voltage sense amplifier. IEEE J Solid-State Circuits, vol. 39, no. 7, pp. 1148–1158, (2004).
13. Xue Lin, Yanzhi Wang and Massoud Pedram.: Stack sizing analysis and optimization for FinFET logic cells and circuits operating in the sub/near-threshold regime. in Proc. IEEE 15th International Symposium on Quality Electronic Design, pp. 341–348, (2014).
14. Nagateja, T., Rao, T. V., Srinivasulu, A.: Low Voltage, High Speed FinFET Based 1-BIT BBL-PT Full Adders. in proc. IEEE International Conference on Communication and Signal Processing (IEEE ICCSP' 15), Melmaruvathur, Tamilnadu, India, April 2-4, (2015), pp. 1247–1251.
15. Chandankhede, R. D., Acharya, D. P., Patra, P.: Design of high speed Sense amplifier for SRAM. in proc. of International Conference on Advanced Communication Control and Computing Technologies (ICACCCT), 8-10 May 2014, pp. 340–343.

Design of QCA-Based D Flip Flop and Memory Cell Using Rotated Majority Gate

Trailokya Nath Sasamal, Ashutosh Kumar Singh
and Umesh Ghanekar

Abstract Quantum-dot cellular automata (QCA) are one of the promising technologies that enable nanoscale circuit design with high-performance and low-power consumption features. This work presents a rotated structure of conventional 3-input majority gate in QCA, which exhibits a symmetric structure that is suitable for a compact implementation of coplanar QCA digital circuits. To show the novelty of this structure, D flip flops and memory cell are proposed. The result shows proposed D flip flops are more superior over the existing designs. In addition, proposed memory cell is 33, 79, and 20% more effective in terms of cell counts, area, and latency, respectively, over the best design in this segment using conventional 3-input majority gate. Designs are realized and evaluated using QCADesigner 2.0.3.

Keywords Quantum-dot cellular automata (QCA) · Memory cell
D flip flop · Majority gate · Digital design

1 Introduction

With the exponential decrease in feature size in CMOS technology, devices are more prone to high leakage current and power consumption [1]. This encourages researchers to come up with some alternative technologies. In this aspect, QCA can be the promising candidate in nanotechnology since it enables device operation at high-frequency with low-power consumption and high device density [2]. QCA is based on confinement of two free electrons within a four dots cell. Due to mutual repulsion of these electrons, two possible states are available depending upon the position of electrons across the diagonal dots. QCA also offers a new horizon in

T. N. Sasamal (✉) · U. Ghanekar
Department of Electronics & Communication, NIT, Kurukshetra 136119, India
e-mail: tnsasamal.ece@nitkkr.ac.in

A. K. Singh
Department of Computer Applications, NIT, Kurukshetra 136119, India

© Springer Nature Singapore Pte Ltd. 2019
B. K. Panigrahi et al. (eds.), *Smart Innovations in Communication and
Computational Sciences*, Advances in Intelligent Systems and Computing 670,
https://doi.org/10.1007/978-981-10-8971-8_22

233

information computation. The above attributes enable binary information to be encoded as charges instead of current unlike the CMOS technology.

Different QCA-based circuits design and structures are discussed previously; it includes the design of ALU [3], adders [4, 5], array divider [6], and sequential circuits [7]. Flip flops and memory cells are inherent blocks of any digital circuit systems. QCA-based flip flop and memory cells have been studied in [8–16]. Authors of Shamsabadi et al. [8] have proposed two novel methods for implementation of low-latency edge-triggered D flip flop that demonstrates independent clock input to relax timing constraints and simplifies the design of sequential circuits in QCA. A new serial memory architecture for QCA implementation has been proposed in [9], which is based on the utilization of basic building blocks referred to as tiles. In [10], authors have presented the "SQUARES" layout rules to simplifying the process of QCA circuit design. In most of the work, RAM cell is loop-based as it needs lesser clock zones. However, the designs are not well optimized.

This work presents a modified majority gate, which is a rotated version of the traditional 3-input majority gate and its applicability in low complexity level and edge trigger D flip flops. Result shows the presented rotated structure is more flexible and generates compact design with respect to existing 3-input majority gate. In addition, we also proposed a coplanar memory cell with set/reset capability using the majority gate structure, which is superior in terms of cell count, area occupation, and delay over the prior designs. The paper is organized as follows: Sect. 2 gives a review of QCA logic. Followed by a detailed analysis of proposed D flip flops as an application of rotated 3-input, majority gate is given in Sect. 3. In Sect. 4, a memory cell is implemented using rotated 3-input majority gate. Simulated results of proposed designs are presented in Sect. 5, and finally paper is concluded in Sect. 6.

2 QCA Background

2.1 QCA Gates and Clocking

QCA technology offers new device architecture at nanoscale. In addition, it enables a new method of information computation and transformation. QCA cells are the basic blocks of QCA technology. Each square-shaped cell has four quantum dots located at the four corners. Logic states of the cells are computed based on the position of the two electrons in the quantum dots. These electrons can be tunneled through the dots and occupy the diagonal dots due to columbic repulsion (achieve minimum energy). In QCA, binary information is encoded by these electrons. They make two possible polarizations as $P = +1$ (logic 1) and $P = -1$ (logic 0), as shown in Fig. 1a.

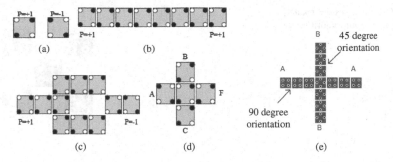

Fig. 1 QCA primitives: **a** cell with two possible polarizations, **b** QCA wire, **c** inverter, **d** traditional majority gate (TMG), **e** coplanar crossover

All QCA-based circuits are built on the basic elements such as the inverter gate and the majority voter (MV) gate. QCA-based wire is formed by cascade of cells, as depicted in Fig. 1b. Figure 1c, d depicts implementation of an inverter and the majority gate, respectively. The majority gate function is as follows (1):

$$MV(A, B, C) = AB + BC + CA \qquad (1)$$

For coplanar wire crossing in QCA, the cells in the two wires must be oriented 90° and 45° as shown in Fig. 1e.

QCA timing is controlled by four-phase clocking scheme. This solves two purposes (a) allow the QCA signal to propagate properly, i.e., control the information flow, (b) supply necessary energy to control the height of inter-dot barrier within the a cell. To drive the input to the desired output, signals need to be passed through four clock zones. Clock signals for each zone are distinct, and each signal clock includes four phases: *Switch, Hold, Release*, and *Relax* as shown in Fig. 2 which are 90° phase shifted [2].

Fig. 2 QCA clocking with four phases

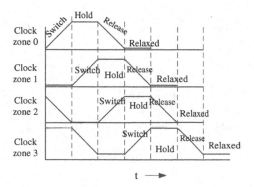

Fig. 3 QCA layout of rotated coplanar 3-input majority gate (RMG)

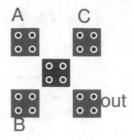

2.2 Rotated 3-Input Majority Gate

Majority gate is the fundamental unit in QCA designs; therefore, the overall optimal design of a digital circuit completely dependents on an optimal orientation of cells in a majority gate. In this section, we introduced a modified 3-input majority gate, which can be used as an alternative to the traditional majority gate as shown in Fig. 3.

3 Design of D Flip Flops

3.1 Level Trigger D Flip Flop

A compact 2:1 multiplexer schematic is presented in Fig. 4a. Proposed multiplexer requires 3 MV (proposed in the previous section) and 1 inverter in QCA implementation. This structure comprises only 17 quantum cells as depicted in Fig. 4b. Further, Fig. 5 shows the QCA realization of the proposed level sensitive D flip flop. In this structure, MVs in first level implement two AND gates that are driven by clock zone 0. The outputs of these MVs are fed to second-level MV which is positioned in the clocking zone 1. The behavior of the presented D flip flop is formulated in Table 1. The input data will propagate to the output by setting clock to 1, otherwise it results the previous output.

(a) **(b)**

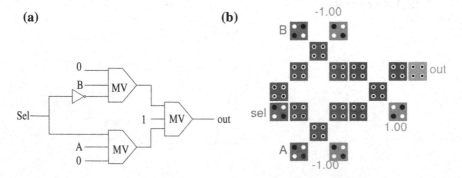

Fig. 4 a Schematic of 2:1 multiplexer **b** proposed QCA layout of 2:1 multiplexer

Fig. 5 Proposed level trigger
D flip flop in QCA

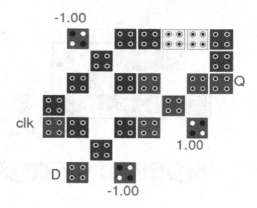

Table 1 Operation of
QCA-based level trigger D
flip flop

Clock (clk)	D	Q (output)
0	X	Q(t − 1)
0	X	Q(t − 1)
1	0	0 (input)
1	1	1 (input)

3.2 Positive Edge Trigger D Flip Flop

Work by Yang et al. [12] first introduced a QCA-based structure for edge trigger
flip flops. It uses 1 majority and 1 inverter. The majority gate works as an AND
gate, which operates on delayed version and inverted version of clock input
(clk) and results an intermediate output I.

The working of the proposed rising edge trigger D flip flop is presented in
Table 2. For instance, during the rising edge of the clock signal, the value at the
output of majority gate is 1. This allows the input to be transmitted to the output.
QCA layout of the proposed rising edge D flip flop is shown in Fig. 6. Output cell
is driven by clock zone 3, and it is clearly perceived that proposed design works
correctly with lesser clock zones, which leads a reduction in total input to output
delay.

Table 2 Operation of QCA-based positive edge trigger D flip flop

clk(t)	clk(t − 1)	$\overline{clk(t-1)}$	Intermediate output (I)	Output (Q(t))
0	0	1	0	Q(t − 1)
0	1	0	0	Q(t − 1)
1	0	1	1	D (input)
1	1	0	0	Q(t − 1)

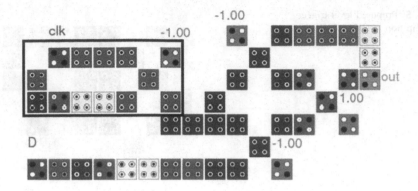

Fig. 6 QCA layout of proposed positive edge trigger D flip flop

3.3 Negative Edge Trigger D Flip Flop

QCA structure of the proposed falling edge D flip flop is shown in Fig. 7. The working of the presented falling edge trigger D flip flop is demonstrated in Table 3. The majority gate works as an AND gate, which operates on $\overline{clk(t)}$ and $clk(t-1)$ and results an intermediate output I. During falling edge of the clock signal, the value at the output of majority gate is 1. This allows the input to be propagated to the output. In this proposed structure, all the signals are properly synchronized, and clock zone 3 is taken for the output cell.

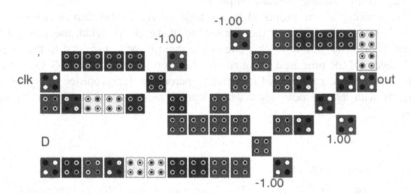

Fig. 7 Proposed negative edge trigger D flip flop in QCA layout

Table 3 Operation of QCA-based negative edge trigger D flip flop

clk(t)	$\overline{\text{clk}(t)}$	clk(t − 1)	Intermediate output (I)	Q(t) (output)
0	1	1	1	D (input)
0	1	0	0	Q(t − 1)
1	0	1	0	Q(t − 1)
1	0	0	0	Q(t − 1)

3.4 Dual Edge Trigger D Flip Flop

QCA structure of the proposed dual edge D flip flop is shown in Fig. 8. This constitutes both rising and falling edge trigger structures. Working of the presented dual edge trigger D flip flop is presented in Table 4. Majority gates at first level works as an AND gate, which operates on delayed version and inverted version of clock input (clk). Output of these AND gates (intermediate signal I_1, I_2) feed to second-level majority gate, which results in intermediate signal I_3. A rising or falling edge of the clock signal (clk) makes $I_3 = 1$. This allows the input to be propagated to the output.

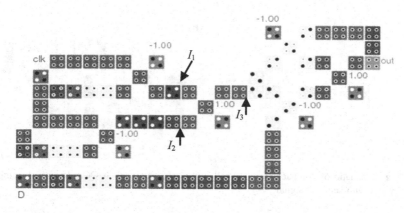

Fig. 8 Proposed dual edge trigger D flip flop QCA

Table 4 Operation of QCA-based dual edge trigger D flip flop

clk(t)	$\overline{\text{clk}(t-1)}$	Intermediate output (I_1)	$\overline{\text{clk}(t)}$	clk (t − 1)	Intermediate output (I_2)	Intermediate output (I_3 = MV (I_1, I_2, 1))	Q(t) (output)
0	0	0	1	1	1	1	D (input)
0	1	0	1	0	0	0	Q(t − 1)
1	0	0	0	1	0	0	Q(t − 1)
1	1	1	0	0	0	1	D (input)

4 Memory Cell

RAM cell is one of the essential components in many digital systems. Its performance is varied upon the complexity and input to output latency in QCA implementation. So optimization can be done in the selection of majority gates and proper clocking zone. In this section, an efficient RAM cell is presented as shown in Fig. 9a, with set and reset ability. QCA layout of the proposed RAM cell is demonstrated in Fig. 9b. It builds upon six majority gates (three majority gates for each 2:1 mux). Signal sel is used as select line for the mux1, and signal R/W (Read/Write) is used as select line for mux2, as shown in Fig. 9. Mux2 operates on output of mux1 and output signal out of the memory cell. A clear view of the RAM cell operation is defined in Table 5. For example, when R/W= 0, the output of the circuit remains unchanged; i.e., out(t) = out($t - 1$). Output of the circuit sees the effect of input and set/reset signals by setting R/W= 1. Further, setting R/W = 1 and sel = 1 allow an input signal to be transmitted to the output. Similarly, for select line (sel) = 0, the output of the circuit gets changed according to set/reset signal. Output cell is driven by clock zone 0, and a valid output can be achieved after 7 clock phases.

(a)   **(b)**

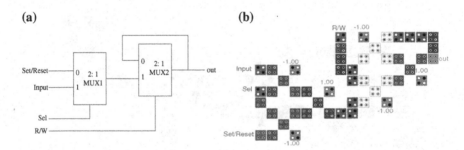

Fig. 9 **a** Schematic of compact memory cell with set/reset signal, **b** QCA layout of compact memory cell with set/reset signal

Table 5 Operation of QCA-based memory cell with set/reset ability	Read/write (R/W)	Select (sel)	Set/reset	Out(t) (output)
	1	1	X	Input
	1	0	0	0 (reset)
	1	0	1	1 (set)
	0	X	X	Out($t - 1$)

5 Simulation Results

Simulation of proposed designs is verified using the simulator QCADesigner-2.0.3 [17]. Coherence vector simulation engine type is used with default parameter values. Also, similar outputs are generated by setting simulation engine to bistable approximation type.

The simulation results of level trigger D flip flop are depicted in Fig. 10a. It is inferred that the presented design works correctly. For instance, when clk = 1, the input gets transferred to the output line after 0.5 clock cycle, otherwise no change in the output. Figure 10b shows the input and output waveform to investigate the operations of rising edge trigger D flip flop. Simulation results show during rising edge of clk the input get transferred to the output line and register a valid output after 1 clock cycle, otherwise no change in the output. Simulation results for falling edge trigger D flip flop is shown in Fig. 10c. This indicates correct operation of proposed structure and a valid output after 1 clock cycle during falling edge of signal clk. Figure 10d shows simulation results for proposed dual edge trigger D flip flop. It allows inputs to be transmitted at both positive and negative edges of signal clk after 1.5 clock cycles. In addition, Fig. 11 shows input and output waveforms of efficient RAM cell with set/reset ability. During normal mode, input data are propagated to the output after 1.25 clock cycles by activating both select and read/write line to +1. Similarly, by fixing set/reset line to 0 or 1, output line can be set or reset, respectively, after a delay of 1.25 clock cycles. So Fig. 11 confirms that presented RAM cell provides expected outputs after first falling edge of clock 0.

5.1 Comparison Results

All the structures are less complex and faster in comparison to prior designs. Also, the new structures can be easily implemented on QCA without any crossover wire. The proposed D flip flop structures are compared with prior designs in terms of cell counts, area occupation, input to output delay, area–delay product (ADP), and the results are highlighted in Table 6. Also, it is worth noticing that the proposed D flip flops are fastest among the all existing structures. The detailed comparison summaries are provided below.

In the presented level trigger flip flop, cell count and area occupation are 52% and 60% less than the most compact existing design, respectively. Positive and negative edge trigger D flip flops are 55% more efficient in terms of area and 36% faster than earlier structure [13]. Dual edge trigger D flip flop surpasses previous best design [12] in terms of area and faster at least 25%.

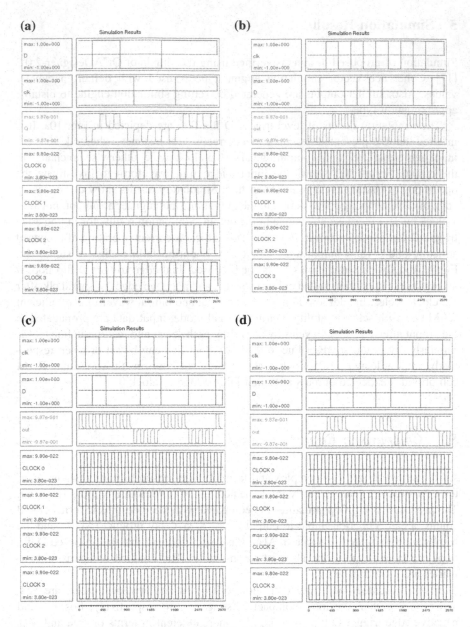

Fig. 10 Input/output waveforms of proposed D flip flops **a** level trigger D flip flop, **b** positive edge trigger D flip flop, **c** negative edge trigger D flip flop, **d** dual edge trigger D flip flop

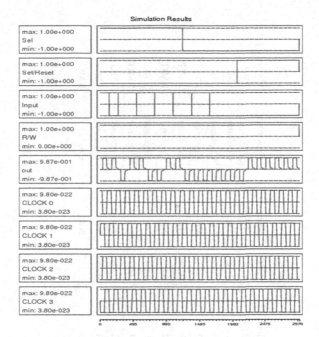

Fig. 11 Input/output waveforms of memory cell with set/reset ability

All the designs achieve the minimum delay and spread over an area lesser than the existing designs in the literature. Hence, it delivers the best area–delay tradeoff. From Table 7, it is clearly perceived that the proposed memory cell with set/reset ability can be a suitable candidate for further complexity and latency analysis in future as a benchmark structure. This structure has outperformed all design [13–16] in terms of cell count, input to output delay, and area occupation.

Table 6 Comparison of QCA D flip flops structure

		Area (μm²)	Cell count	Delay	Clock phases	ADP	Set/reset feature	Crossover type	Majority gate types
QCA D flip flop (level trigger)	[11]	0.08	66	1.5	6	0.12	No	Coplanar	TMG[a]
	[13]	0.05	48	1	4	0.05	No	Coplanar	TMG[a]
	Proposed	0.02	23	0.5	2	0.01	No	Coplanar	RMG[b]
Positive edge trigger QCA D flip flop	[13]	0.09	84	2.75	11	0.247	No	Coplanar	TMG[a]
	Proposed	0.04	47	1.75	7	0.07	No	Coplanar	RMG[b]
Negative edge trigger QCA D flip flop	[13]	0.09	84	2.75	11	0.247	No	Coplanar	TMG[a]
	Proposed	0.04	47	1.75	7	0.07	No	Coplanar	RMG[b]
QCA dual edge D flip flop	[12]	0.16	116	3	12	0.48	No	Coplanar	TMG[a]
	[13]	0.14	120	3.25	13	0.455	No	Coplanar	TMG[a]
	Proposed	0.1	81	2.25	9	0.225	No	Coplanar	RMG[b]

[a]TMG traditional 3-input majority gate
[b]RMG rotated majority gate

Table 7 Comparison of RAM cell structure

Memory cell	Coplanar crossover	Set/reset ability	Area (μm^2)	Cell count	Delay	Clock phases	ADP	Majority gate types
[15]	Yes	No	0.16	158	2	8	0.32	TMG[a]
[14]	Yes	No	0.11	100	3	12	0.33	TMG[a]
	No	No	0.07	63	2	8	0.14	
[13]	No	Yes	0.13	109	1.75	7	0.2275	TMG[a]
[16]	No	Yes	0.08	88	1.5	6	0.12	TMG[a], 5-input majority gate
Proposed	No	Yes	0.06	49	1.25	5	0.075	RMG[b]

[a]*TMG* traditional 3-input majority gate
[b]*RMG* rotated majority gate

6 Conclusion

In this work, a rotated 3-input majority gate structure has been considered. It is observed that this single layer gate is flexible and can be used to implement more complex QCA designs. As memory cell and flip flops are rudimentary for most of the digital circuits, having a high speed and less complex memory cell is significantly important. To showcase the efficacy of the rotated structure, a new memory cell and D flip flop structures were introduced. All the designs enjoy coplanar non-crossover wires with enhanced QCA layout. Simulations using QCADesigner confirm that the presented designs using rotated structure have outperformed all prior designs and show significant improvements. Due to optimized structures, the proposed modules can be used to design high-performance QCA sequential circuits and systems in the future. Further extensions such as fault analysis, power analysis, and realistic clocking scheme could be investigated.

References

1. Lent, C. S., Tougaw, P. D., Porod, W., Bernstein, G. H.: Quantum cellular automata. Nanotechnology 4, 49–57 (1993).
2. Lent, C. S., Tougaw, P. D.: A device architecture for computing with quantum dots. Proc. IEEE 85(4), 541–557 (1997).
3. Sasamal, T. N., Singh, A. K., Mohan, A.: Efficient design of reversible alu in quantum-dot cellular automata. Optik 127(15), 6172–6182 (2016).
4. Sasamal, T. N., Singh, A. K., Mohan, A.: An optimal design of full adder based on 5-input majority gate in coplanar quantum-dot cellular automata. Optik 127(20), 8576–8591 (2016).
5. Navi, K., Farazkish, R., Sayedsalehi, S., Azghadi, M. R.: A new quantum-dot cellular automata full-adder. Microelectron. J. 41, 820–826 (2010).
6. Sasamal, T. N., Singh, A. K., Ghanekar, U.: Design of non-restoring binary array divider in majority logic-based QCA. Electronics Letters 52(24), 2001–2003 (2016).
7. Roohi, A., Khademolhosseini, H., Sayedsalehi, S., Navi, K.: A symmetric quantum-dot cellular automata design for 5-input majority gate. J. Comput. Electron. 13, 701–708 (2014).
8. Shamsabadi, A. S., Ghahfarokhi, B. S., Zamanifar, K., Vafaei, A.: Applying inherent capabilities of quantum-dot cellular automata to design: D flip-flop case study. J. Syst.Archit. 55, 180–187 (2009).
9. Vankamamidi, V., Ottavi, M., Lombardi, F.: A serial memory by quantum-dot cellular automata (QCA). IEEE Trans. Comput. 57, 606–618 (2008).
10. Berzon, D. Fountain, T. J.: A memory design in QCAs using the squares formalism. In Proceedings of the Great Lakes Symposium. VLSI, pp. 166–169 (1999).
11. Vetteth, A., Walus, K., Dimitrov, V. S., Jullien, G. A.: Quantum-Dot Cellular Automata of Flip-Flops. ATIPS Laboratory 2500 University Drive, N.W., Calgary, Alberta, Canada T2N 1N4, 2003.
12. Yang, X., Cai, L., Zhao, X.: Low power dual-edge triggered flip-flop structure in quantum dot cellular automata. Electron. Lett. 46, 825–626 (2010).
13. Hashemi, S., Navi, K.: New robust QCA D flip flop and memory structures. Microelectronics Journal 43, 929–940 (2012).

14. Dehkordi, M. A., Shamsabadi, A. S., Ghahfarokhi, B. S., Vafaei, A.: Novel RAM Cell designs based on inherent capabilities of quantum dot cellular automata. Microelectronics Journal 42, 701–708 (2011).
15. Walus, K., Vetteth, A., Jullien, G. A., Dimitrov, V. S.: RAM design using quantum-dot cellular automata. In Proc. Nanotechnology Conference and Trade Show, 2, pp. 160–163 (2003).
16. Angizi, S., Sarmadi, S., Sayedsalehi, S., Navi, K.: Design and evaluation of new majority gate-based RAM cell in quantum-dot cellular automata. Microelectronics Journal 46, 43–51 (2015).
17. Walus, K., Dysart, T. J., Jullien, G. A., Budiman, R. A.: QCAdesigner: a rapid design and simulation tool for quantum-dot cellular automata. IEEE Trans. Nanotechnol. 3, 26–31 (2004).

On a Hardware Selection Model for Analysis of VoIP-Based Real-Time Applications

Shubhani Aggarwal, Gurjot Kaur, Jasleen Kaur, Nitish Mahajan, Naresh Kumar and Makhan Singh

Abstract Voice over IP (VoIP) is a methodology for transmitting data, voice, video, messaging, and chat services over Internet Protocol. The quality of VoIP is heavily dependent on the type of hardware used for call manager. Hardware calibration is a mechanism for selecting a suitable hardware (processor) which can match the processing requirements of call manager. The aim of this paper is to propose a hardware selection model that can handle VoIP-based real-time applications by considering parameters like CPU utilization and RAM utilization. The proposed algorithm takes into consideration input arrival rate, type of applications (peer-to-peer or back-to-back), codec, and call holding time (CHT) corresponding to which a hardware recommendation is generated. The experiments are conducted for different types of hardware, and results show that the proposed model can help VoIP service providers to select the appropriate hardware.

Keywords IP PBX · Voice over IP · SIP · FreeSWITCH · Codec

1 Introduction

VoIP is a technology that carries the voice communication over IP network. Instead of using circuit-switched network, we use packet-switched network where audio is transmitted in the form of digital packets [1]. VoIP call includes the call setup signaling protocol, call admission control, transmission of RTP payload. A large n of factors are involved in making a high-quality VoIP call like codec selection, packetization, packet loss, delay, filter to provide QoS [2]. VoIP implementation

S. Aggarwal (✉) · J. Kaur · N. Mahajan · M. Singh
Computer Science and Engineering, University Institute of Engineering and Technology,
Panjab University, Chandigarh 160014, India
e-mail: shubhaniaggarwal529@gmail.com

G. Kaur · N. Kumar
Electronics and Communication Engineering, University Institute of Engineering and
Technology, Panjab University, Chandigarh 160014, India

© Springer Nature Singapore Pte Ltd. 2019
B. K. Panigrahi et al. (eds.), *Smart Innovations in Communication and
Computational Sciences*, Advances in Intelligent Systems and Computing 670,
https://doi.org/10.1007/978-981-10-8971-8_23

uses hard IP phones and soft IP phones and does not rely on a traditional PBX [3]. An IP PBX is a private branch exchange that switches calls between VoIP users on local lines while allowing all users to share a certain number of external phone lines. The typical IP PBX can also switch calls between a VoIP user and a traditional telephone user. Call servers perform IP phone registration and coordinate call signaling. In VoIP, call servers are also known as softswitch [4]. RTCP is a protocol that controls end-to-end information about the session of RTP packets [5]. SIP is an application layer protocol used for initiating, establishing, modifying, and terminating the session [6]. SIP is responsible for finding the location of the end users on the basis of URL. This can be done with the help of DNS server [7].

User Agent Client (UAC) sends requests to SIP server, and SIP server sends back responses. The SIP server acts as a proxy server and sends requests to some another server on the favor of a client [8]. The SIP server also acts as a registrar and checks that UAC is authorized or not.

FreeSWITCH is an open-source IP-based platform that can be customized to route the calls and interconnect different communication protocols by using audio, text, and video [9]. Its design uses central core which is stable and has modules for specific functionality [10]. A session is initiated by the connection between Free-SWITCH and a SIP protocol. It combines with other IETF protocols like RTP, RTCP, and gateway control protocol (GCP) for establishing the entire multimedia communication [11]. FreeSWITCH can handle the calls in two modes: peer-to peer (P2P) mode in which it handles registration and signaling load only; i.e., SIP packets flow through the PBX or FreeSWITCH, but voice in the form of RTP packets travels directly between endpoints or clients. In other mode, i.e., back-to-back (BTB) configuration, all the SIP and RTP packets flow through the server; i.e., FreeSWITCH will act as an intermediary between its clients. This provides the PBX much better control over the voice or RTP packets when operated in BTB mode as compared to P2P.

A codec in VoIP compresses VoIP packets for faster transmission and then decodes back by decompressing them. There are different types of codecs based on the selected data rate, sampling rate, and compression rate [12]. Thus, to improve the quality of VoIP applications, two factors should be considered: codec selection and possible hardware to handle real-time application load [13, 14]. As per rigorous literature review, till date, no work has been done for selection of optimized hardware to handle different VoIP-based applications. In this paper, we have used StarTrinity to generate load on call manager which takes input as desired load, i.e., CPS or BHCA, and gives output as whichever hardware best suits for the given input. We also analyzed the output of our proposed scheme by considering different codecs like G.711, G.729 that measure the quality of the VoIP calls on network for varying input load.

The remainder of the paper is organized as follows: Sect. 2 introduces the proposed work. Section 3 presents an experimental study to evaluate the performance of VoIP applications. Section 4 concludes the research paper by giving future directions.

2 Proposed Work

In this section, a model for hardware calibration is presented. Based on our simulation, we will select an appropriate codec which consumes less CPU utilization and RAM utilization for every hardware. RAM utilization and CPU utilization are the parameters that are being considered while hardware selection. We have done performance testing several numbers of times and established a relationship based on actual simulation results using regression analysis. Regression equation is used in making prediction of CPU utilization and RAM utilization based on captured data for desired calls per second (CPS) as shown in Eq. (1).

$$y = ax + b \tag{1}$$

Here a, b are constants as shown in Eqs. (2) and (3), respectively

$$a = \frac{(\sum y)(\sum x^2) - (\sum x)(\sum xy)}{n(\sum x^2) - (\sum x)^2} \tag{2}$$

$$b = \frac{n(\sum xy) - (\sum x)(\sum xy)}{n(\sum x^2) - (\sum x)^2} \tag{3}$$

The detailed algorithm is shown in Fig. 1. It takes as input Busy Hour Call Attempt (BHCA) or CPS or maximum concurrent calls. Maximum concurrent calls are the highest number of concurrent ongoing calls achieved in a given simulation.

The proposed algorithm works for different applications that call manager can handle like registration load, peer-to-peer load, back-to-back load, audio recording. The proposed algorithm also considered different codecs for analyzing performance of call manager.

So, with the help of this model, we can get as output whichever is the suitable hardware based on observed CPU utilization and RAM utilization. Although we performed the simulation on two processors, this algorithm is applicable for any number of processors. For testing of any kind of processor, we can also hire cloud services and test call load on it by using StarTrinity SIP tester as load generator for desired results. Further, based on this model, we have conducted different performance tests that are presented in Sect. 3.

3 Analysis and Results

In this section, we have presented simulation results and analysis of the proposed model discussed in previous section. We have defined and executed performance tests that allow us to measure resource usage accurately for selection of hardware at

Algorithm 1: Proposed algorithm for VoIP calls.

INPUT: a) BHCA or CPS or maximum concurrent calls
b) Call load: P2P or BTB
If BTB, then audio recording enabled or disabled
c) Codec selection: G.711 or G.729
d) Call Holding Time (CHT)
PROCESSING
Case 1. P2P calls for selected codec and selected CHT
        If CPS <= α or maximum concurrent calls < y
        then processor P1
        Else processor P2.
Case 2. BTB calls for selected codec and selected CHT
    a) Without audio recording
        If CPS< β
        then processor P1
        Else processor P2.
    b) With audio recording
        If CPS< μ
        then processor P1
        Else processor P2.
    where, α, β, μ are the particular thresholds of maximum load that
        a particular processor can handle for different scenarios.
        y indicates maximum concurrent calls count that a
        particular processor can handle for different scenarios.
        P1, P2 are the processors.

Fig. 1 Proposed algorithm

different call scenarios: P2P calls, BTB calls, with audio recording, without audio recording, and transcoding with codecs G.711 and G.729.

3.1 Simulation Parameters

StarTrinity as a simulation framework is used to generate different VoIP calls. StarTrinity is an emulator used for load testing and performance of SIP servers [15]. It also monitors the VoIP quality of live IP network servers and simulates thousands of simultaneously outgoing and incoming SIP calls. In StarTrinity, blocking factor is set to zero. In our simulation study, we have considered four PCs such that two PCs behave as call originators and two PCs are used to receive the incoming calls (receiver). For this simulation study, only two processors have been considered, i.e., i7 and low-end Raspberry Pi. Total 1000 users were registered on incoming end. An input to call manager is in the form of calls per second (CPS) which is deduced from BHCA. Simulation has been carried out for one hour.

For all tests, two important parameters are taken into consideration, i.e., CPU utilization and RAM utilization. CPU utilization means usage of CPU for processing resources, i.e., amount of work handled by CPU at a particular time. It depends on the type and amount of load being processed. It must not exceed 80%. Similarly, RAM usage also increases as the load on the system increases. It must be less than 70%. We have performed simulation 10 times, and average values have been taken for analysis purpose as presented in next section.

3.2 Result Analysis

Case 1. This section presents analysis of results under different cases. In this case, results are presented for P2P calls for two call manager's—i7 and Raspberry Pi. A comparison is made for two systems when we gradually increase the load in terms of CPS that can be determined by call manager parameters: CPU utilization and RAM utilization. The data are presented with different codecs G.711 and G.729 for call holding time (CHT) of 1 and 2 min to evaluate distinctly the effect of varying CHT on performance parameters.

(A) Using codec G.711

See Tables 1 and 2.

Table 1 Simulation results for P2P calls with call duration using G.711 codec = 1 min

I7			Raspberry Pi		
CPS	CPU utilization (%)	RAM utilization (%)	CPS	CPU utilization (%)	RAM utilization (%)
10	10	33.9	1	9	17.08
20	25	45	5	45	24
25	32	52	7	66.5	34.7
30	38	60	8	70.75 (unstable)	36.7

Table 2 Simulation results for P2P calls with call duration using G.711 codec = 2 min

I7			Raspberry Pi		
CPS	CPU utilization (%)	RAM utilization (%)	CPS	CPU utilization (%)	RAM utilization (%)
10	20	38.6	2	18	21.6
13	25	43	3	28	29.62
15	27	48.3	4	43.75	33.5
20	40	52.4	5	52.5 (unstable)	38.45

(B) Using codec G.729

See Tables 3 and 4.

In P2P calls, server handles only the signaling part, i.e., SIP packets and not the RTP media. Hence, i7 performs well even up to 30 and 20 CPS for CHT of 1 and 2 min, respectively, for both codecs as shown in Tables 1, 2, 3, and 4. The longer call duration results in greater number of concurrent calls because the previous sessions still persist, while new sessions build up every second.

Raspberry Pi is able to handle maximum of 500 concurrent calls in this case. Beyond this, CPU utilization of the call manager considerably increases and calls begin to drop. This load corresponds to 7 and 4 CPS for CHT 1 and 2 min, respectively, for both codecs G.711 and G.729 as shown in Tables 1, 2, 3, and 4.

Different codecs mainly result in non-similar RTP payload size due to distinct compression techniques employed, which further results in varied load on the Call Manager. There is no significant difference in the performance of the call manager when we used two different codecs at the endpoints. This is because, in P2P calls, call manager only handles SIP signaling which is similar in all cases.

Table 3 Simulation results for P2P calls with call duration using G.729 codec = 1 min

I7			Raspberry Pi		
CPS	CPU utilization (%)	RAM utilization (%)	CPS	CPU utilization (%)	RAM utilization (%)
10	10	15	1	9.25	16.2
20	20	39.4	5	52	27.4
25	30	48.2	7	54.75	31.67
30	42.5	62.4	8	59 (unstable)	32.9

Table 4 Simulation results for P2P calls with call duration using G.729 codec = 2 min

I7			Raspberry Pi		
CPS	CPU utilization (%)	RAM utilization (%)	CPS	CPU utilization (%)	RAM utilization (%)
10	15	38.4	2	19.5	22.5
14	18.5	42.6	3	33	28
15	20	44.7	4	45	31.35
20	30	49.8	5	56 (unstable)	32.43

Case 2. In this case, the results for BTB calls are shown for two processors, i.e., i7 and Raspberry Pi. The CPU utilization of call manager is considerable when it is involved in managing the RTP media as more CPU cycles are required to handle full load. With longer CHT, it increases further due to greater number of concurrent ongoing calls. The results are also provided for the case when audio recording of the BTB calls is done to observe its impact on call manager's performance.

Without audio recording. I7 can handle full load up to 17 and 13 CPS for CHT 1 and 2 min, respectively, as shown in Tables 5 and 6. Beyond this value, CPU overshoots to the unacceptable range, and hence, the systems slow down and cannot process further load.

Raspberry Pi can handle up to 2 and 1 CPS for CHT 1 and 2 min, respectively, as shown in Tables 5 and 6. Afterward, CPU usage becomes 98%.

With audio recording. With even greater CPU cycles required for audio recording, i7 can handle only up to 12 and 8 CPS for CHT 1 and 2 min, respectively, as shown in Tables 7 and 8.

Raspberry Pi handles up to 1 CPS for CHT 1 min and 2 min as shown in Tables 7 and 8. As observed from Tables 7 and 8, RAM utilization is also considerably higher when audio recording of the calls is done because the streams are first transferred to RAM before recording them to the hard disk.

Case 3. Transcoding is done to set a channel between two endpoints with different codecs. Following simulations were performed to evaluate the effect of transcoding on call manager's performance. The codecs selected for the simulation were chosen to be different for both the ends. The calls were generated in a burst with CHT of 2 min, and the results were captured. For the purpose of comparative analysis, same simulation was performed but with similar codecs on both the ends. The results capture the estimated effect of transcoding.

Table 5 Simulation results for BTB calls with call duration = 1 min (without recording)

I7			Raspberry Pi		
CPS	CPU utilization (%)	RAM utilization (%)	CPS	CPU utilization (%)	RAM utilization (%)
10	43	42	1	22.5	17.2
17	70	45.3	2	62.5	24.3
18	80 (hanged)	48.4	3	98	41.4

Table 6 Simulation results for BTB calls with call duration = 2 min (without recording)

I7			Raspberry Pi		
CPS	CPU utilization (%)	RAM utilization (%)	CPS	CPU utilization (%)	RAM utilization (%)
10	55	45	1	60	20.5
13	74	53	2	95	34.05
14	80 (hanged)	57			

From Tables 9 and 10, we observed that transcoding requires greater CPU utilization and RAM utilization as compared to using same codecs at both ends. As shown in Table 10, G.729 utilizes more CPU and RAM than G.711 as it incorporates greater compression. Hence, more number of CPU cycles and RAM are needed for compression. Thus, from various results, it can be concluded that with the help of proposed algorithm, suitable processors can be selected as call manager in a cost-effective manner.

Thus, the proposed model can help VoIP service providers to select the appropriate hardware and provide better QoS for real-time-based VoIP applications like voice, video, data, chat services.

Table 7 Simulation results for BTB calls with call duration = 1 min (with recording)

I7			Raspberry Pi		
CPS	CPU utilization (%)	RAM utilization (%)	CPS	CPU utilization (%)	RAM utilization (%)
10	45	54.6	1	51.6	18.8
12	60	58.5	2	96	46.37
13	80	65.9			

Table 8 Simulation results for BTB calls with call duration = 2 min (with recording)

I7			Raspberry Pi		
CPS	CPU utilization (%)	RAM utilization (%)	CPS	CPU utilization (%)	RAM utilization (%)
6	45	46.9	1	71.2	33.18
8	65	51	2	100	88.64
9	80	63.9			

Table 9 Simulation results for transcoding with different codec at both ends

G.729–G.711			G.711–G.729		
CPS	CPU utilization (%)	RAM utilization (%)	CPS	CPU utilization (%)	RAM utilization (%)
10	10	14.8	10	8.8	12.4
20	12.75	15	20	11.7	12.7
30	15.2	15.2	30	15	13

Table 10 Simulation results for transcoding with same codec at both ends

G.729–G.729			G.711–G.711		
CPS	CPU utilization (%)	RAM utilization (%)	CPS	CPU utilization (%)	RAM utilization (%)
10	3.5	12.4	10	2	12.5
20	5	13	20	3	12.7
30	7.5	13.7	30	4.5	12.9

4 Conclusion

In this paper, we have proposed an algorithm for selecting a suitable hardware from a set of processors with different codecs by considering call manager parameters like CPU utilization and RAM utilization. We have evaluated its performance by conducting several simulation runs for estimating CPU utilization and RAM utilization for different processors and presented an equation for estimating load based on captured results. Experimental results prove that our proposed work is able to calibrate the appropriate hardware among all available hardware given any call load in form of CPS or BHCA. This would help VoIP service providers to provide efficient services optimally by providing better QoS for different VoIP services like voice, video, data, chat services in VoIP network. In future work, we will consider load balancing and clustering of different SIP services.

References

1. Yu, J. and Al Ajarmeh, I.: Design and Traffic Engineering of VoIP for Enterprise and Carrier Networks. International Journal On Advances in Telecommunications, vol. 1 (2008)
2. Goode, B. (2002) 'Voice over internet protocol (VoIP)', Proceedings of the IEEE, 90(9) (pp. 1495–1517)
3. Dantu, R., Fahmy, S., Schulzrinne, H. and Cangussu, J.: Issues and challenges in securing VoIP, computers & security 28.8 743–753 (2009)
4. Shunyi, R. X. Z. (2001) 'Next generation network architecture based on softswitch', Telecommunications Science, Vol. 8, pp. 25–31
5. Mäkelä, M.: Signaling in Session Initiation Protocol (2003)
6. Schulzrinne, H. and Wedlund, E.: Application-layer mobility using SIP. Mobile Computing and Communications Review, 4(3), 47–57 (2000)
7. An Sari, A. M., Nehal, M. F., & Qadeer, M. A.: SIP-based Interactive Voice (2013)
8. Rosenberg, J., Schulzrinne, H., Camarillo, G., Johnston, A., Peterson, J., Sparks, R. and Schooler, E.: SIP: session initiation protocol (2002)
9. FreeSwitch: http://www.freeswitch.org, accessed on April (2013)
10. Minessale, A. and Schreiber, D.: FreeSWITCH 1.0.6. Birmingham, UK: Packt Publishing (2010)
11. Ott, J., Wenger, S., Sato, N., Burmeister, C., and Rey, J.: Extended RTP profile for real-time transport control protocol (RTCP)-based feedback (RTP/AVPF) (2006)
12. K. Kim and Y. J. Choi, "Performance Comparison of Various VoIP Codec in Wireless Environments," in Proceedings of ACM International Conference on Ubiquitous Information Management and Communication (ICUIMC 11), Seoul, Korea, pp. 1–10, 21–23 Feb, 2011
13. Dougherty, B., White, J., Thompson, C. and Schmidt, D. C.: Automating hardware and software evolution analysis. Engineering of Computer Based Systems, ECBS 2009. 16th Annual IEEE International Conference and Workshop IEEE (2009)
14. Queenette, U. I. and Jerome, I. O.: Selection Criteria for Computer Software and Hardware, A Case Study of Six University Libraries in Nigeria, Chinese Librarianship: an International Electronic Journal 32 (2011)
15. www.StarTrinity.com

Trajectory Planning and Gait Analysis for the Dynamic Stability of a Quadruped Robot

Mayuresh S. Maradkar and P. V. Manivannan

Abstract Trajectory planning of robot's center of gravity (CoG) is the main concern when a legged robot is walking. The trajectory of the robot should be framed such that the center of pressure (CoP) of the robot should lie within supporting polygon at all time. This paper deals with the study of support polygon and graphical analysis to find the location of CoG, where the robot has high chances to go to instability. The quadrilateral supporting phases are utilized to avoid these instability locations. Further, the analysis is done to find the timely sequence of lift and touchdown of legs (lift and touch are called as events of legs). Based on the sequence of events and the support polygon analysis, trajectory of the robot is defined, which can produce smooth, steady, and stable robot motion. Though the robot gains static stability by trajectory planning, its dynamic stability should also be verified. This is done using zero moment point (ZMP) method. The analysis done in this paper is for the unswaying robot, walking on flat terrain.

Keywords Quadruped · Support polygon · Margin of stability
Gait · Zero moment point (ZMP)

1 Introduction

Robots are primarily classified in three categories: wheeled, tracked, and legged. Wheeled and tracked robots have access to less than 50% of earth surface, while on the other hand, legged animals have greater capability to traverse on off road, uneven terrain [1]. Gaining the inspired from the animals, research in legged robots

M. S. Maradkar (✉)
Indian Institute of Technology Madras, Chennai 600036, India
e-mail: mayuresh.maradkar87@gmail.com

P. V. Manivannan
Department of Mechanical Engineering, Indian Institute of Technology Madras,
Chennai 600036, India
e-mail: pvm@iitm.ac.in

© Springer Nature Singapore Pte Ltd. 2019
B. K. Panigrahi et al. (eds.), *Smart Innovations in Communication and Computational Sciences*, Advances in Intelligent Systems and Computing 670,
https://doi.org/10.1007/978-981-10-8971-8_24

started. Unlike the wheeled and the tracked robots, legged robots take discrete steps for producing motion, making them agiler providing ability to easily avoid obstacles [2]. With the high agility of legged robots, they have demerits too, including more actuators requirement, high energy consumption, complex locomotive mechanism. But more important than them, they need complex control system [3].

Motion in the wheel and tracked robots is controlled by the varying the speed of the wheels alone but in the quadruped (legged) robot motion is defined by the gaits it follows. Gait is a sequence of the motion produced by legs, harmonized with the motion of the body [4]. Based on the periodicity, gaits are classified into periodic and non-periodic. Periodic gaits are primarily used for motion on the even terrains [5]. Static walk, trot, pace, gallop are some of the periodic gaits. Non-periodic gaits are produced for motion on uneven terrains [6]. During the static walk, quadruped robot takes discrete steps such that three or four legs are supporting the body while the remaining legs will be in swing phase such that the duty cycle (β) of the stance will be: $1 > \beta > 0.75$ [7, 8]. For the static walk, one needs to plan the trajectory of the robot such that the CoG of the robot will always lie in support polygon [9]. Legged robots can be in two different postures: (1) erect and (2) crawl [10]. Robots in erect posture have greater ability to avoid the ground obstacles, whereas robots in crawl have higher ability to avoid overhanging obstacles [11].

The robot presented in this paper has the ability to transform itself from the erect to sprawl and vice versa depending on the requirement. The analysis done in the paper deals with the trajectory generation and stability analysis to produce the smooth, steady, and stable static motion of this presented quadruped robot in an erect posture. Study of support polygon and the analysis of the timely sequence of events (events here are lift and touchdown of legs) is done for planning the foot placement [12]. The further trajectory of the robot is developed and joint angular displacements required to achieve the desired motion are obtained. Zero moment point method (ZMP) is utilized for verifying the dynamic stability of the robot.

2   Design of Quadruped Robot

The quadruped robot presented in this paper has a capability to walk in both erect as well as Sprawl posture. Figure 1 shows the CAD model of this robot in an erect posture with the frontal and parasagittal planes, respectively. Each leg of this robot has four degrees of freedom (DOF), all provided by revolute joints. Out of these four DOF, two are provided at the hip by two independent joints (joint 1 and joint 2). Axis of joint 1 is normal to the frontal plane while the axis of joint 2 is normal to the joint 1. The remaining two DOF are provided at the knee by joint 3 and joint 4. Figure 2 shows the joint arrangements and the axis of joints in a leg. During erect posture, the axis of the joint 3 is normal to the parasagittal plane while the axis of the joint 4 is normal to the frontal plane. Thus, the robot has in all 16 DOF. Each DOF is controlled by digital servo motor CLS406MD.

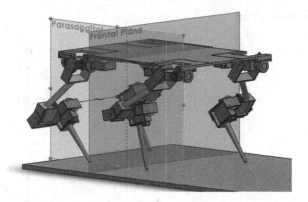

Fig. 1 CAD model of the robot

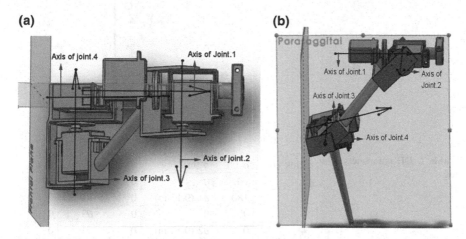

Fig. 2 Joint arrangements in a leg of quadruped Robot. **a** Top view of the leg, **b** side view of the leg

The details of dimensions and the mass properties of linkages of legs are given in Fig. 3 and Tables 1 and 2. Figure 3 shows the line diagram of a leg of this robot. Here, lines represent the linkages of legs, solid dots represents the origin 'O^i' of the local frame of references F^i (F^{i-1} is the local frame of reference of link i), and hollow dots represents the center of mass (C^i) of links i. Denavit Hartenberg (DH) parameters of the leg mechanism of this robot are given in Table 1 where, α, a, d, θ, and θ_h represent the twist angle, link length, joint offset, joint angle, and home joint angle, respectively. Table 2 gives the mass and locations of the CoG of the links with respect to their corresponding local frame of references.

Fig. 3 Line diagram of a leg

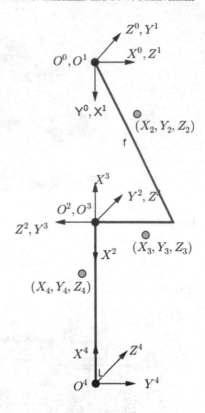

Table 1 DH parameters of a leg

Link no.	α (°)	a (m)	d (m)	θ (°)	θ_h (°)
1	90	0	0	$\theta 1$	90
2	180	$a1$ (0.15 m)	0	$\theta 2$	0
3	90	0	0	$\theta 3$	0
4	0	$a2$ (0.15 m)	0	$\theta 4$	0

Table 2 Mass and the location of CoG of links w.r.t. their local reference frames

Link no.	X (mm)	Y (mm)	Z (mm)	Mass (g)
1	0	0	0	125
2	115	0	49	118
3	4	0	−30.5	23
4	15	0	10	100

3 Gait Analysis

The stability of the robot depends on the trajectory of body and support polygon it is following [13]. Thus, the stability analysis demands the study of the support polygon, trajectory planning, and the timely sequence of events (together called as gait analysis). The study done here is for the robot walking on the flat terrain.

3.1 Support Polygon

The support polygon defines the polygonal area formed by the robot's feet resting on the ground. Support polygon required for the smooth and effective walk for this quadruped robot is shown in the Fig. 4, where the honeycombed triangle represents the common area for support polygon $1^1 2^1 3^1$ and $1^1 3^1 4^1$ (where $a^i b^j c^k$ is a support polygon formed by leg a, leg b, and leg c and is similar to the polygon formed by joining points a^i, b^j, and c^k). Similarly, the cross-hatched triangle represents the common area for support polygon $1^1 2^2 4^1$ and $2^2 3^2 4^1$. Sl is the stroke length of robot, L is hip-to-hip distance of legs in parasagittal plane, 'w' is hip-to-hip distance between frontal legs (or the rear legs), The symbol 'A' denotes the intersection of the midline of robot and line joining $3^1 2^1$, while 'B' denotes the intersection point of midline of robot and line joining $4^1 1^1$. The 'd' is a constant, for which CoG of robot lies at S (midpoint of 'A' and 'B') during the transition of the supporting phase from $1^1 2^1 3^1$ to $1^1 3^1 4^1$. The sequence of support polygons followed by the robot is given by Eq. (1). The $a^i b^j c^k$ are triangular support polygon while $a^i b^j c^k d^l$ are quadrilateral support polygons.

$$3^1 4^0 2^1,\ 3^1 4^0 2^1 1^1,\ 1^1 2^1 3^1,\ 1^1 2^1 3^1 4^1,\ 1^1 3^1 4^1,\ 1^1 2^2 3^1 4^1,\ 1^1 2^2 4^1,\ 1^1 2^2 3^2 4^1,\ 2^2 3^2 4^1,\ 1^2 2^2 3^2 4^1,\ 1^2 2^2 3^2,\ 1^2 2^2 3^2 4^2, \ldots$$

$$(1)$$

The duration for which the robot lies in the quadrilateral support affects the speed of the robot. Longer the normalized duration ($2\Delta t_q$, the ratio of the duration of the event and the cycle time) of the robot on quadrilateral supporting phase, lesser is its speed. Hence for enhancing the robot's speed, the Δt_q should tend to zero. Thus, the sequence of support polygon turns out to:

$$3^1 4^0 2^1,\ 1^1 2^1 3^1,\ 1^1 3^1 4^1,\ 1^1 2^2 4^1,\ 2^2 3^2 4^1,\ 1^2 2^2 3^2, \ldots \qquad (2)$$

From Fig. 4, it may be noted that the polygons $3^1 4^0 2^1$ and $1^1 2^1 3^1$ have no common area. Hence for an unswaying robot, during the end of supporting phase of $3^1 4^0 2^1$, the CoG should lie within support polygon $3^1 4^0 2^1$, but it (CoG) should tend toward 'A'. Similarly, during the start of supporting phase of $1^1 2^1 3^1$, the CoG should lie within support polygon $1^1 2^1 3^1$, but it (CoG) should tend away from 'A'. Thus, the transition from support polygon $3^1 4^0 2^1$ to $1^1 2^1 3^1$ has very high chances of

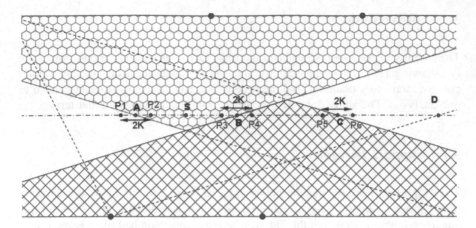

Fig. 4 Support polygon based on the dimensions of the robot

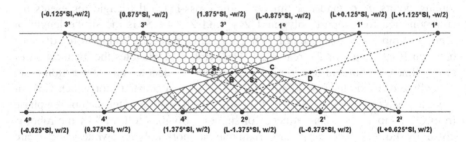

Fig. 5 Detailed view of the support polygons where the robot has high ability to get unstable

producing the instability in robot motion. However, the stability can be maintained, if the CoG is made to lie at 'A' during the transition of support polygons. Similarly, during the transition from $1^1 3^1 4^1$ to $1^1 2^2 4^1$, 'B' is the critical point, and for the transition from $2^2 3^2 4^1$ to $1^2 2^2 3^2$, 'C' is the critical point. The stability at 'A', 'B', 'C' (as shown in Figs. 4 and 5) is maintained by quadrilateral supports $3^1 4^0 2^1 1^1$, $1^1 2^2 3^1 4^1$, $3^2 4^1 2^2 1^2$, respectively.

3.2 Timing of Events

In legged robot's motion analysis, the event is defined as the lift or the touchdown of the leg. For quadruped robots, there are eight events in all; four correspond to the lift and the remaining four corresponding to the touchdown legs. The sequence of the events that the robot has to follow for the stable walk is given by Eq. (3), where α_i and γ_i represent the touchdown and lift of leg i, respectively. The mathematical representation of γ and α is given by Eqs. (4) and (5).

$$\alpha_4, \gamma_2, \alpha_2, \gamma_3, \alpha_3, \gamma_1, \alpha_1, \gamma_1, \alpha_4 \tag{3}$$

$$\gamma = \{0.75 - \Delta t_q, 0.25 - \Delta t_q, 0.5, 0\} * nT \tag{4}$$

$$\alpha = \{0.5, 0, 0.25 + \Delta t_q, 0.75 + \Delta t_q\} * nT \tag{5}$$

From the support polygon analysis done in the previous section, it can be observed that robots have a high probability of becoming unstable, during diagonal legs swapping their phases. At such instances, if somehow the robot is made to support itself on the quadrilateral support polygon, this instability issue can be overcome. Equation (6) gives the relation between the duty cycle of the robot and the normalized time of robot's quadrilateral phase.

$$\beta = 0.75 + \Delta t_q \tag{6}$$

where, β is the duty cycle of the leg, T is the cycle time of stance, and n is the integral value. Figure 6 shows the timing diagram of the events.

3.3 Trajectory Planning

From support polygon analysis done in above section, it has been found that in a complete motion cycle, the robot has to undergo six different supporting phases. Out of these 6, two are quadrilateral supporting phases, (1) for stability at A (provided by $1^1 2^1 3^1 4^0$) and (2) for stability at B (provided by $1^1 2^2 3^1 4^1$). The remaining four supporting phases are triangular support phases ($1^1 2^1 3^1$, $1^1 3^1 4^1$, $1^1 2^2 4^1$, $2^2 3^2 4^1$). Thus for simplifying the motion analysis, robot's trajectory is segmented into six sub-trajectories. Each sub-trajectory corresponds to one supporting phase. At the start and end of each supporting phase, at least one leg is either getting lifted or touched down. Hence, for the smooth transition of the robot from one supporting phase to other, each sub-trajectory function must satisfy the following conditions:

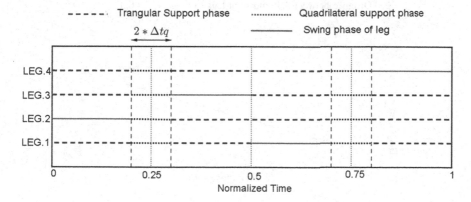

Fig. 6 Timing diagram of events

$$\frac{df_k}{dt}\bigg|_{t_s^k} = \frac{df_k}{dt}\bigg|_{t_e^k} = 0 \tag{6}$$

$$\frac{d^2 f_k}{dt^2}\bigg|_{t_s^k} = \frac{d^2 f_k}{dt^2}\bigg|_{t_e^k} = 0 \tag{7}$$

$$f_{k-1}\big|_{t_e^{k-1}} = f_k\big|_{t_s^k} \tag{8}$$

$$f_k\big|_{t_e^k} = f_{k+1}\big|_{t_s^{k+1}} \tag{9}$$

where, f_k, t_s^k and t_e^k represents the sub-trajectory function, the time of start and time of the end of supporting phase k, respectively. If the above conditions are not satisfied, the joint motors generate jerky motion, which leads to dynamic instability of robot. Thus for stable motion, each sub-trajectory has to satisfy in all six conditions. The best solution obtained for f_k is a polynomial equation with a degree at least equals to 5. For analysis, the considered degree is 5. Thus, the sub-trajectory function is given by Eq. (10). Table 3 gives the considered motion parameters while Table 4 gives the mathematical expression for the initial displacements of the robot during the start and end of the supporting phases where, N is the integral quotient of time (t) and cycle time (T). The trajectory given by Eq. (10) is shown in Fig. 7.

Table 3 Assumptions considered during analysis

S. No.	Parameter	Considered value
1	Average velocity (v)	0.2 m/s
2	Stroke length (Sl)	0.2 m
3	Ground height (h)	0.25 m
4	Duty cycle	0.8
5	Distance between front and rear legs (L)	0.7
6	Distance between left and right legs (w)	0.3

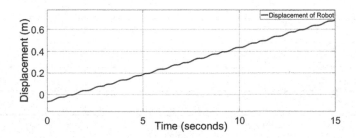

Fig. 7 Displacement profile of robot along its heading direction

$$f_k = \prod_{i \equiv 0}^{5} a_{i,k} t^i \qquad (10)$$

Figures 8 and 9 show the required relative displacement of feet w.r.t. their corresponding F^0 to generate the trajectory shown in Fig. 7. From Fig. 8, it is clear that the trajectory of leg 3 is similar to leg 4 with a time delay of $T/2$. Figure 9 shows that trajectory of leg 1 is similar to the trajectory of leg 2 with a time delay of $T/2$.

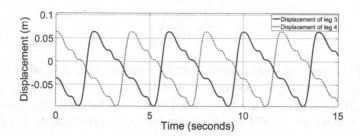

Fig. 8 Displacement profile of feet w.r.t. reference frame F^0 attached at the corresponding hips of legs

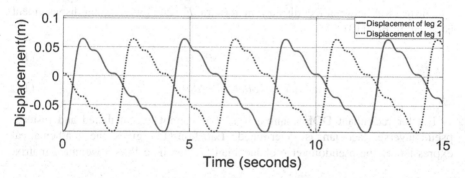

Fig. 9 Displacement profile of feet w.r.t. reference frame F^0 attached at the corresponding hips of legs

Table 4 Displacements at the start and the end of supporting phase

Support phase no	Supporting phase	Displacement at start	Displacement at end
1	$1^1 2^2 3^1 4^1$	$N * Sl$	$N * Sl + 2m$
2	$1^1 2^1 3^1$	$N * Sl + 2m$	$N * Sl + m + Sl/4$
3	$1^1 3^1 4^1$	$N * Sl + m + Sl/4$	$N * Sl + Sl/2$
4	$1^1 2^1 3^1 4^0$	$N * Sl + Sl/2$	$N * Sl + 3 * Sl/4m$
5	$1^1 2^2 4^1$	$N * Sl + 3 * Sl/4m$	$(N + 1) * Sl-2m$
6	$2^2 3^2 4^1$	$(N + 1) * Sl-2m$	$(N + 1) * Sl$

4   Inverse Kinematics of Robot

Digital servo motors (Model: ClS406MD) are used as actuators in the presented robot. These digital servos use joint angular displacement as operational signals. Thus, to produce the robot's desired motion, joint angular displacement needs to be feed to the corresponding joint motors. Hence, finding the joint angle profiles is crucial for the motion control of the robot. The Jacobian inverse method is used to obtain these angular profiles. The Jacobian matrix of the manipulator (here, it is leg) is given by Eq. (11).

$$J = \left(\frac{\partial p_i}{\partial \theta_j}\right)_{i,j} \tag{11}$$

where 'P' is a relative displacement function between feet and F^0 while, 'i' is the direction of displacement (coordinate axes of F^0, i.e., $i \in \{X^0, Y^0, Z^0\}$, j is the joint number. This p is obtained using Eq. (12), where 'n' is DOF of the manipulator, k is the link number and M_k DH transformation matrix corresponding to link k.

$$p = \left(\prod_{k \equiv 1}^{n} M_k\right)[0 \quad 0 \quad 0 \quad 1]^T \tag{12}$$

The relation between feet displacement (w.r.t. F^0) and joint angular displacement is given by Eq. (13).

$$\Delta S_t = J\Delta\theta_t \tag{13}$$

$$\Rightarrow \Delta\theta_t = J^{-1}\Delta S_t \tag{14}$$

But for redundant DOF manipulators, J^{-1} cannot be calculated and instead, pseudoinverse Jacobian is determined. Equation (15) gives the mathematical expression of the pseudoinverse of Jacobian (J^+), where W is a weighted matrix.

$$J^+ = W^{-1}J^T \left(JW^{-1}J^T\right)^{-1} \tag{15}$$

$$\Delta\theta_t = J^+\Delta S_t \tag{16}$$

$$\theta_{t+1} = \theta_t + \Delta\theta_t \tag{17}$$

Figures 10 and 11 show the obtained joint angle profile of leg 4 and 2, respectively. The joint angle profiles of the leg 3 are similar to leg 4 (Fig. 10), but with a time delay of $T/2$. Here, T is the cycle time. Similarly, the joint angle profiles of leg 1 are similar to leg 2 (Fig. 11), with the time delay of $T/2$.

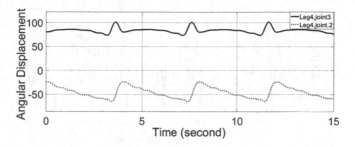

Fig. 10 Joint angle profiles in leg 4

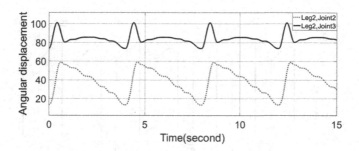

Fig. 11 Joint angle profiles of leg 2

5 Dynamic Stability Analysis: Zero Moment Point (ZMP)

In the previous section, we developed the robot's displacement trajectory capable of providing static stability but the dynamic stability of the robot is still to be verified. This dynamic stability is validated using ZMP method. However, as the mass of links of this robot is very less compared to its body mass, inertial effects produced by the body alone are considered. Equation (18) gives the mathematical expression of ZMP, relative to CoG of the body.

$$X_{zmp} = y_{CG} * \tan^{-1}\left(\frac{-x''_{CG}}{g - y''_{CG}}\right) \tag{18}$$

where, X_{zmp} is the ZMP along the heading direction of the robot. y_{CG} is the ground height of CoG of the body; x''_{CG} is the acceleration of CoG along heading direction and is given in Fig. 12. y''_{CG} is the acceleration along an upward direction which is zero for this trajectory. The location of ZMP must lie inside the supporting polygon for the stable motion of the robot. Figure 13 shows the location of ZMP inside the support polygons where solid line curve represents the distance of the frontal edge of support polygon from ZMP, while dotted line curve represents the distance of ZMP from the rear edge of support polygon.

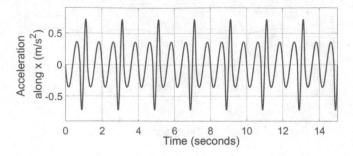

Fig. 12 Acceleration of the body of the robot along the heading direction

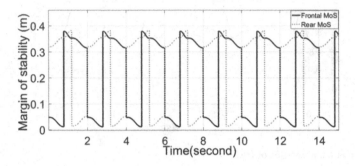

Fig. 13 Marginal stability of robot

6 Conclusion

In this work, using the support polygon analysis, we were able to find the position of the CoG that leads to instability of robot's motion. During the diagonal legs swapping their phases (i.e., support to lift or vice versa), the robot loses its stability. At these instances, the stability of robot can be maintained by the quadrilateral support. Based on the study of support polygons and the timing diagram of legs, the robot's trajectory was planned. The acceleration of the robot following this trajectory lies in the range of -0.7 to 0.7 m/s^2 showing the gradual motion of the robot. ZMP method gave the location of center of pressure. This center of pressure is found to be always lying inside the supporting polygon. We were able to obtain a minimum margin of stability margin is 14.8 mm, which is close to the desired minimum margin of stability (15 mm). Thus, the dynamic stability of the robot following the planned trajectory is verified.

References

1. Anon.: Logistical Vehicle Off-Road Mobility. Project TCCO 62–5; U. S. Army Transportation Combat Developments Agency; Fort Eustis, Va (1967).
2. D. Vishal, P. V. Mannvannan.: Design and analysis of active vibration and stability control of a bio-inspired quadruped robot. Int. Jou. Computational Vision and Robotics (2016).
3. Boston Dynamics Corp, Big Dog Overview [online]. Big Dog: An Overview. http://www. bostondyanmics.com/img/bigdog/overview.
4. D. Vishal, P. V. Manivannan.: Multibody dynamics simulation and gait pattern analysis of a bio-inspired quadruped robot for unstructured terrain using adaptive stroke length, Artificial Life and Robotics-Springer (2016).
5. Min-Hsiung Hung, Fan-Tien Cheng, Hao-Lun Lee & David E. Orin.: Increasing the stability margin of multilegged vehicles through body sway. Journal of the Chinese Institute of Engineers, Vol. 28, No. 1, pp. 39–54 (2005).
6. A. A. Frank.: Automatic Control System for Legged Locomotion Machines," USC Rept., p. 273 (1969).
7. McGhee, R. B., and Frank, A. A.: On the Stability of Quadruped Creeping Gaits. Mathematical Biosciences, Vol. 3, No. 3/4, pp. 331–353.
8. Hironori Adachi, Noriho Koyachi, and Eiji Nakano.: Mechanism and Control of a Quadruped Walking Robot, IEEE Control Systems Magazine.
9. Reza Yazdani, Vahid. Johari Majd, and Reza Oftadeh.: Dynamically Stable Trajectory Planning for a Quadruped Robot. 20th Iranian Conference on Electrical Engineering, (ICEE2012), Tehran, Iran (2012).
10. W. Scott Persons, Victoria Arbour, Jessica Edwards, Matthew Vavrek, Philip Currie, and Eva Koppelhus.: DINO101- Moving Around, University of Alberta.
11. Mayuresh Maradkar, Aaditya Chandramouli, P. V. Manivannan.: Bio-inspired Reconfigurable Robot: Conceptual Design of an All-terrain Robot Capable of Transforming from an Erect to Sprawling Posture. 3rd Int. Conf. Mechatronics and Robotic Engineering (ICMRE), Paris, France (2017).
12. Mayuresh Maradkar, P. V. Manivannan.: Kinematics and Multi-body Dynamics of a Bio-inspired Quadruped Robot with Nine Linked Closed Chain Legs. International Conference on Robotics: Current Trends and Future Challenges (RCTFC), Thanjavur, India, (2016).
13. Shuaishuai Zhang, Xuewen Rong, Yibin Li and Bin Li.: A Composite COG Trajectory Planning Method for the Quadruped Robot Walking on Rough Terrain. International Journal of Control and Automation Vol. 8, No. 9, pp. 101–118, (2015).

Application of Evolutionary Reinforcement Learning (ERL) Approach in Control Domain: A Review

Parul Goyal, Hasmat Malik and Rajneesh Sharma

Abstract Evolutionary algorithms have come to take a centre stage in diverse areas spanning multiple applications. Reinforcement learning is a novel paradigm that has recently evolved as a major control technique. This paper presents a concise review on implementing reinforcement learning with evolutionary algorithms, e.g. genetic algorithm (GA), particle swarm optimization (PSO), ant colony optimization (ACO), to several benchmark control problems, e.g. inverted pendulum, cart–pole problem, mobile robots. Some techniques have combined Q-Learning with evolutionary approaches to improve their performance. Others have used knowledge acquisition to obtain optimal fuzzy rule set and genetic reinforcement learning (GRL) for designing consequent parts of fuzzy systems. We also propose a Q-value-based GRL for fuzzy controller (QGRF) where evolution is performed after each trial in contrast to GA where many trials are required to be performed before evolution.

Keywords Reinforcement learning · ERL · GA · PSO · QGRF

1 Introduction

Reinforcement learning is a subpart of the larger machine learning framework in computer science. The basic idea is to learn to behave optimally based on experiential knowledge gained during interactions with the system we intend to control. The resulting controllers are optimal yet intelligent and have been applied by a variety of researchers in several fields. Reinforcement learning has emerged as a viable and efficient method for designing adaptive controllers.

A major breakthrough in RL came when function approximation was incorporated into the RL framework. This allowed extension of RL to very high-dimensional or

P. Goyal (✉) · H. Malik · R. Sharma
Department of Instrumentation and Control Engineering, Netaji Subhas Institute
of Technology, New Delhi 110078, India
e-mail: parulthaparian@gmail.com

© Springer Nature Singapore Pte Ltd. 2019   273
B. K. Panigrahi et al. (eds.), *Smart Innovations in Communication and
Computational Sciences*, Advances in Intelligent Systems and Computing 670,
https://doi.org/10.1007/978-981-10-8971-8_25

continuous state space domains. Function approximation has been used in RL using neural networks, fuzzy systems, wavelets and, recently, using genetic algorithms or a combination of them.

In order to solve RL problems, EAs represent an interesting alternative which offers potential advantages for scaling to real applications. In particular, ERL systems can address challenges faced in RL implementation which include incomplete state information, large state spaces and non-stationary environments. EA methods depend more on the net results of the decision sequences rather than on ambiguous sensor information. There are two ways in which ERL policy representations can address large state space problems: generalization and selectivity. It specifies policy at abstraction level which is better than observed states to actions explicit mapping as used in pure RL. In rule-based representations, rule language permits conditions to match state sets thus reducing storage greatly. ERL systems use selective representations of policies as they eliminate explicit information related to less desirable actions. Policies that result in bad decisions are discouraged by EAs and gradually eliminated from population. Its advantage is that focus remains on pro-table actions only thus reducing policies space requirements. If environment of agent changes over time, optimal policy turns into a moving target making RL problem becomes more difficult. Simple modifications in standard EAs can track non-stationary environments and provide an encouraging RL approach for these difficult cases. This paper attempts to sum up the contributions to the RL field by these evolutionary techniques.

GA, a powerful random search method, handles optimization problems. This is particularly useful in solving complex optimization problems especially where great number of parameters makes analysis of global solution difficult. It has been successfully applied in diverse areas such as neural network parameter tuning [1], fuzzy control [2–4], load forecasting, eBook applications. PSO relates particle movement over search space, which is adjusted according to particle and its proximate member's flying experiences. The particle flies towards better search area over course of search process. The study has shown that PSO convergence rate is notably high than other techniques like GA [5]. These techniques when combined with RL or Q-Learning can accelerate the results. Like ACO, an optimization technique, when incorporated with Q-Learning gives high performance than ACO alone [6]. In this paper, we present a review of various combinations of EA and RL in search of better results. Every technique has its own advantages and drawbacks in different applications.

2 Literature Review

Applications of ERL in different domains of engineering as well as other fields have been presented in Fig. 1.

Fig. 1 Applications of ERL in different research domains

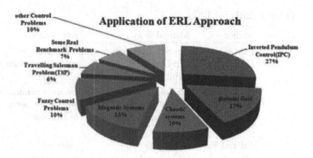

2.1 ERL for Inverted Pendulum Control (IPC)

Lam et al. [7] have used GA to design gains of fuzzy model-based nonlinear controller and to determine the stability conditions based on Lyapunov's stability theory. The cart–pole inverted pendulum has been stabilized using TSK fuzzy plant model. Dadios and Williams [1] have used an elastic pole in inverted pendulum problem with higher complexity in dynamics; e.g. additional load of 0.35 kg is attached to tip of pole which increased its time period. A comparison has been made between the neural network, fuzzy logic and fuzzy-genetic approaches, and the fuzzy genetic is found to have the highest accuracy.

Pal and Pal [8] have proposed a GA-based self-organized scheme (SOGARG) for a fuzzy logic controller which extracts optimal rule set. It is applied on truck back (23 rules) with ITAE 71.42 and inverted pendulum problem (average 16.6 rules) with ITAE 0.1019. Kumar et al. [5] have designed a PID controller using particle swarm optimization (PSO) to balance an inverted pendulum. This method tunes the controller parameters more efficiently than GA and compared with GA-PID controller. It has less computation burden as it has one variable, i.e. velocity.

Lin and Chen [9] have proposed a hybrid evolutionary algorithm cultural cooperative PSO (CCPSO) along with RL for cart–pole and ball–beam system. It increases global search capacity with the use of belief space (information repository). This technique automatically adjusts the self-evolving neuro-fuzzy inference network (SENFIN) parameters and performs well in RL problems.

Prado et al. [10] proposed PSO to obtain fuzzy rules for knowledge acquisition. It is simple and converges faster when used on inverted pendulum. Results obtained with this strategy are better than other RL and GA-based classical methods. Chiang et al. [11] propose a self-learning fuzzy controller genetic reinforcement fuzzy learning controller (GRFLC). It can predict the physical system behaviour using action-evaluation network; it can be applied to unknown analytical process models. Simulations show that membership functions and control rules can be generated without expert knowledge.

Hsu and Lin [12] proposed a hybrid reinforcement evolutionary learning algorithm, i.e. R-HELA to construct a TSK-type neuro-fuzzy controller, i.e. TNFC. Population is evolved using Compact GA (CGA) and modified variable-length GA

Table 1 ERL for IPC

S. No.	Author	Methodology	Application	Result
1	Lam et al. [7]	Genetic algorithm and Lyapunov stability theory	Cart–pole type inverted pendulum	GA used to formulate task of solving both gains and stability in a single problem
2	Dadios and Williams [1]	Neural network fuzzy control system fuzzy-genetic approach	Inverted pendulum with more degrees of freedom	GA optimizes fuzzy logic controller automatically with a decent convergence speed and high accuracy
3	Pal and Pal [8]	SOGARG	Inverted pendulum truck back	SOGARG is insensitive to changes in GA and system parameters
4	Kumar et al. [5]	PSO	Inverted pendulum	Convergence rate of PSO is high over other methods like GA
5	Lin and Chen [9]	SENFIN-REL and CCPSO	Cart–pole and ball–beam system	It performs well than other RL problem-solving methods
6	Prado et al. [10]	Knowledge acquisition using PSO	Inverted pendulum	Final results prove that it outperforms other classical learning methods
7	Chiang et al. [11]	GRFLC	Cart–pole problem	It is a combination of neural network and fuzzy logic controller in order to achieve great learning capability
8	Hsu and Lin [12]	R-HELA hybrid reinforcement evolutionary learning algorithm	Cart-inverted pendulum problem	On comparison with R-SE, R-GA, CQGAF, TDGAR, etc., it was found that it performs better in terms of CPU time, pole-balance trials and successfully controls pendulum in 30 runs

(MVGA). This method has been compared with R-GA and R-SE (reinforcement symbiotic evolution) where they manage to balance pole at 149th and 80th generation, respectively; R-HELA controls the pendulum in 30 runs successfully. Comparison os CQGAF and TDGAR, it is found that this technique is better in terms of CPU time and pole-balance trials.

A brief summary of ERL for IPC is presented in Table 1.

2.2 ERL in Robotics

Bhargava et al. [13] applied strong optimization and rapid convergence capability of GA in path-planning domains for mobile robots. Experiment is done in two environments: one less cluttered and other complex. Maximum unknown environment has to be explored without bumping into obstacles, while minimum velocity requirement has to be maintained. Controllers cooperate to save energy and achieve goal by using problem-solving capability of GA.

Shah and Gopal [14] proposed design of a neuro-fuzzy model (DENFIS) using RL. Its potential has been shown on a two-link robot manipulator. Its performance is better than Q-Learning-based RL system in terms of low tracking errors, low absolute error value and control effort. Training time of this approach is higher, but it does not affect its performance.

Zhou [15] proposes fuzzy reinforcement learning which is based on genetic algorithm (GAFRL) to accelerate robot learning. GA can evaluate candidate solutions without feedback from environment thus performing very effective global search. It can handle local minimum problem by using global optimization property of GA. A robot can start with heuristic knowledge and can refine its behaviour with GAFRL.

Das et al. [16] have proposed Q-Learning-based improved PSO (QIPSO-DV) algorithm for optimal trajectory path-planning of multi-robots in clutter environment for improving convergence. It reduces both space and time complexities as compared to classical Q-Learning technique. It minimizes path length, arrival time and turning angle of robots.

Kamio and Iba [17] integrated RL with genetic programming so that a robot can adapt its actions in real environment. RL is executed on a real robot and can learn effective actions. Experimentally, it was found that box moving task was done by the robots effectively and the performance was better using this technique. A brief summary of ERL in robotics is presented in Table 2.

Table 2 ERL in robotics

S. No.	Author	Methodology	Application	Result
1	Bhargava et al. [13]	Genetic algorithm and reinforcement learning	Mobile robot	Bootstrap problem is removed considerably and simulation time is reduced
2	Shah and Gopal [14]	DENFIS approach to RL	Two-link robot manipulator	DENFIS has high growth ability and ensures robust structure for real-time control applications
3	Zhou [15]	GAFRL	Physical biped robot	It can solve local minima problem
4	Das et al. [16]	QIPSO-DV	Multi-mobile robot navigation	It shows high optimization as compared to other algorithms such as IPSO, PSO
5	Kamio and Iba [17]	RL and genetic programming	Real robot	Its performance is higher than traditional Q-Learning method

2.3 ERL in Chaotic Systems

Juang [18] has combined Q-value-based GA with online clustering. This clustering-based approach designs the precondition part, and Q-value-based GRL designs the consequent part of a fuzzy system. Fuzzy rules generated are smaller when compared to grid type.

Lin and Jou [19] have applied TDGAR to chaotic systems where it produces small perturbations and converts chaotic oscillations into regular ones. It does not require knowing the property or mathematical model of chaotic control system. It makes controller designs practical and more feasible for real-world applications.

Juang and Hsu [20] have proposed Q-value-aided ACO for online rule generation. Consequent parts are selected based on this technique using Q-values and pheromone trails which are updated according to reinforcement signals. It is efficient, robust to noise and eases design effort. A brief summary of ERL for chaotic systems is shown in Table 3.

2.4 ERL in Magnetic Systems

Juang and Lu [6] have proposed ACO with fuzzy Q-Learning and RL. Q-values and pheromone levels select consequent values best combination. Comparatively, it performs better than other genetic fuzzy systems.

Juang et al. [21] have used GA with symbiotic evolution (GRL) which reduces CPU time and control trials considerably as compared to traditional GA method. It partitions input space flexibly, and no knowledge of mathematical model is required.

Lin and Jou [22] have integrated GA, temporal difference and gradient descent methods to form one learning system (TDGAR). It accelerates GA learning and achieves satisfactory results in controlling the magnetic bearing system. Juang and Lu [23] have combined fuzzy Q-Learning with ACO for magnetic systems. It finds

Table 3 Application of ERL for chaotic systems

S. No.	Author	Methodology	Application	Result
1	Juang [18]	CQGAF	chaotic system, magnetic levitation and cart–pole problem	This is very effective and efficient technique
2	Lin and Jou [19]	TDGAR	Chaotic dynamic control systems	It can be used in real-world directly to control physical chaotic problem
3	Juang and Hsu [20]	ORGQACO	Magnetic levitation, chaotic system and truck-backing control	It is more effective and efficient

Table 4 ERL in magnetic systems

S. No.	Author	Methodology	Application	Result
1	Juang and Lu [6]	ACO-FQ	Magnetic levitation, truck backup and water—bath temperature control	This technique's performance is better than ACO
2	Juang et al. [21]	GA and SEFC	Magnetic levitation system, cart–pole and temp control of water bath	It is superior and efficient as compared to other control problems
3	Lin and Jou [22]	TDGAR	Real magnetic bearing control system	This technique is more practical and feasible
4	Juang and Lu [23]	ACO-FQ	Magnetic levitation, truck backup water-bath temperature control	It performs much better than ACO

best combinations of actions suggested collectively by pheromone and Q-values. Its convergence quickly, and performance is better than ACO. A brief summary of ERL in magnetic systems is shown in Table 4.

2.5 ERL in Fuzzy Control

Hsu et al. [2] have used evolutionary reinforcement learning for solving control problems. In TFC model, the number of fuzzy rules is determined by SA (self-adaptive) method. Fuzzy rules are evaluated by multi-group symbiotic evolution (MSGE). R-ELDMA identifies groups for chromosome selection, both suitable and unsuitable, through data mining selection strategy (DSS). It tunes TFC model parameters efficiently. This technique has been applied to control a chaotic system and a cart–pole problem, and the results have been compared with other methods like R-SA, R-HELA, R-GA. It performs better in terms of shorter CPU time and faster convergence.

Lin and Xu [3] have proposed a comparative study of reinforcement sequential search-based GA (R-SSGA) and traditional GA method to tune fuzzy controller. Success or failure is indicated by reinforcement signal. It performs efficient mutation, minimizes population size, converges quickly, has small angular deviation and makes the TSK-type fuzzy controller design more practical. According to comparative analysis on IPC and water-bath temperature control systems, this method has shorter CPU time, better disturbance rejection capabilities and good control. Pourpanah et al. [4] have used fuzzy classifier with Q-Learning (QFAM) for data classification and GA for rule extraction or a two-stage hybrid model (QFAM-GA). Here, RL selects best winning nodes and improves FAM performance. This method is tested with UCI data sets and yields good results in noisy environments. A brief summary of ERL in fuzzy control problems is shown in Table 5.

Table 5 ERL in fuzzy control problems

S. No.	Author	Methodology	Application	Result
1	Hsu et al. [2]	R-ELDMA	TSK fuzzy controller-2 simulations discussed cart–pole balance and chaotic system control	It performs better and converges quickly
2	Lin and Xu [3]	Genetic reinforcement learning (R-SSGA)	Nonlinear fuzzy control problems—cart–pole balance and water-bath temperature control	It converges quickly, minimizes population size and produces smaller angular deviations
3	Pourpanah et al. [4]	Hybrid QFAM-GA model	Rule extraction and data classification	It yields high accuracy and minimizes complexity of the model

2.6 ERL in Travelling Salesman Problem (TSP)

Liu and Zeng [24] have proposed reinforcement mutation with improved genetic algorithm (RMGA) to solve TSP problem. Reinforcement learning is used as mutation operator (RL-M) and genetic algorithm as framework (H-EAX crossover operator) thus making it an evolutionary algorithm. Convergence rate of H-EAX is slow but with the local optimization of RL-M, convergence rate can be improved. Experimental results show that every time, RMGA can get an optimum tour in a reasonable time thus outperforming EAX-GA and LKH in terms of running time and quality of solutions.

Jiang and Liu [25] have proposed exploration–exploitation strategy based ACO for RL. This strategy generates adaptive set of straggled ants (SA) and elitist ants (EA) which helps it to converge to optimal solution quickly. Comparisons have been made with ACS and Ant-Q (Original Ant Model), and this strategy is found to have higher efficiency. Percentage of learned optimal states is more for AACO-RL.

A brief summary of ERL in travelling salesman problem is shown in Table 6.

2.7 ERL in Some Other Benchmark Problems

Mabu et al. [26] proposed a combination of reinforcement learning and evolution (genetic network programming) technique, i.e. GNP-RL to simulate stock trading models: tile-world problem and Khepera robot's wall following behaviour. This technique can adjust to changes in environment during the occurrence of sensor

Table 6 ERL in TSP

S. No.	Author	Methodology	Application	Result
1	Liu and Zeng [24]	Reinforcement mutation and genetic algorithm (RMGA)	Travelling salesman problem (TSP)	Results on TSP show that RMGA outperforms EAX-GA. Running rate of RL-M can be improved
2	Jiang and Liu [25]	Adaptive ACO and RL (AACO-RL)	TSP	AACO-RL converges faster to optimal solution

faults using WEBTOS robot simulator. Determination of number of sub-nodes has to be done appropriately as they affect adaptability and fitness. GNP-RL can change programs automatically using alternative actions, and it works well even if wrong information acquisition by the sensor.

Samma et al. [27] have integrated PSO with local search method and introduced RL-based memetic PSO (RLMPSO) model. Each particle is subjected to low-jump, high-jump, convergence, fine-tuning and exploration in accordance with the actions generated by RL algorithm. This model is evaluated on uni-modal, multi-modal, composite, shifted and rotated benchmark problems. RLMPSO produces good results with little penalty values, high delay value and low-cost value. It starts with slow convergence speed but converges rapidly if global optima regions are identified. As per the indication by 95% confidence intervals, it outperformed other methods significantly.

A brief summary of ERL on some benchmark problems is shown in Table 7.

Table 7 ERL in some benchmark problems

S. No.	Author	Methodology	Application	Result
1	Mabu et al. [26]	GNP-RL, i.e. genetic network programming and reinforcement learning	Elevator group control problem, tile-world problem, stock trading models and Khepera robot's wall following behaviour	GNP-RL with two sub-nodes performs the best as more sub-nodes means more search space thus lower fitness
2	Samma et al. [27]	RL and PSO (RLMPSO)	Uni-modal, multi-modal, composite, shifted, rotated and real benchmark problems like gear design, pressure vessel design	RLPMSO performed very well as compared to other methods except Ackley benchmark problem. It has slower convergence speed than other methods

2.8 ERL in Other Real-Life Problems

Bora et al. [28] incorporated a parameter-free self-tuning GA and RL approach (NSGA-RL). It is applied to satellite coverage problem and is more robust than others. Though it is comparatively slower, it works great for broadband reflector antenna satellites.

Houli et al. [29] have proposed multi-objective reinforcement learning to control traffic signals. It is efficient and less time consuming than traditional methods. Multi-RL is a car-based algorithm which prevents queue spillovers and avoids large-scale traffic jams. Building a model more closely to real traffic system is the future work.

Daneshfar and Bevrani [30] have proposed multi-agent RL based on optimization of GA for LFC in large-scale power system. It performs comparatively well due to its flexibility and independence in specification of control objective leading to high scalability. A brief summary of ERL in real-life problems is shown in Table 8.

2.9 Research Gap Findings

Following are the research gaps we found, after a careful critical analysis of the papers listed above.

- GA is faster than ACO and has been employed for global optimization successfully, but degradation in its efficiency is a major concern and can be overcome by more using PSO.
- GA needs variance in data to converge to the objective optimally.
- ERL technique can solve many problems which could not be tackled by EAs and RL working independently; e.g., initial population value and mutation point

Table 8 ERL in real-life problems

S. No.	Author	Methodology	Application	Result
1	Bora et al. [28]	NSGA-RL	Satellite coverage problem	It has advantage of complex parameter tuning and no-time spending
2	Houli et al. [29]	Multi-objective RL	Traffic signal control	This method is more efficient than the others
3	Daneshfar and Bevrani [30]	MARL and GA	Load frequency control	It is very scalable to realistic problems

were randomly generated, and constant range generated mutational value and population sizes.

• Although ERL is promising and has successful applications, a number of challenges still remain like handling learning rate where learning of the agent happens directly from experiences in operational environments.

3 Suggested Approach

In our view, we can use EA to find most suitable rules in a fuzzy Q-Learning setup. Our motivation for this novel approach is based on the fact that an evolutionary algorithms application has very few requirements:

1. An appropriate mapping between search space and space of chromosomes.
2. An appropriate fitness function.

The user has to decide on large number parameters, i.e. recombination rates, population size, parent selection rules, mutation rates, but research suggests that EAs are comparatively robust over a large range of control-parameter settings. In Table 9 we give a pseudo-code of EAs.

Next step is to build an RL framework that fits EAs, and there are two major concerns:

• Firstly, how can space of policies be depicted by chromosomes in EA?
• Secondly, how can fitness of population be estimated?

Answers to these queries depend on how the user integrates EA into RL. We give a pseudo-code for ERL in Table 10.

Once fitness of all policies in population has been decided, a new population is generated according to steps in usual EA (Table 9).

Table 9 Pseudo-code for evolutionary algorithm

```
procedure EA begin
    t = 0; initialize M(t);
    evaluate structures in M(t);
    while termination condition not satisfied do begin
        t = t + 1;
        select M(t) from M(t-1); alter structures in
    M(t); evaluate structures in M(t);
        end
end.
```

Table 10 Pseudo-code for evolutionary

```
procedure ERL begin
  t = 0;
  initialize a population of
  policies, M(t);
   evaluate policies in M(t);
while termination condition not satisfied do begin
      t = t + 1;
      select high-payoff policies, M(t), from policies in M(t-1);
      update policies in M(t);
      evaluate policies in M(t);
      end
  end
```

First, for reproduction, parents are selected, probabilistically on relative fitness:

$$P_r(p_i) = \frac{\text{Fitness}(p_i)}{\sum_{j=1}^{n}\text{Fitness}(p_j)} \quad (1)$$

where n is total number of individuals and p_i represents individual i. Using this selection rule, policy's fitness is proportional to expected number of offspring for given policy.

In our proposed work, we will be using GA-based optimization of the fuzzy rules that take part in generating optimal solution in a fuzzy RL controller. In conventional fuzzy Q-controller, all the rules take part in generating the optimal controller action, irrespective of their firing strengths. In our proposed approach, we will use GA to prune these rules so that only the fittest of these take part in producing an optimal action. These will also be used in the RL-based tuning of fuzzy inference system. This would allow only the fittest of the rules to participate in the control action generation, thereby cutting down the redundancy in the FQL-based controller design.

4 Proposed Model

We propose QGRF approach as shown in Fig. 2. Fuzzy system has two parts—precondition and consequent part. Precondition part can be decided by performing grid-type partition on input variables, and consequent part is outlined by GA. In GA, candidate consequent part population is generated. Here, each individual encodes consequent part and has corresponding Q-value. Utility of Q-value is to evaluate action recommended by individuals. High Q-value means higher reward

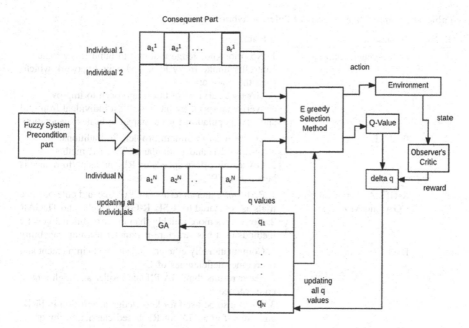

Fig. 2 Proposed architecture of QGRF

will be achieved. At every time step, depending on the Q-value of an individual, E - greedy method is used to choose one of the individuals. With the GA selected individuals, an action and system Q-value is evaluated. This action is applied to the environment with a consequent reward. Depending on this reward, Q-value of every individual is updated using temporal difference algorithm of RL. After every trial, these new Q-values are used as fitness values for evolution of GA. New populations can be generated by using GA and lower Q-value individuals can be replaced by newly generated ones. This process is repeated until success and forms the backbone of QGRF.

5 Conclusions

GA has been combined with RL or Q-Learning for many applications with decent convergence speed, and high-accuracy albeit PSO has better convergence than GA. Combination of RL and PSO performs better than other methods except that it has slower convergence speed. Similarly, RMGA outperforms other methods but still its running time can be improved. Various EA combinations with RL can lead to high-performance techniques as compared to GA, FL or EA as standalone as depicted in Table 11.

Table 11 Comparative analysis of different hybrid techniques

S. No.	Technique	Advantages
1	Fuzzy-genetic approach	1. No prior knowledge required to build fuzzy logic controller unlike fuzzy logic and neural network which need these apriori 2. GA uses survival of fittest approach to improve convergence speed by picking fittest individual from old and new population for participation in next generation
2	GAFRL	1. Solves local minimum problem in traditional actor-critic RL and is similar to TDGAR method 2. GA learning dose speeds up RL and its performance is better than FRL
3	R-HELA (combination of CGA and MVGA)	1. Performs better in terms of CPU time and pole-balance trials as compared to R-SE, R-GA, CQGAF and TDGAR 2. Converges more quickly than other traditional genetic methods as it has structure-parameter learning capability
4	PSO	1. Computationally efficient and simple to implement and overcomes deficiencies of GA 2 Better results than GA-PID controller and high rate of convergence 3. It can also be used for knowledge acquisition in FRBs and outperforms GA- or RL-based classical techniques
5	CCPSO	1. In PSO, performance degrades if search space dimensionality increases but CPSO uses multiple swarms to improve this drawback 2. CPSO may find suboptimal solution; hence, we use CCPSO as it increases global search capacity, avoids getting trapped in suboptimal solution
6	QIPSO-DV	1. This hybrid model improves path-planning performance of multi-robots. 2. In classical PSO, convergence is premature thus global optimum cannot be discovered but this model improves its performance and removes this flaw. 3. This method is efficient in number of steps needed to reach at destination, avoids local optima problem, achieves less errors and has fast convergence rate as compared to IPSO-DV, PSO and CQL
7	RLMPSO	1. It overcomes high computational cost and premature convergence problems of PSO 2. It has slow convergence rate in beginning due to small population size as compared to other models
8	ACO-FQ	1. ACO finds local or global optimal solution and ACO-FQ accelerates the learning process 2. ACO-FQ performs better than GA, ACO, CQGAF and fuzzy Q-Learning in terms of CPU time and trial number

(continued)

Table 11 (continued)

S. No.	Technique	Advantages
9	R-ELDMA	1. It identifies near-optimal solutions and reduces number of evolutionary generations 3. It is effective, feasible and performs better in terms of shorter CPU time and convergence than R-SA, R-HELA, R-GA methods
10	R-SSGA	1. It minimizes population size and converges quickly. 2. It gives small errors and shorter CPU time as compared to SEFC and TGFC methods

References

1. Elmer P. Dadios and David J. Williams, "Nonconventional Control of the Flexible Pole–CartBalancing Problem: Experimental Results", IEEE Transactions On Systems, Man, And Cybernetics—Part B: Cybernetics, Vol. 28, No. 6, 1998.
2. Chi-Yao Hsu, Yung-Chi Hsu and Sheng-Fuu Lin, "Reinforcement evolutionary learning using data mining algorithm with TSK-type fuzzy controllers", Applied Soft Computing 11 (2011) 3247–3259.
3. Cheng-Jian Lin and Yong-Ji Xu, "Anovel genetic reinforcement learning for nonlinear fuzzy control problems", Neurocomputing 69(2006) 2078–2089.
4. Farhad Pourpanah, Chee Peng Lim and Junita Mohamad Saleh, "A hybrid model of fuzzy ARTMAP and genetic algorithm for data classification and rule extraction", Expert Systems With Applications 49 (2016) 74–85.
5. Deepak Kumar, Brijesh Dhakar and Rajpati Yadav, "Tuning a PID controller using Evolutionary Algorithms for an Non-linear Inverted Pendulum on the Cart System", International Conference on Advanced Developments in Engineering and Technology (ICADET-14), India, Vol. 4(1), 2014.
6. Chia-Feng Juang and Chun-Ming Lu, "Ant Colony Optimization Incorporated With Fuzzy Q-Learning for Reinforcement Fuzzy Control", IEEE Transactions On Systems, Man, And Cybernetics—Part A: Systems And Humans, Vol. 39(3), 2009.
7. H. K. Lam, Frank H. Leung and Peter K. S. Tam, "Design and Stability Analysis of Fuzzy Model-Based Nonlinear Controller for Nonlinear Systems Using Genetic Algorithm," IEEE Transactions On Systems, Man, And Cybernetics—Part B: Cybernetics, Vol. 33, No. 2, 2003.
8. Tandra Pal and Nikhil R. Pal, "SOGARG: A Self-Organized Genetic Algorithm-Based Rule Generation Scheme for Fuzzy Controllers", IEEE Transactions On Evolutionary Computation, Vol. 7, No. 4, 2003.
9. Cheng-Jian Lin and Cheng-Hung Chen, "Nonlinear system control using self-evolving neural fuzzy inference networks with reinforcement evolutionary learning", Applied Soft Computing 11 (2011) 5463–5476.
10. R. P. Prado, S. Garc´ıa-Gal´an, J. E. Mu˜noz Exp´osito and A. J. Yuste, "Knowledge Acquisition in Fuzzy-Rule-Based Systems with Particle-Swarm Optimization", IEEE Transactions On Fuzzy Systems, Vol. 18(6), 2010.
11. Chih-Kuan Chiang, Hung-Yuan Chung and Jin-Jye Lin, "A Self-Learning Fuzzy Logic Controller Using Genetic Algorithms with Reinforcements", IEEE Transactions On Fuzzy Systems, Vol.5(3), 1997.
12. Yung-Chi Hsu and Sheng-Fuu Lin, "Reinforcement Hybrid Evolutionary Learning for TSK-type Neuro-Fuzzy Controller Design", Proceedings of the 17th World Congress The International Federation of Automatic Control Seoul, Korea, July 6–11, 2008.

13. Yesoda Bhargava, Anupam Shukla and Laxmidhar Behera, "Improved Approach to Area Exploration in an Unknown Environment by Mobile Robot using Genetic Algorithm, Real time Reinforcement Learning and Co-operation among the Controllers", Third International Conference on Advances in Control and Optimization of Dynamical Systems March 13–15, 2014. Kanpur, India.

14. Hitesh Shah and M. Gopal, "A Reinforcement Learning Algorithm with Evolving Fuzzy Neural Networks", Third International Conference on Advances in Control and Optimization of Dynamical Systems March 13–15, 2014. Kanpur, India.

15. Changjiu Zhou, "Robot learning with GA-based fuzzy reinforcement learning agents", Information Sciences 145 (2002) 45–68.

16. P. K. Das, H. S. Behera and B. K. Panigrahi, "Intelligent-based multi-robot path planning inspired by improved classical Q-learning and improved particle swarm optimization with perturbed velocity", Engineering Science and Technology, an International Journal 19 (2016) 651–669.

17. Shotaro Kamio and Hitoshi Iba, "Adaptation Technique for Integrating Genetic Programming and Reinforcement Learning for Real Robots', IEEE Transactions On Evolutionary Computation, Vol. 9 (3), 2005.

18. Chia-Feng Juang, "Combination of Online Clustering and Q-Value Based GA for Reinforcement Fuzzy System Design", IEEE Transactions On Fuzzy Systems, Vol. 13(3), 2005.

19. Chin-Teng Lin and Chong-Ping Jou, "Controlling Chaos by GA-Based Reinforcement Learning Neural Network", IEEE Transactions On Neural Networks, Vol. 10(4), 1999.

20. Chia-Feng Juang and Chia-Hung Hsu, "Reinforcement Interval Type-2 Fuzzy Controller Design by Online Rule Generation and Q-Value-Aided Ant Colony Optimization", IEEE Transactions On Systems, Man, And Cybernetics—Part B: Cybernetics, Vol. 39(6), 2009.

21. Chia-Feng Juang, Jiann-Yow Lin and Chin-Teng Lin, "Genetic Reinforcement Learning through Symbiotic Evolution for Fuzzy Controller Design", IEEE Transactions On Systems, Man, And Cybernetics—Part B: Cybernetics, Vol. 30(2), 2000.

22. Chin-Teng Lin and Chong-Ping Jou, "GA-Based Fuzzy Reinforcement Learning for Control of a Magnetic Bearing System", IEEE Transactions On Systems, Man, And Cybernetics—Part B: Cybernetics, Vol. 30, No. 2, 2000.

23. Chia-Feng Juang and Chun-Ming Lu, "Ant Colony Optimization Incorporated With Fuzzy Q-Learning for Reinforcement Fuzzy Control", IEEE Transactions On Systems, Man, And Cybernetics—Part A: Systems And Humans, Vol. 39(3), 2009.

24. Fei Liu and Guangzhou Zeng, "Study of genetic algorithm with reinforcement learning to solve the TSP", Expert Systems with Applications 36 (2009) 6995–7001.

25. Tanfei Jiang and Zhijng Liu, "An Adaptive Ant Colony Optimization Algorithm Approach to Reinforcement Learning", International Symposium on Computational Intelligence and Design, 2008.

26. Shingo Mabu, Andre Tjahjadi and Kotaro Hirasawa, "Adaptability analysis of genetic network programming with reinforcement learning in dynamically changing environments", Expert Systems with Applications 39 (2012) 12349–12357.

27. Hussein Samma, Chee Peng Lim and Junita Mohamad Saleh, "A new Reinforcement Learning-based Memetic Particle Swarm *Optimizer*", Applied Soft Computing 43 (2016) 276–297.

28. Teodoro C. Bora, Luiz Lebensztajn and Leandro Dos S. Coelho, "Non-Dominated Sorting Genetic Algorithm Based on Reinforcement Learning to Optimization of Broad-Band Reflector Antennas Satellite", IEEE Transactions On Magnetics, Vol. 48(2), 2012.

29. Duan Houli, Li Zhiheng and Zhang Yi, "Multiobjective Reinforcement Learning for Traffic Signal Control Using Vehicular Ad Hoc Network", Hindawi Publishing Corporation EURASIP Journal on Advances in Signal Processing Volume 2010, Article ID 724035, 7 pages.

30. F. Daneshfar and H. Bevrani, "Load–frequency control: A GA-based multi-agent reinforcement learning", Published in IET Generation, Transmission & Distribution.

Fingerprint-Based Support Vector Machine for Indoor Positioning System

A. Christy Jeba Malar and Govardhan Kousalya

Abstract The position of a movable object is required in an indoor environment for providing various business interest services and for emergency services. The techniques implemented on WLAN (802.11b Wireless LANs) endow with more ubiquitous (Feng et al. in IEEE Trans Mob Comput 12(12), 2012, [1]) within the environment and the requirement for additional hardware is not necessary, thereby reducing infrastructure cost and enhancing the value of wireless data network. The received signal strength (RSS) from various reference points (RP) were recorded by a tool and fingerprint radio map is constructed. The signal property of a fingerprint will differ in each point. The location can be found by comparing the current signal strength with already collected radio maps. Almost all indoor environments are equipped with Wi-Fi devices. No additional hardware is required for the setup. In this paper, we introduce SVM classifier (Roos et al. in IEEE Trans Mob Comput 1 (1), 59–69, 2002 [2]) as a methodology with minimum cost and without scarifying accuracy. The obtained results show minimal location error and accurate location of the object.

Keywords Pervasive computing · Received signal strength · Indoor positioning Support vector machine

1 Introduction

Nowadays, Global Navigation Satellite Systems (GNSSs) play a dominant role in locating objects and it is intended for covering globally. It is difficult to decode global positioning system (GPS) signals in indoor. Since it is attenuated by

A. Christy Jeba Malar (✉)
Sri Krishna College of Technology, Coimbatore, India
e-mail: a.christyjebamalar@skct.edu.in

G. Kousalya
Coimbatore Institute of Technology, Coimbatore, India
e-mail: kousir@gmail.com

© Springer Nature Singapore Pte Ltd. 2019
B. K. Panigrahi et al. (eds.), *Smart Innovations in Communication and Computational Sciences*, Advances in Intelligent Systems and Computing 670,
https://doi.org/10.1007/978-981-10-8971-8_26

buildings, it is not suitable for indoor environment; indoor positioning systems have engrossed much attention in recent years and have been extensively studied. Figure 1 shows the region of interest in indoor grid environment. Location awareness in indoor promises a new business market and it encompasses tracking of objects, security services, etc. Because of the multifaceted nature of indoor environments, the development of an indoor positioning technique is always facing some challenges like smaller dimensions, non-line of sight (NLOS), impact of obstacles like walls, furniture, movement of objects, doors, and other factors. Many indoor positioning systems have been implemented based on Bluetooth technique [3], Radio waves, Infrared Ray and Cellular network [4]. Even though these systems are giving accurate results, additional hardware is required for this environmental setup. Location awareness in indoor rely upon different types of parameters such as angle of arrival (AOA), time of arrival (TOA), time difference of arrival (TDOA), and received signal strength (RSS) in wireless local area network (WLAN). Standard IEEE 802.11 termed as Wi-Fi has become de facto; thereby, (Wi-Fi) WLAN RSS techniques are widely used.

The WLAN RSS techniques are economical solutions as many indoor environment areas are usually equipped with a Wi-Fi network. These Wi-Fi networks act as a part of the communication infrastructure, avoiding expensive and time-consuming deployment of wireless network infrastructures. In addition, nowadays, all personal laptops and mobile devices are equipped with WLAN capability and are simple to be communicated from server system. Thereby, the RSS measurements based on WLAN take advantage on flexibility, mobility, and easy deployment. WLAN RSSI techniques are deployed based on fingerprinting or propagation channel modeling. Figure 2 shows that the number of models has been proposed for indoor propagation analysis.

Fig. 1 Region of interest and access points

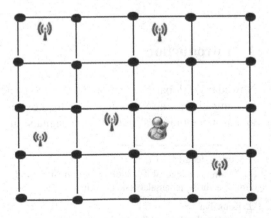

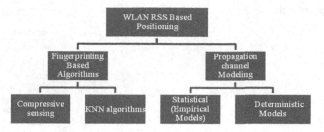

Fig. 2 Positioning algorithms

2 Related Work

RFID-based methods have made a major impact in the field of indoor positioning systems [5]. RF-enabled devices are used in the logistics to communicate with each other to achieve the desired goal that has to define new algorithms in that hardware. Radar operates [6] on signal strength information, recorded at multiple base stations. Empirical measurements and signal strength propagation are combined to determine position of interest and thereby provides location-aware services. Various machine learning and probabilistic-based methods are applied in the location estimation domain [7] without taking physical properties of signal propagation. Kushki et al. [7] proposed a kernel-based measurement for evaluation of similarity among the observed RSS vector and the trained RSS records. The Horus system [8] uses clustering of location to minimize the computational cost of the algorithm. In deterministic methods of calculation [2, 6], the received signal strength at a reference point is measured by average of RSS values, and its uses some direct methods to estimate the position of interest. In the radar system [6], the nearest neighbor algorithm has been used to infer the location of interest. But the probabilistic techniques [9] concern about the signal strength distributions from the reference points while constructing radio map in the offline phase [10]. In the online phase, probabilistic techniques are used to estimate the location of interest. IOT-based indoor positioning is discussed as an evolving approach [11]. Statistical-based measurement like support vector machine is used in locating nodes in wireless sensor networks (WSNs) [12]. Indoor navigation is done by Least Square SVM classifier [13] by combining physical motion with wireless positioning. This system is based on measuring received signal strength at the reference points from various Wi-Fi access points in indoor campus. The received signal strength around the Wi-Fi access point is measured.

Figure 3 shows the signal strength variation around a wireless access point in different radius. Because of the nonlinear behavior of the RSS propagation within indoor campus, the majority of the received signal strength-based indoor localization makes use of RSS radio map. The system comprises of two steps. In the offline phase, RSS radio map is constructed from the received signal strength measured at reference points. Second phase is online phase, and there the location of interest is calculated by receiving RSS values of that position.

Fig. 3 Signal strength
variation

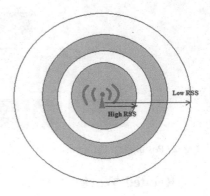

3   Radio Map Construction

Because of nonlinear nature of environmental received signal strength (RSS),
measurement at reference points from Wi-Fi APs is heavily dependent on the
environmental obstacles. In accordance with path loss and multipath fading effect,
the RSS value P_{ss} (dBm) at a distance d from a particular AP can be designed as

$$P_{ss} = PL + k + \gamma \log(d/d_0) \tag{1}$$

In Eq. (1), PL represents the transmitted power (dBm) of the transmitter, and d_0
is the reference distance from the antenna area. In the offline phase, RSS readings
from various reference points are recorded. In the grid view environment, the (x_i, y_i)
positions along with the RSS readings from all APs are recorded. The RSS radio
map is a vector R consists of the RSS values from all access points (APs) with AP
id, reference location $L(x_i, y_i)$. The radio map is constructed by all vectors, for a grid
of locations in the indoor area. For positioning, a MS obtains a sample RSS vector
$P = (p_1, p_2, \ldots, p_N)$. The unknown location is then estimated by the random
samples collected at the location. Unknown Mobile User (MU) location is calculated by the SVM classification algorithm at the online phase. Accuracy error is
minimized by taking repeated reading and by calculating the average. Figure 4
shows the architecture diagram of this system.

4   Support Vector Machine

The support vector machines are based on supervised and strong mathematical
foundations. It is a useful technique for prediction on binary classification. SVM
has become popular in various applications like engineering, medicine, and pattern
recognition with excellent performance. SVMs, very well working for binary
classification, make use of hyperplanes that separate the data into two different

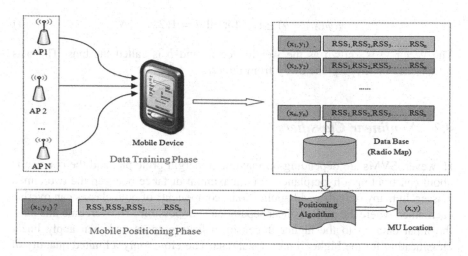

Fig. 4 System architecture

classes. SVMs are able to handle both simple, linear classification tasks, as well as more complex, i.e., nonlinear, classification problems. Training of SVM is based on the parameters, like preprocessing the data, choosing appropriate kernel.

4.1 Linear Classifiers

SVMs are performing binary classification of data. The data for binary classification problem consists of elements labeled with any one of two labels of the respective classes; for expediency, the labels are taken as +1 (for positive class) or −1 (for negative class). The component x_i will belong to the ith vector in a sample set $\{(x_i, y_i)\}_{i=1}^{n}$ where $y_i \in +1, -1$ the labels in association with x_i. The data points for x_1 to x_i are called examples or fingerprint components. The components are assumed as vectors. After setting up the kernels, this hypothesis will be removed, and the components could become any continuous or discrete component. Linear classifier is taken as the dot product between two vector points, and represented as an inner product, formulated as $W^T x = \sum_i w_i x_i$. A linear classification is referred by the linear discriminate function,

$$W^T \phi(x_i) + b \geq +1, \, y_i = +1 \tag{2}$$

$$W^T \phi(x_i) + b \leq -1, \, y_i = -1 \tag{3}$$

which is represented as,

$$y_i(\phi(x_i) + b) \geq 1, \quad \text{for all } i = 1, 2, \ldots, N \qquad (4)$$

The vector w is known as the weight vector, and b is called the bias. The bias b translates the hyperplane away from the origin.

4.2 Nonlinear Classifiers

However, SVMs are depending on maximum margin principle, and the concern is about constructing a hyperplane with a maximum distance between the two aimed classes. In many of real-world applications, components of both classes can overlap each other, where the perfect linear separation is impossible. To overcome this map, the input data on to the higher dimensional feature space and then apply linear classification for the higher dimensional data. Therefore, only a limited quantity of misclassifications might be accepted around the margins. Nonlinear classifier provides better accuracy. Figure 5 depicts the mapping of overlapping data to a new feature space by increasing the dimensions.

4.3 Kernel Functions

Radial Basis Function (RBF) kernel is a function of the Euclidean distance between the points, whereas all other kernels are functions of inner product of the points. In case, two points are close to the origin, but on opposite sides, then the kernels based on inner product assign the low pair value, but Euclidean distance-based kernels assign the high pair value. For some real-life applications, inner product of points is sometimes more preferable than the RBF kernel, where it is concerned for direction rather than the count of data points.

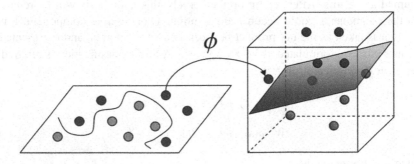

Fig. 5 Mapping of data to a new feature space

Linear kernel function is represented as,

$$K(s_i, s_j) = s_i^T s_j \tag{5}$$

Polynomial kernel function is represented as,

$$K(s_i, s_j) = (s_i^T s_j + 1)^{\gamma} \tag{6}$$

RBF kernel function is taken Eq. (7) the following form:

$$K(s_i, s_j) = e^{-\frac{\|s_i - s_j\|^2}{2\sigma^2}} = \prod_{q=1}^{N} e^{-\frac{(s_{iq} - s_{jq})^2}{2\sigma^2}} \tag{7}$$

where σ is a user-specified parameter.
First-order ANOVA RBF kernel is given by Eq. (8),

$$K(s_i, s_j) = \sum_{q=1}^{N} e^{-\frac{(s_{iq} \cdot s_{jq})^2}{2\sigma^2}} \tag{8}$$

The ANOVA kernel represented above is useful for developing new kernel functions using the ANOVA trick [14].

A nonlinear transformation from input vector R^n to feature vector F represented by average of n signal strength readings $\{R_i\}_{i=1}^{n}$ recorded at the reference point at a particular time at different angle is stored in a column-wise input vector and is mentioned by R. For n sets of input vectors, R_i is representing the ith vector and R_{iq} is representing its qth record taken as the qth feature. And, the target location of interest is simply the coordinate $(x_i; y_i)$. The resultant classifier is working on the grid space and is given by

$$f(x_i, y_i) = \text{sign}\left(\sum_{i=1}^{N} \alpha_i s_i R(x, x_i) + b\right) \tag{9}$$

where R is the kernel matrix. Equation (9) shows the estimated location.

5 Experimental Setup

Normally, all public campus like, universities, hospitals, shopping malls are already equipped with Wi-Fi access points. The proposed positioning system is designed for indoor environment where the buildings are already equipped with Wi-Fi access points (APs) [15]. The environment is taken as a grid-based environment.

In addition to the Wi-Fi access points fixed already, a number of reference points are considered. At a regular time interval, consistent RSS readings have reported the server to construct the radio maps. Together with the 8 deployed access points, 10 reference points are considered across the region. The mobile device records RSS data from all detectable APs at specific RPs at same orientation collected by the device at a time interval Δt and transmits these data to the server, which has constraint on the device's network card and other hardware limitations. At the online phase, RSS readings from unknown location can be denoted as $R(t) = [R_1(t), R_2(t), ..., R_n(t)]$, $t = 0, 1, 2, ...$, where $R_n(t)$ refers to the RSS value from RP n at time t. Then, the proposed positioning system uses $R(t)$ to compute the position estimate, by applying the support vector machine algorithm.

6 Results and Discussion

In the experimental setup, random RSS data samples are collected from all surveyors. It is taken as training data and the RSS data collected during online phase are used for testing. This experiment is showing how the support vector machines are handling the large amount of variations in the RSS data, recorded at regular time intervals.

The error rate from the figure supports that the algorithm is suitable for tracking humans and objects in the region of interest (ROI) with high accuracy given in Fig. 6.

Figure 7 illustrates the comparison of the proposed support vector algorithm-based calculation with the existing algorithm. Figure 8 illustrates the number of iterations by proposed model and existing model. The complexity analysis comparison proves that the proposed algorithm outperforms the existing methodology.

Fig. 6 Location sensing of proposed algorithm

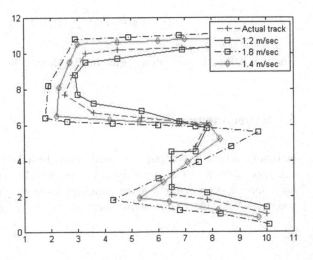

Fig. 7 Error value

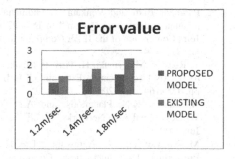

Fig. 8 Number of iterations

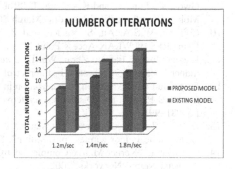

7   Conclusions

The proposed algorithm proves to be an efficient key for the indoor location tracking applications. The proposed algorithm performs better on the existing algorithm [12] by error and count of iterations. Results deduced prove that the algorithm works efficiently for reduced speed. However, the error rate in increased speed is also less when compared to the existing protocol. The future work covers, increase in ROI size and tracking of location with hindrances.

References

1. Chen Feng, Wain Sy Anthea Au, Shahrokh Valaee, Zhenhui Tan.: "Received-Signal-Strength-Based Indoor Positioning Using Compressive Sensing", IEEE Transactions on mobile computing. 2012, vol. 11, no. 12.
2. Roos, T., Myllymaki, P., and Tirri, H. "A Statistical Modelling Approach to Location Estimation", IEEE Transactions on Mobile Computing 1, 1 (January–March 2002), 59–69
3. R. Bruno and F. Delmastro, "Design and analysis of a bluetooth-based indoor localization system," in Proceedings of IEEE International Conference on Personal Wireless Communications, vol. 27, pp. 711–725, Venive, Italy, September 2003.
4. H. Liu, H. Darabi, P. Banerjee, and J. Liu, "Survey of wireless indoor positioning techniques and systems," IEEE Transactions on Systems, Man, and Cybernetics, Part C, vol. 37, pp. 1067–1080, November 2007.

5. Paramvir Bahl and Venkata N. Padmanabhan, "RADAR: an in-building RF-based user location and tracking system," in INFOCOM 2000 Proceedings of the Nineteenth Annual Joint Conference of the IEEE Computer and Communications Societies, March 2000, vol. 2, pp. 775–784

6. T. Roos, P. Myllymaki, H. Tirri, P. Misikangas, and J. Sievanen, "A Probabilistic Approach to WLAN User Location Estimation," Int'l J. Wireless Information Networks, vol. 9, no. 3, pp. 155–164, July 2002.

7. A. Kushki, K.N. Plataniotis, and A.N. Venetsanopoulos, "Kernel-Based Positioning in Wireless Local Area Networks," IEEE Trans. Mobile Computing, vol. 6, no. 6, pp. 689–705, June 2007

8. M. Youssef and A. Agrawala, "The Horus WLAN Location Determination System," Proceedings Third International Conference of Mobile Systems, Applications, and Services, pp. 205–218, 2005.

9. Gwon, Y., Jain, R., and Kawahara, T. "Robust Indoor Location Estimation of Stationary and Mobile Users", In IEEE Infocom (March 2004).

10. C. Feng, W.S.A. Au, S. Valaee, and Z. Tan.: "Compressive Sensing Based Positioning Using RSS of WLAN Access Points", pp. 1–9, Proc. IEEE INFOCOM, 2010.

11. Haojun Huang. Jianguo Zhou. Wei Li. Juanbao Zhang. Xu Zhang. Guolin Hou: "Wearable indoor Localisation approach in Internet of Things", IET Networks, (Jul. 2016), pp. 1–5.

12. D. A. Tran and T. Nguyen, "Localization in wireless sensor networks based on support vector machines Parallel and Distributed Systems", IEEE Transactions on, vol. 19, no. 7, (2008), pp. 981–994.

13. L. Pei, J. Liu, R. Guinness, Y. Chen, et al., "Using LS-SVM based Motion recognition for smart phone indoor wireless positioning", Sensors, vol. 12, no. 5, (2012), pp. 6155–6175.

14. T. Graepel, "Kernel Matrix Completion by Semi-Definite Programming" Proc. Int'l Conf. Artificial Neural Networks, 2002.

15. G.Kousalya, P Narayanasamy, Jong Hyuk Park, Tai-hoon Kim.: "Predictive handoff mechanism with real-time mobility tracking in a campus wide wireless network considering", ITS Computer Communications, 2008, vol. 31, no. 12, pp. 2781–2789

GPU Approach for Handwritten Devanagari Document Binarization

Sandhya Arora, Sunita Jahirabadkar and Anagha Kulkarni

Abstract The optical character recognition (OCR) is the process of converting scanned images of machine printed or handwritten text, numerals, letters, and symbols into a computer processable format such as ASCII. For creating OCR's paperless application, a system of high speed and of better accuracy is required. Parallelization of algorithm using graphics processing unit (GPU) along with CPU can be used to speed up the processing. In GPU computing, the compute-intensive operations are performed on GPU while serial code still runs on CPU. Binarization is one of the most fundamental preprocessing techniques in the area of image processing and pattern recognition. This paper proposes an adaptive threshold binarization algorithm for GPU. The aim of this research work is to speed up binarization process that eventually will help to accelerate the processing of document recognition. The algorithm implementation is done using Compute Unified Device Architecture (CUDA) software interface by NVIDIA. An average speedup of $2\times$ is achieved on GPU GeForce 210 having 16 CUDA cores and 1.2 compute level, over the serial implementation.

Keywords CUDA · GPU · OCR · Binarization · Parallelization
Pattern recognition

1 Introduction

Devanagari is the script of Hindi, which is the official languages of India and which is used by majority of people [1, 2] and other languages like Marathi, Bengali, Gujarati, Konkani, Sanskrit, and other north Indian languages. Devanagari script

S. Arora (✉) · S. Jahirabadkar
Department of Computer Engineering, Cummins College of Engineering for Women,
Pune, India
e-mail: sandhya.arora@cumminscollege.in

A. Kulkarni
IT Department, Cummins College of Engineering for Women, Pune, India

© Springer Nature Singapore Pte Ltd. 2019
B. K. Panigrahi et al. (eds.), *Smart Innovations in Communication and Computational Sciences*, Advances in Intelligent Systems and Computing 670,
https://doi.org/10.1007/978-981-10-8971-8_27

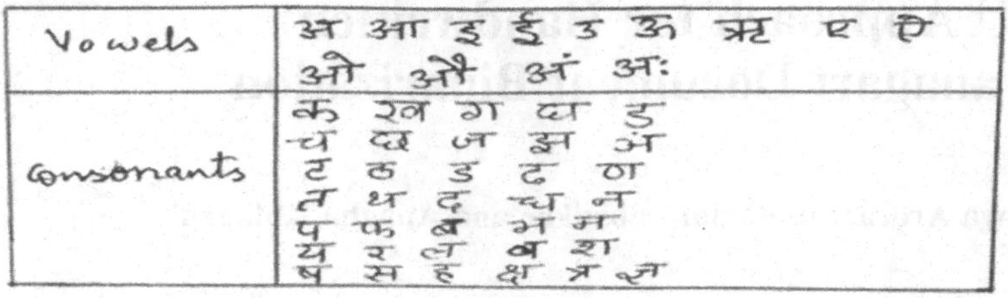

| Vowels | अ आ इ ई उ ऊ ऋ ए ऐ ओ औ अं अः |
| Consonants | क ख ग घ ङ च छ ज झ ञ ट ठ ड ढ ण त थ द ध न प फ ब भ म य र ल व श ष स ह क्ष त्र ज्ञ |

Fig. 1 Vowels and consonants of Devnagari script

consists of thirteen vowel symbols called matras and thirty-six consonants called vyanjans as shown in Fig. 1. Devanagari character recognition is complicated because of the variations and similarity in shape of characters, in writing styles due to various factors, its complex character grapheme structure, etc. [3].

Optical character recognition (OCR) [4] is the process that recognizes text (handwritten or printed) on document pages and converts them to a machine-understandable text format. Binarization plays a very important role in the OCR tool chain and in all image processing, pattern recognition-based applications. Binarization is the first fundamental step of image preprocessing. Binarization process also called as thresholding converts all gray-levels of gray image into 0 or 1 level, i.e., in two-tone image [5]. Accuracy of character recognition impacts the OCR quality [5]. In real-time applications such as in banks, both accuracy and high speed of OCR are required [6]. For example, in the banking cheque reading [6], mail sorting [7], very fast OCR systems are required with the acceptable error rate close to zero. Hence, fast and accurate algorithms are needed at all stages of optical character recognition (OCR). Graphics processing unit (GPU), a general purpose computation hardware, can be used to speed up the processing. GPU is used for parallel programming, and its low cost makes it productive. A GPU has massively parallel architecture of thousands of cores as compared to CPU, which contains only few cores for processing. GPU cores are designed for handling large blocks of data processing in parallel. In GPU computing, the compute-intensive operations are performed on GPU device while serial code still runs on CPU. From user's perspective, applications simply run significantly faster. Most of the OCR research work is implemented using single processor (CPU) only [1–3, 8–14].Graphics processing unit (GPU) computation is yet to be fully implemented in OCR. Some studies of GPU implementation are in [11, 15–21, 22–24]. Some research work [21, 25–28] gives the details of general purpose computation's parallel implementations on GPU to reduce the computation of numerical problems. Some parallel implementations of handwritten character recognition on GPUs were proposed in [28, 29–32]. In [33], the matrix multiplication using neural network was implemented on GPU to enhance the time performance. Neural network based color images text localization was proposed by Jung [34]. A profiling-based segmentation algorithm's parallel implementation on GPU for Devanagari character recognition was proposed by Singh et al. [35].

This paper presents a very simple adaptive binarization GPU-based technique which can be applied to handwritten Devanagari characters/documents. GPU architecture is discussed in Sect. 2. Adaptive binarization algorithms for CPU and GPU are detailed in Sects. 3 and 4, respectively. Results of the sequential (on CPU) and parallel (on GPU) implemented algorithms on handwritten Devanagari document images are shown in Sect. 5.

2 NVIDIA CUDA/GPU Architecture

Graphics processing unit (GPU) is a heterogeneous chip multiprocessor that can perform rapid mathematical calculations, primarily for the purpose of efficiently manipulating computer graphics by using its highly parallel structure. CUDA is a parallel computing platform and programming model created by NVIDIA and implemented by graphics processing units (GPUs). CUDA stands for Compute Unified Device Architecture.

CPU has one control unit, one cache, and few arithmetic logic units, while GPU contains arrays of ALUs, and for each array, there is one control unit and cache. In both CPU and GPU, there is one dynamic random access memory (DRAM). GPU hardware mainly consists of three blocks: memory, streaming multiprocessors (SM), and streaming processors (SP). The memory can be global memory, shared memory, read-only constant and texture memory as shown in Fig. 2. The overall throughput of GPU is largely determined by number of SPs present, the bandwidth of global memory, and how well the programmer can make use of his programming skills to fully make use of available highly parallel structure. GPU has array of SMs, each of which consists of multiple SPs. Each SP can be considered as a core. Each SM has access to register file which is just a chunk of memory. Registers run at very fast speed; i.e., their speed is almost equal to the speed of SP. The register file stores the registers in use for the threads running on SPs. Shared memory block is accessible to the individual SM [33]. This memory block acts as programmer controller cache. Each SM has accessibility to constant memory, texture memory, and global memory through separate bus. The texture memory is used for data where there is interpolation. Constant and texture memory are used for read-only data. The multiprocessors communicate through the global or device memory.

At the software level, CUDA programming model organizes parallel computations using groups of threads, blocks, and grids. CUDA is an extension of C language that allows GPU code to be written in regular C language. The single code is written to execute on host processor (CPU) and on device processor (GPU). The host processor can spawn the multithreaded tasks to GPU device. Host processor identifies the GPU code (multithreaded task) and sends that code along with the data to execute on GPU, and rest of the code gets executed on CPU only. CUDA kernel function is the basic building block of GPU computing. The prefix—global —keyword to the function differentiates that it is a GPU code not CPU code. The kernel function is invoked as kernel function ≪num-of-blocks, num-of-threads≫ (parameters of the function). This kernel function is called as a task. When launching kernel function, developer specifies how many copies of it to run in by

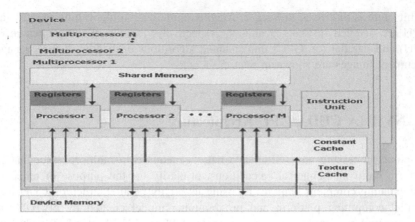

Fig. 2 GPU architecture [36]

num-of-blocks and num-of-threads. These tasks run in parallel in blocks and threads. A kernel can be executed by a one-dimensional or two-dimensional thread blocks as shown in Fig. 3. A thread block can be arranged in (x, y, z), (x, y), or (x)-dimensional group of threads using dim3 attribute in CUDA programming. All threads in same block can share data using shared memory, but different block threads cannot use that shared memory. Grid contains blocks, arranged in rows and columns in one, two, or three dimensions as shown in Fig. 3. And within each block, there are threads to form a wrap. For data transfer from CPU to GPU, we need to allocate memory space on the host, transfer the data to the device using built-in API, retrieve the data, transfer the data back to the host, and finally free the memory. Built-in APIs are available to perform all these tasks on host.

Fig. 3 CUDA programming model [36]

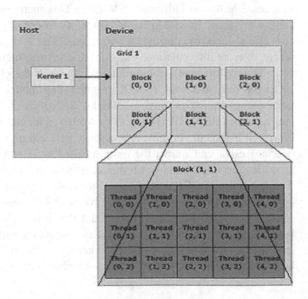

3   Binarization

Binarization of images is the first most fundamental preprocessing step of scanned images. In the process of binarization, a threshold value is computed, which is used to differentiate foreground (object pixels) and background pixels of image. Color image and gray level image processing requires more computing power of the system as compared to binary image processing. Ideally, an input character image should have two tones, i.e., black and white pixels (represented by 1 and 0). Image binarization converts a gray-level image containing up to 256 gray levels into a two-level containing 0 or 1 image.

A thresholded image $g(m, n)$ is defined as:

$$g(m, n) = \begin{cases} 1, & f(m, n) > T \\ 0, & f(m, n) \leq T \end{cases}$$

where '1' represents the object/foreground, '0' represents background, and $f(m, n)$ is the gray level at pixel position (m, n) in the original image. The goal of binarization is to identify a threshold value dynamically that would help to distinguish between image pixels representing the character pattern and those representing background. The threshold value identified would be completely dependent on the nature of the documents, the contrast level, and the difference of contrast between the foreground and background pixels, and their concentration in the document. The methodology applied to gray-level image (range of pixel values 0–255) is as follows.

3.1   Adaptive Threshold() Algorithm

1. An initial threshold (T) is chosen randomly. This can be done using any other desired method desired. Let it be 128 (midway between 0 and 255).
2. The image is segmented into object and background pixels, creating two sets $G1$ and $G2$:

   $G1$ = Contains all background pixels $\{f(m, n) : f(m, n) > T\}$
   $G2$ = Contains all object pixels $\{f(m, n) : f(m, n) \leq T\}$

   (where $f(m, n)$ represents the gray-level value of the pixel at mth column, nth row position)
3. The average gray-level value of each set is computed.

   $m1$ = average gray-level value of the pixels in $G1$
   $m2$ = average gray-level value of the pixels in $G2$

4. A new threshold is computed as the average of $m1$ and $m2$

$$T = (m1 + m2)/2$$

5. If the difference between this new threshold value and the old one is less than 2% the gray-level range (0–255), then the new threshold is taken as final else repeat the steps 2, 3, and 4 with this new threshold value.

4 GPU Algorithm for Binarization

Above algorithm is parallelized using CUDA platform. All the experiments are carried out using the GeForce 210 GPU card, with 1 GB DDR3, 1.2 compute level, 16 CUDA cores, RAM 1 GB, 520 MHz core clock, 1066 MHz memory clock. In CUDA, both host and device maintain their own memory. Memory is allocated at both host (CPU) and device (GPU) using malloc and cudaMalloc functions, respectively. Each CUDA thread has a unique id, and each thread is assigned some computation, and all thread managed in block or grids computes in parallel. The following pseudo-code outlines the parallel structure of above method (Fig. 4).

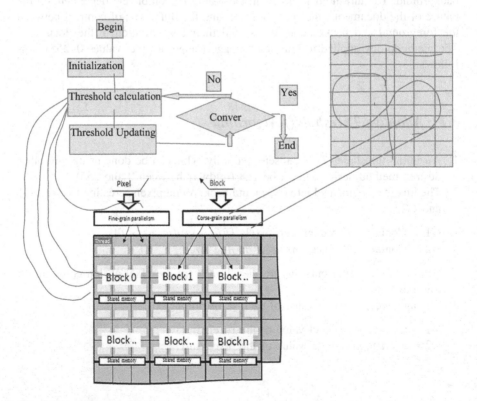

Fig. 4 Binarization on GPU

Kernel_Threshold(Threshold) Algorithm:
for each thread in block
   if (current-Pixel-Value > Threshold)
     current-Pixel-Value = $\{f(m, n) : f(m, n) > T\}$ (background pixels)
   else current-Pixel-Value = $\{f(m, n) : f(m, n) < T\}$ (object pixels)
end for

Global_Threshold_Calculation():
   $m1$ = average gray level value of the background pixels
   $m2$ = average gray level value of the object pixels
   Threshold = $(m1 + m2)/2$

Kernel_Thresholded_Image (Threshold):
For every thread inside the block
        If (current-Pixel-Value > Threshold)
              current-Pixel-Value = WHITE

        Else current-Pixel-Value = BLACK

End for

Main ()

{

    If ((newthreshold-oldthreshold) > 2% of object pixels)
    {

      Define block and grid
      Kernal_Threshold (Threshold) // Computed on GPU
    Global_Threshold_Calculation() // Computed on CPU

    }

    Kernel_Thresholded_Image (Threshold) // On GPU

}

5   Results

The proposed binarization method is evaluated for different block sizes on GPU as well as on CPU. The comparison of the processing time between GPU and CPU shows the GPU acceleration. It is observed that as we increase the block size, processing time decreases in CPU implementation and increases in GPU implementation. And the speedup of GPU-implemented binarization algorithm decreases as we increase the block size. The execution on the GPU is faster when the number of pixels in the image blocks is close to the physical number of threads in the GPU blocks. In this case, with higher block size, each thread processes one pixel directly. Otherwise, that pixel's thread follows a queue and does not start execution until a thread gets free (Fig. 5 and Table 1).

Fig. 5 Binarized images (input images from WWW)

Table 1 Execution time comparison of CPU and GPU implementation

Blocks	Device	Time (ms)	Speedup
8 × 8	CPU	2416	6 times
	GPU	398	
16 × 16	CPU	2246	4.5 times
	GPU	489	
32 × 32	CPU	2112	2.4 times
	GPU	880	

6   Conclusion

The binarization is a very important component in the OCR. For some OCR applications, high speed and accurate system are required. In this context, the parallel implementation of adaptive threshold algorithm for binarization on graphics processing unit (GPU) has been proposed. According to experiment, it offers a high speedup. This proposed algorithm has been tested on handwritten Devanagari document images. Results show effective speedup on different block sizes compared to CPU implemented algorithm. Speedup of binarization algorithm for OCR tool chain will eventually speed up the document recognition process.

References

1. I.K. Sethi and B. Chatterjee, "Machine Recognition of constrained Handprinted Devnagari", Pattern Recognition, Vol. 9, pp. 69–75, 1977.
2. Pal, U. and Chaudhuri, B.B., "Indian script character recognition: a survey". Pattern Recognition 37, 1887–1899, 2004.
3. S. Arora, "Studies on some Soft Computing Techniques: A Case Study for Constrained Handwritten Devnagari Characters and Numerals", International Journal of Scientific & Engineering Research, Volume 4, Issue 7, 2013.
4. Xiu, P., Baird, H.S.: Whole-book recognition. IEEE Trans. Pattern Anal. Mach. Intell. 34(12), 2467–2480, 2012.
5. Kae, A., Huang, G.B., Doersch, C., Learned-Miller, E.G., "Improving state-of-the-art OCR through high-precision document-specific modeling" In: CVPR, pp. 1935–1942,2010.
6. Eikvil, L. "OCR-Optical Character Recognition", 1993, http://www.nr.no/*eikvil/OCR.pdf.
7. Gaceb, D., Eglin, V., Lebourgeois, F. "A new mixed binarization method used in real time application of automatic business document and postal mail sorting" Int. Arab J. Inf. Technol. 10(2), 179–188, 2013.
8. M. Hanmandlu and O.V. Ramana Murthy, "Fuzzy Model Based Recognition of Handwritten Hindi Numerals", In Proc. Intn Conference on Cognition and Recognition, 490–496, 05.
9. M. Hanmandlu, O.V. Ramana Murthy, Vamsi Krishna Madasu, "Fuzzy Model based recognition of Handwritten Hindi characters", IEEE Computer society, Digital Image Computing Techniques and Applications, 2007.
10. Reena Bajaj, Lipika Dey, and S. Chaudhury, "Devnagari numeral recognition by combining decision of multiple connectionist classifiers", Sadhana, Vol. 27, part. 1, pp.-59–72, 02.
11. S. Kumar, C. Singh, "A Study of Zernike Moments and its use in Devnagari Handwritten Character Recognition", Intl. Conf. On Cognition and Recognition, 514–520, 05.
12. U. Bhattacharya, B. B. Chaudhuri, R. Ghosh and M. Ghosh, "On Recognition of Handwritten Devnagari Numerals", In Proc. of the Workshop on Learning Algorithms for Pattern Recognition, Sydney, pp. 1–7, 2005.
13. N. Sharma, U. Pal, F. Kimura, S. pal, "Recognition of Off Line Handwritten Devnagari Characters using Quadratic Classifier", ICCGIP 2006, LNCS 4338, pp 805–816, 2006.
14. S. Arora, D. Bhattacharya, M. Nasipuri, " Recognition of Non-Compound Handwritten Devnagari Characters using a Combination of MLP and Minimum Edit Distance", International Journal of Computer Science and Security (IJCSS), Volume (4): Issue (1), 2010.
15. Brij Mohan Singh, Rahul Sharma, Ankush Mittal, Debashish Ghosh, "Parallel Implementation of Otsu's Binarization Approach on GPU", International Journal of Computer Applications (0975 – 8887) Volume 32–No. 2, October 2011.

16. Sauvola, J. and Pietikainen, M., "Adaptive document image binarization" Pattern Recognition, Vol. 33, 225–236, 2000.
17. Kim, I.K., Jung, D.W. and Park, R.H., "Document image binarization based on topographic analysis using a water flow model", Pattern Recognition, Vol. 35, 265–277, 2002.
18. Gatos, B., Pratikakis, I. and Perantonis, S. J. "Adaptive degraded document image binarization", Pattern Recognition, Vol. 39, 317–327, 2006.
19. Chang, Y.F., Pai, Y.T. and Ruan, S.J. "An efficient thresholding algorithm for degraded document images based on intelligent block detection", IEEE Intern Conf on Systems, Man, and Cybernetics, 667–672, 08.
20. Valizadeh, M., Komeili, M., Armanfard, N. and Kabir, E., "Degraded document image binarization based on combination of two complementary algorithms", International Conf. of Advances in Computational Tools for Engineering Applications, IEEE, 595–599, 2009.
21. Owens, J. D., Luebke, D., Govindaraju, N., Harris, M., Kruger, J., Lefohn, A. E. and Purcell, T. J., "A survey of general-purpose computation on graphics hardware", In proceeding of Eurographics, State of the Art Reports, 21–51, 2005.
22. M. Soua, R. Kachouri, M. Akil, " GPU parallel implementation of the new hybrid binarization based on Kmeans method HBK", Journal of Real-Time Image Processing, 14.
23. Yi, Y., Lai, C., Petrov, S.: Efficient parallel CKY parsing using GPUs. J. Log. Comput. 24(2), 375–393, 2014.
24. George Cherian Panappally, M.S. Dhanesh, "Design of graphics processing unit for image processing", 1st Intern Conf on Computational Systems & Communications, 2014.
25. Larsen, E. S., McAllister, D.," Fast Matrix Multiplies using Graphics Hardware", International Conference for High Performance Computing and Communications, 159–168, 2001.
26. Trendall C. and Stewart, A. J., "General calculations using graphics hardware with applications to interactive caustics Rendering Techniques", 11th Eurographics Workshop on Rendering, 287–298, 2000.
27. Li, Wei, Xiaoming, A. & Kaufman, "Implementing lattice boltzmann computation on graphics h/w," Intern. Conf. for High Performance Computing and Communins, 2001.
28. Kruger, J. and Westermann, R., "Linear operators for GPU implementation of numerical algorithms", In Proceedings of SIGGRAPH, San Diego, 908–916, 2003.
29. Mizukami, Y., Koga, K. and Torioka, T., "A handwritten character recognition system using hierarchical extraction of displacement", IEICE, J77-D-II(12): 2390–2393, 1994.
30. Steinkraus, D., Buck, I., and Simard, P. Y., "GPUs for machine learning algorithms", International Conference of Document Analysis and Recognition, 1115–1120, 2005.
31. Ilie, "A. Optical character recognition on graphics hardware", www.cs.unc.edu/~adyilie/IP/Final.pdf.
32. Lazzara, G., Graud, T., "Efficient multiscale Sauvola's binarization", Int. Journal of Document Analysis and Recognition, 2(14), 105–123,2014.
33. Oh, K.S. and Jung, K., "GPU implementation of neural networks", Pattern Recognition, Elsevier, 1311–1314, 2004.
34. Jung, K. "Neural Network-based text localization in color image",. Pattern Recognition Letters, Vol. 22, (4), 1503–1515, 2001.
35. Singh, B.M., Mittal A. and Ghosh, D., "Parallel implementation of Devanagari text line and word segmentation approach on GPU", IJCA 24(9): 7–14, 2011.
36. NVIDIA Corporation: NVIDIA CUDA programming guide. Jan 2007, http://developer.download.nvidia.com/compute/cuda/2_0/docs/NVIDIA_CUDA_Programming_Guide_2.0.pdf.

On an Improved *K*-Best Algorithm with High Performance and Low Complexity for MIMO Systems

Jia-lin Yang

Abstract Multiple-input multiple-output (MIMO) techniques are significantly advanced in contemporary high-rate wireless communications. The computational complexity and bit-error-rate (BER) performance are main issues in MIMO systems. An algorithm is proposed based on a traditional *K*-Best algorithm, coupling with the fast *QR* decomposition algorithm with an optimal detection order for the channel decomposition, the Schnorr–Euchner strategy for solving the zero floating-point and sorting all the branches' partial Euclidean distance, the sphere decoding algorithm for reducing the search space. The improved *K*-Best algorithm proposed in this paper has the following characteristics: (i) The searching space for the closest point to a region is smaller compared to that of the traditional *K*-Best algorithm in each dimension; (ii) it can eliminate the survival candidates at early stages, and (iii) it obtains better performance in the BER and the computational complexity.

Keywords MIMO · *K*-Best algorithm · Fast *QR* decomposition
BER performance · Computational complexity

1 Introduction

The bit-error-rate (BER) performance of multiple-input multiple-output (MIMO) systems will be improved by using spatial diversity or the channel capacity via spatial multiplexing. However, the hardware complexity of the data detector will be increased by the spatial multiplexing technique in an MIMO system. Thus, the maximum-likelihood (ML) decoder as the basic optimal method is infeasible, for a straightforward exhaustive search implementation suffering from prohibitive complexity, particularly in high-dimensional MIMO communication systems [1]. So as to reduce the computational complexity of the ML algorithm, a *K*-Best algorithm shortening the search space is presented [2].

J. Yang (✉)
20th Research Institute of Electronic Technology Corporation, Xi'an, China
e-mail: 17791290017@163.com

© Springer Nature Singapore Pte Ltd. 2019
B. K. Panigrahi et al. (eds.), *Smart Innovations in Communication and Computational Sciences*, Advances in Intelligent Systems and Computing 670,
https://doi.org/10.1007/978-981-10-8971-8_28

A large number of modified K-Best algorithms have been presented in the literature, aiming to lower the computational complexity or to obtain a better BER performance. The computational complexity is always a key issue in a real MIMO system. Mondal et al. [3] take the advantage of the quadratic amplitude modulation constellation structure to accelerate the detection procedure and achieve 50% lower computational complexity. The sorting-free K-Best algorithm [4] combining with the distributed K-Best algorithm obtains a good performance in computational complexity. Kim and Park [5] propose a modified K-Best algorithm that reduces the complexity via tree expansion and sorting, and they also design a hardware implementation scheme of the efficient pipeline scheduling to reduce the overall processing latency. Although the aforementioned studies have reduced the computational complexity, their BER performance has not been improved remarkably, i.e., those algorithms can only keep or even lower the BER performance compared with the traditional K-Best algorithm. In order to improve BER performance of the MIMO systems, some modified K-Best algorithms are accordingly reported, but they are at the expense of their complexity. Chen et al. [6] present an MIMO detector design solution by using implementation-oriented breadth-first tree search instead of conventional breadth-first tree search, which achieves near optimum detection performance with the same computational complexity as the traditional K-Best algorithm does. El-Mashed et al. [7] propose an ordered successive interference cancellation (OSIC), where the MIMO detector is divided into the small blocks with the small dimension. The result shows the BER of the OSIC based on K-Best algorithm is only 5×10^{-3} at the condition that the values of SNR and K are 30 and 5, respectively, at 16 QAM. However, there are few algorithms that not only reduce computational complexity but also obtain better BER performance, as far as the authors know.

The objective in the present study is to improve the existing K-Best algorithms such that high BER performance and low complexity for MIMO systems are achieved. An improved algorithm is presented, coupling the fast QR decomposition, the Schnorr-Euchner strategy (SE), and the sphere decoding algorithm (SDA). The fast QR decomposition algorithm applies to the channel decomposition. The SE solves the zero floating-point and sorts all the branches' partial Euclidean distance (PED), and the SDA aggressively discards non-promising treed nodes.

2   System Model

Considering a symbol synchronized and coded MIMO system with transmitting antennas and receiving antennas, the model [7] is

$$y = Hs + n \qquad (1)$$

where $y = [y_1, y_2, \ldots, y_i, \ldots, y_{N_r}]^\mathrm{T}$ is the N_r-dimensional received symbol vector, $s = [s_1, s_2, \ldots, s_i, \ldots, s_{N_t}]^\mathrm{T}$ denotes the N_t-dimensional pre-inverse fast Fourier transform (IFFT) symbol vector, each vector from a common square QAM constellation with equal power across all symbols and the constellation points are M_c, H is the $N_r \times N_t$ frequency-domain channel response matrix that generally contains correlated elements, and $n = [n_1, n_2, \ldots n_r]^\mathrm{T}$ represents the zero-mean complex circular-symmetric Gaussian noise component observed on the N_r-dimensional independent identically distributed receive antennas, each vector with power σ^2.

3 Proposed K-Best Algorithm

We propose a modified K-Best algorithm for an MIMO system. It exploits the fast QR decomposition algorithm with an optimal defection order, the SE strategy, and the SDA algorithm. The purpose of this study is to obtain the better BER performance and lower computational complexity.

First, we briefly outline the principle of the basic fast QR decomposition algorithm [8]. The rank value is directly obtained according to the results of QR decomposition in the fast QR decomposition algorithm, while the diagonal elements of R are calculated just in the opposite order in a standard QR algorithm, which makes the optimal order of detection difficult to be found. The channel matrix H with the fast QR decomposition algorithm can be expressed as:

$$H = QR = (q_1, q_2, \ldots, q_{N_t}) \begin{pmatrix} r_{1,1} & r_{1,2} & \cdots & r_{1,N_t} \\ & r_{2,2} & \ddots & r_{2,N_t} \\ & & \ddots & \vdots \\ & & & r_{N_t,N_t} \end{pmatrix} \tag{2}$$

where R is an $N_t \times N_t$ upper triangular matrix and Q is the orthogonal matrix.

The transmitting power at the each ith layer is assumed as 1 as in Ref. [8]. Signal-to-interference-plus-noise ratio at ith layer (SINR$_i$) can be expressed to obtain the detection order and the orthonormal column matrix (P):

$$\mathrm{SINR}_i = \frac{1}{\sigma^2 \left((QR)^H (QR) \right)_{i,i}^{-1}} = \frac{1}{\sigma^2 \left[R^{-1} (R^{-1})^H \right]_{i,i}}$$

$$= \frac{1}{\sigma^2 \left(R^{-1} (R^{-1})^H \right)_{i,i}} = \frac{1}{\sigma^2 \left\| (R^{-1})_{i,i} \right\|_2} \tag{3}$$

where $(QR)^H$ is the conjugate transposed matrix of QR, $\left[R^{-1} (R^{-1})^H \right]_{i,i}$ is the ith row and column of $R^{-1} (R^{-1})^H$ matrix, $\left\| (R^{-1})_{i,i} \right\|_2$ is the second vector norm, and

σ is the additive Gaussian white noise. It can be noted that the rank values by SINR_i are equal to the rank based on the R^{-1} module value. The signal can be more easily detected with the smaller R^{-1} module value. R^{-1} is also an upper triangular matrix since R is an upper triangular matrix. The R^{-1} can be expressed as:

$$R^{-1} = \begin{pmatrix} r''_{1,1} & r''_{1,2} & \cdots & r''_{1,N_t} \\ & r''_{2,2} & \cdots & r''_{2,N_t} \\ & & \ddots & \vdots \\ & & & r''_{N_t,N_t} \end{pmatrix} \tag{4}$$

where the values of R^{-1} can be obtained as:

$$\begin{cases} r''_{ii} = 1/r_{ii} \\ r''_{i,i+1} = -r_{i,i+1} r''_{i+1} r''_{ii} \\ \quad\vdots \\ r''_{i,i+k} = -r''_{ii} \sum_{j=1}^{k} r_{i,i+1} r''_{i+j,i+k} \end{cases} \tag{5}$$

Finally, the optimal detection order and the P can be obtained by $\text{SINR}_{Nt} < \text{SINR}_{Nt-1} < \ldots < \text{SINR}_1$. RP can be obtained according to the P. The calculation of the QR decomposition once again is described as

$$RP = Q_2 \begin{pmatrix} r'_{1,1} & r'_{1,2} & \cdots & r'_{1,N_t} \\ & r'_{2,2} & \cdots & r'_{2,N_t} \\ & & \ddots & \vdots \\ & & & r'_{N_t,N_t} \end{pmatrix} = Q_2 R' \tag{6}$$

where Q_2 is the orthogonal matrix and R' is the upper triangular matrix, respectively. The final optimal detection order can be expressed as

$$H' = HP = QRP = QQ_2 R' = Q'R' \tag{7}$$

where H' is the optimal defection order channel matrix.

Then, the SE strategy that can search the closest lattice point in the nearest hyperplane and enhance the tree search is adopted to solve the zero floating-point in partial Euclidean distance. According to the work in Ref. [9], the nearest reference center is calculated as

$$\hat{s}_i = \left(y_i - \sum_{j=i+1}^{2N_r} r'_{ij} s_j \right) / r'_{ii} \tag{8}$$

where $\hat{s}_i$ is the expansion center of the parent node. Then, the nearest constellation point related to $\hat{s}_i$ is confirmed to obtain the expanding child nodes at every layer, which just need to calculate all the PEDs one time and K times comparison and selection each other than K times PED calculations as the conventional *K*-Best algorithm does. The SE algorithm solves the zero floating-point and sorts all the branches' partial Euclidean distance (PED); then the SDA aggressively discards non-promising tree nodes earlier.

The SDA is utilized further to simplify the searching branches sharply, as given in Eqs. (9) and (10)

$$d(\hat{s}_i) = \sum_{l=i}^{2N_r} e^l(\hat{s}_l) = \sum_{i=2N_r}^{1} \left| y_i - \sum_{j=i}^{2N_r} r'_{ij}\hat{s}_j \right|^2 \leq c_o \qquad (9)$$

$$e^i(\hat{s}_i) = \left| y_i - \sum_{j=1}^{2N_r} r'_{ij}\hat{s}_j \right| \qquad (10)$$

where c_o is the squared radius of the sphere, $e^i(\hat{s}_i)$ is the branch cost function associated with nodes at the ith layer, and $d(\hat{s}_i)$ is the partial sum of $\hat{s}_i$. The decoding process can be regarded as descending down in a tree where each node has $\hat{s}_i$ branches. If $d(\hat{s}_i)$ exceeds c_o, the entire ith branch and all its descendants will be pruned. Finally, the minimum value of $d(\hat{s}_i)$ is chosen as the best way to search the next layer. The loop stops until reaching the bottom of the tree, i.e., $i = 1$. This algorithm is implemented as shown in Fig. 1.

The computational complexity of the traditional and proposed *K*-Best algorithms is discussed as follows and only the multiplications are considered in this paper. The operation of the traditional *K*-Best algorithm mainly includes the generation of the candidate complex vectors, the calculation of squared PEDs and the generation of routes. The traditional *K*-Best algorithm keeps the best K nodes that have the smallest metrics at each layer instead of expanding every node and the obtainment of the best path need to visit KM_c nodes in the next layer. The computational complexity of the traditional *K*-Best algorithm can be obtained as follows: (i) The generation of the candidate complex vectors is $N_t^2 N_r$, (ii) the generation of routes is $2N_t^2 KM_c$, and (iii) the PED is $KM_c N_t + KM_c$ [11]. Totally, the number of multiplications is

$$2N_t^2 KM_c + KM_c N_t + N_t^2 N_r + KM_c \qquad (11)$$

The calculations of the computational complexity of SE and SDA-Kest algorithm can be found in Refs. [10, 11], respectively. For the proposed *K*-Best algorithm, the signal channel is decomposed by the fast *QR* decomposition twice, and the computational complexity of the whole fast *QR* decomposition is $1/2N_t^3 + 1/4N_t^2 N_r$ according to Eqs. (2) and (5) [8]. The SE strategy is used to solve the zero floating-point and sort the branches' PED described in Eq. (8), in

Fig. 1 Flowchart of the
proposed K-Best algorithm

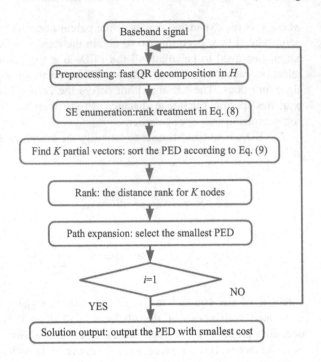

which the multiplications of $R'_{ij}s_j$ are replaced by shift-and-add operations since the
floating-point is a simple integer result. For the entire searching tree with N_t levels,
the complexity for the SE strategy can be represented as $1/2KN_t^2 + N_tN_r + KM_cN_t + 1/2KN_t$. The SDA searches the smallest PED value described in Eqs. (9)
and (10), where the PEDs need to be computed once and the reuse can be applied to
the remaining layers which have only K survival nodes. At each layer, the square
operations will be calculated $M_cK + (M_c - 1)K$ times [12]. The total complexity of
the SDA is KM_cN_t. Finally, the complexity is

$$1/2N_t^3 + 1/4N_t^2N_r + 1/2KN_t^2 + N_tN_r + 2KM_cN_t + 1/2KN_t \qquad (12)$$

The comparison of the computational complexity of the traditional and proposed
algorithms under different K's is shown in Table 1, where the quadratic amplitude
modulation constellation is 16 QAM, i.e., $M_c = 16$.

The process of reducing the searching complexity is shown in Table 1, we can
observe that the proposed algorithm successfully achieves lower computational
complexity compared with the traditional K-Best algorithm, SE [10] algorithm and
SDA [11] algorithm. Specifically, the complexity of the proposed method reduces
by 74.0–76.2% with different K's (4, 8, 16, respectively) compared to the traditional
K-Best algorithm.

Table 1 Comparison of multiplication times on the two algorithms (16 QAM 4 × 4)

Algorithms	Computational complexity	$K = 4$	$K = 8$	$K = 16$
Traditional method [6]	$2N_t^2 KM_c + KM_c N_t + N_t^2 N_r + KM_c$	2432	4800	9536
SE [10]	$2N_t^2 KM_c + 2N_t K - N_t M_c + N_t^2 N_r + KM_c$	2144	4288	8576
SDA [11]	$2N_t^2 KM_c + KM_c N_t + N_t^2 N_r$	2368	4672	9280
Proposed method	$1/2N_t^3 + 1/4N_t^2 N_r + 1/2KN_t^2 + N_t N_r + 2KM_c N_t + 1/2KN_t$	632	1168	2272

4 Performance Evaluation

The performance of the proposed K-Best algorithm is evaluated by numerical simulations. Throughout the simulation, we assume that the channel is channel which is evaluated at $4T \times 2R$ (the overdetermined system), $4T \times 4R$ (static flat Rayleigh fading channel), respectively, which are typical system in the MIMO system. The modulation system is 16 QAM and 64 QAM, respectively. The result comparison of the BER performance of the proposed K-Best algorithm versus several K-Best based algorithms is shown in Figs. 2 and 3.

The proposed algorithm has lower BER than that of the traditional K-Best algorithm at all SNRs at $4T \times 4R$ and 16 QAM in Fig. 2. The performance of the two algorithms significantly decreases with the increase of SNRs. There is a slight gap in BER between the two algorithms before SNR = 5. When the SNR exceeds 5, the BER of the new K-Best algorithm is obviously lower compared to that of the traditional algorithm. For example, when $K = 4$ and SNR = 30, the BER of the new algorithm is improved by 67.6% compared with the traditional algorithm.

Figure 3a–d shows the BER performance for the QR [8], SDA [11], SE [10], and the proposed algorithm under the conditions of $K = 4$, 8, and 16 at $4T \times 2R$, $4T \times 4R$ and 16 QAM, 64 QAM, respectively. The BERs decrease as the SNR and K increase. The BER performance of the proposed K-Best algorithm is the best in all algorithms as shown in Fig. 3a–d.

The percent of timesaver ST is defined as

$$ST(\%) = (T_{\text{traditional}} - T_{\text{new}})/T_{\text{traditional}} \times 100\% \tag{13}$$

where $T_{\text{traditional}}$ is the computational cost by the traditional K-Best algorithm and T_{new} is by the proposed K-Best algorithm. Figure 4 depicts the timesaver in running time for the proposed K-Best algorithm compared with the traditional K-Best algorithm at different SNRs at $4T \times 4R$ and 16 QAM under the condition of $K = 4$, 8, 16, respectively.

Fig. 2 BER performance curve of the traditional and proposed algorithms

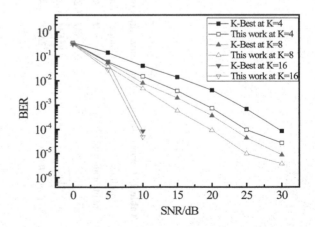

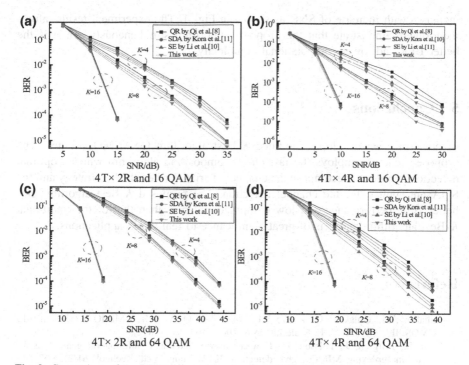

Fig. 3 Comparison of the BER performance of various K-Best algorithms

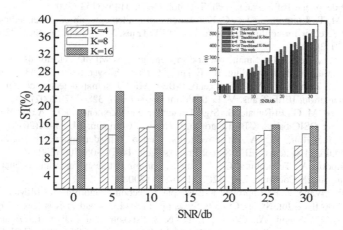

Fig. 4 Timesaver for running time

In Fig. 4, it can be noted that the computational cost in this work is lower than that of the traditional *K*-Best algorithm at $4T \times 4R$ and 16 QAM. The computational cost is reduced by 11.13–23.60% when the SNR ranges from 0 to 30 dB at the calculated cases, and it is more efficient when $K = 16$. The absolute timesaver increases with increase of SNR as shown in Fig. 4. The experiment results afore-mentioned demonstrate that the proposed algorithm simultaneously achieves the better BER and lower computational complexity.

5 Conclusions

In this work, we develop a modified K-Best algorithm for MIMO systems. The proposed scheme employs the fast QR decomposition algorithm with an optimal detection in order to decompose the channel matrix. Further, the SE strategy and the SDA are applied for the channel detection. The proposed K-Best algorithm has better BER performance and low computational cost than that of the traditional K-Best algorithm, and it is of great significance to real-world applications.

References

1. Shabany, M., Gulak, P. G.: A 675 Mbps, 4 × 4 64-QAM K-Best MIMO detector in 0.13 CMOS. IEEE Trans. V. L. Scale Int. Syst. **20**(1), 135–147 (2012)
2. Liu, L., Lofgren, J., Nilsson, P.: Low-complexity likelihood information generation for spatial-multiplexing MIMO signal detection. IEEE Trans. Veh. Technol. **61**(2), 607–617 (2012)
3. Mondal, S., Eltawil, A. M., Salama, K. N.: Architectural optimizations for low-power K-Best MIMO decoders. IEEE Trans. Veh. Technol. **58**(7), 3145–3153 (2009)
4. Shiue, M. T., Long, S. S., Jao, C. K.: Design and implementation of power -efficient K-Best MIMO detector for configurable antennas. IEEE Trans. V. L. Scale Int. Syst. **22**(11), 2418–2422 (2012)
5. Kim, T. H., Park, I. C.: Small-area and low-energy-best MIMO detector using relaxed tree expansion and early forwarding. IEEE Trans. Circuits & Syst. **57**(10), 2753–2761 (2010)
6. Chen, S., Zhang, T., Xin, Y.: Relaxed-Best MIMO signal detector design and VLSI implementation. IEEE Trans. V. L. Scale Int. Syst. **15**(3), 328–337 (2007)
7. El-Mashed, M. G., El-Rabaie, S.: Signal detection enhancement in LTE-A downlink physical layer using OSIC-based K-Best algorithm. Physical Commun. **14**, 24–31(2015)
8. Qi, X. F., Holt, K.: A lattice-reduction -aided soft demapper for high-rate coded MIMO-OFDM systems. IEEE Signal Process. Lett. **14**(5), 305–308 (2007)
9. Ghaderipoor, A., Tellambura, C.: A statistical pruning strategy for schnorr- euchner sphere decoding. IEEE Commun. Lett. **12**(2), 121–123 (2008)
10. Li, Q.W., Wang, Z. F.: Improved K-Best sphere decoding algorithm for MIMO systems. In: Proc. 2006 IEEE Int. Symp. Circuits Syst., pp. 110–113. Island of Kos, Greece (2006)
11. Kora, A. D., Saemi, A., Cances, J. P.: New list sphere decoding (LSD) and iterative synchronization algorithms for MIMO- OFDM detection with LDPC FEC. IEEE Trans. Veh. Technol. **57**(6), 3510–3524 (2008)
12. Amiri, K., Dick, C., Rao, R., Cavallaro, J. R.: Novel sort-free detector with modified real-valued decomposition (M-RVD) ordering in MIMO systems. In: Proc. 2007 IEEE Global Telecommun., pp. 1–5. New Orleans, USA (2007)

Dynamic Testing for RFID Based on Photoelectric Sensing in Internet of Vehicles

Xiaolei Yu, Dongsheng Lu, Donghua Wang, Zhenlu Liu
and Zhimin Zhao

Abstract With the development of application scale of radio frequency identification (RFID) technology in Internet of vehicles (IOV), the scenarios of multi-reader and multi-tag have become increasingly common. Firstly, the relationship between IOV and RFID technology is presented in this paper. Moreover, key of electronic vehicle identification (EVI) is presented, which is the main application of RFID in IOV. Finally, the novel semi-physical verification system based on photoelectric sensing is introduced for dynamic testing of RFID in EVI. The experimental results show that the testing method as well as system is applicable not only for reading range test of single-tag scenario, but also for anti-collision test of multi-tag scenario. The proposed method has good characteristics in terms of speed and accuracy. Thus, the research of this paper is very useful for testing of EVI RFID system performance in a variety of harsh environments.

Keywords Internet of vehicles · Electronic vehicle identification
RFID · Dynamic testing · Reading range

1 Introduction

Internet of vehicles (IOV) originates from the Internet of things (IOT), which is an information interactive network composed of location, speed, and routes. IOV integrates a variety of advanced and sophisticated electronic information

X. Yu · D. Wang (✉)
Jiangsu Institute of Quality and Standardization, Nanjing 210029, China
e-mail: wangdonghua126@126.com

X. Yu (✉) · D. Lu · Z. Liu · Z. Zhao
College of Science, Nanjing University of Aeronautics and Astronautics,
Nanjing 210016, China
e-mail: nuaaxiaoleiyu@126.com

© Springer Nature Singapore Pte Ltd. 2019
B. K. Panigrahi et al. (eds.), *Smart Innovations in Communication and
Computational Sciences*, Advances in Intelligent Systems and Computing 670,
https://doi.org/10.1007/978-981-10-8971-8_29

technology, such as radio frequency identification (RFID) technology. Therefore, it can effectively manage vehicles and road infrastructures [1].

Compared with other countries, Chinese IOV started relatively late, but it developed very quickly in recent years. Chinese Ministry of Industry and Information Technology proposed the "Development and Innovation Action Plan of the Internet of Vehicles (2015–2020)", which aimed to promote the research of IOV. The Chinese national standard "Automotive Electronic Marking General Technical Conditions" drafted by the Ministry of Public Security will also be released [2].

As the main application of RFID in IOV, electronic vehicle identification (EVI) has been researched widely. Marais modeled the key elements of the EVI environment and determined the optimal installation angle of readers [3]. Colella presented a measurement platform for the performance analysis of UHF passive RFID tags, which could effectively guarantee high versatility [4]. Hu presented an evaluation method for EVI adaptability and discussed the feasibility of this technology in urban traffic [5]. Simultaneously, there are some studies on RFID static testing which focused on physical testing and electromagnetic compatibility testing. In practice, the performance of RFID in IOV system is related to dynamic parameters. However, there are few reports of dynamic testing for key parameters such as reading range.

This paper can be separated into three parts. Firstly, the relationship between IOV and RFID is introduced. Then, the EVI is analyzed. Finally, a semi-physical verification technology is proposed, which combines the testing platform with computer simulation. The reading range and anti-collision performance of RFID tags are tested by this new testing technology. The proposed method has good characteristics in terms of speed and accuracy. This paper provides an important technical support for the development and application of RFID technology in IOV.

2   The Role of RFID in IOV

In a typical system of IOV, the effective vehicle communication is the most important technology. It contains the RFID, satellite positioning, cloud computing, and other key technologies. Through the electromagnetic radiation, the system can identify multiple RFID tags in a variety of harsh environments without human intervention. Further, the central processing system can obtain the relevant information.

Firstly, when RFID tags enter the magnetic field, RF signals are received by antennas and the induced current is generated. Secondly, the product information memorized in RFID tags with the energy is obtained by the induced current. RFID tag can also send a signal of a certain frequency proactively. Finally, when RFID reader receives the signal and decodes the information, it transfers the information to central information system for data processing.

3 Main Application of RFID Technology—EVI

Compared with the traditional technology, RFID technology can accurately identify the high-speed objects. Therefore, RFID technology will be widely used in the future intelligent transportation system, and the EVI is main application of RFID technology in IOT.

3.1 Basics of EVI

Compared with ordinary license plate, EVI can communicate with the readers, which are installed on the roadside. A typical RFID tag for EVI is shown in Fig. 1.

When the vehicle with EVI tags turns into the valid reading range of the reader, EVI tags send signal to reader. The reader receives the signal and performs demodulation processing. Lastly, reader gets the information and transfers it to the control system [6]. The whole process is shown in Fig. 2.

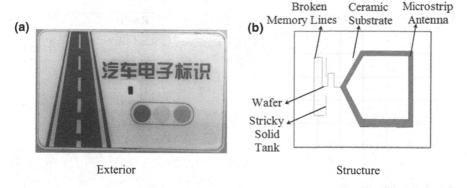

Exterior   Structure

Fig. 1 RFID tag for EVI

Fig. 2 Principle of EVI [6]

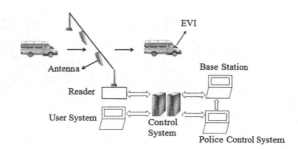

3.2 Dynamic Testing of EVI

EVI is one of the most important applications of RFID technology in IOV, which is very useful in intelligent transportation system. Therefore, dynamic testing of EVI is necessary. In general, the temperature of dynamic testing is from 20 to 26 °C, and the relative humidity is from 40 to 60% [7].

3.2.1 Reading Range Test

The test procedure is as follows:

(a) The test system shall be connected according to the schematic (Fig. 3).
(b) The EVI shall be attached to the vehicle.
(c) The vehicle shall be driven through a RFID gate, as shown in Fig. 4.
(d) The laser ranging sensor shall be used to test the distance between the vehicle and RFID gate, when the reader reads EVI. The distance can be used to calculate the reading range.

Fig. 3 Principle of EVI test system

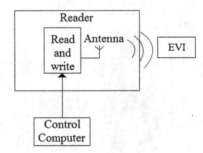

Fig. 4 Reading range test site

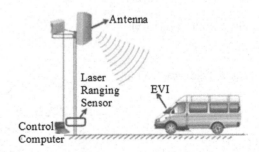

3.2.2 Effect of Vehicle Speed on Reading Performance

The test procedure is as follows:

(a) The test system shall be connected according to Fig. 3.
(b) The relevant parameters shall be set.
(c) The vehicle shall be driven through a RFID gate at different speeds.
(d) The results shall be recorded and analyzed.

4 Semi-physical Dynamic Testing

In practical applications, temperature, noise, and the medium have a great impact on the application of RFID technology. Therefore, photoelectric sensing-based testing technology has become an important technical support in research and application of RFID technology.

4.1 Test Purposes

The test of RFID system includes dynamic testing and static testing. Our main purposes are: (a) to find appropriate vehicle speed and driving trajectory, by carrying out dynamic testing of RFID tags; (b) to explore best antenna angle, by studying the effect of antenna angle on RFID performance; (c) to design the experimental platform, and to complete relevant experiments.

4.2 Test Methods

We propose a semi-physical verification system, which combines a testing platform with computer simulation, as shown in Fig. 5. The synchronous motor is used to drive package with RFID tags on the surface. The RFID antenna is used to read RFID tags, which is on the box. The laser ranging sensor is used to measure the distance between RFID antenna and RFID tags. The results are displayed on the screen of control box.

In vehicle speed test scenarios, RFID tag is attached to the front of vehicle. Firstly, the researcher sets vehicle speed and the transmit power of reader. Then, the results are recorded and analyzed by testing software. Finally, the researcher finds optimal vehicle speed, when the performance of RFID tags is best.

In antenna angle test scenarios, RFID tag is attached to the front of vehicle. Firstly, the researcher sets vehicle speed and the transmit power of reader, and the

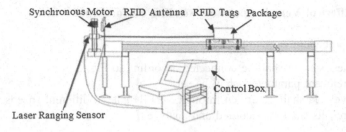

Fig. 5 Testing platform of RFID system

antenna angle is set from 10° to 90° randomly. Then, the results are recorded and analyzed by testing software. Finally, the researcher finds appropriate antenna angle, when the performance of RFID tags is best. Based on the above experiments, the variation of RFID tags can be obtained, which is a support for reading performance.

4.3 Test Results and Analysis

The testing platform is used to test RFID tag's reading performance, which meets the requirements of Chinese national standard GB/T 29768-2013. It is composed of track, object (including RFID tags), laser ranging sensor, RFID antenna, RFID reader, computer, control box, and console. The test steps are shown as follows:

(a) The researcher inputs user name and password, then enters software system.
(b) After entering the system, the researcher clicks the antenna connection button. If the antenna power text box appears data, it indicates that the antenna is connected successfully. Then, the researcher clicks the system connection button, when the buttons of system disconnection, motor up, motor down, and motor reset are from gray to black; it indicates that the device is connected successfully.
(c) After the antenna is connected successfully, the researcher inputs date in antenna power text box and clicks power setting button. At the same time, the researcher inputs the number of tags and the measurement scene, then clicks the parameter confirm button.

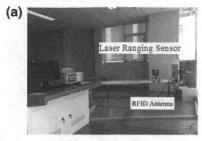

System Structure   Test Results Show

Fig. 6 Single-tag test scenario

4.3.1 Single-Tag Performance Test Scenario

As shown in Fig. 6a, antenna power is set to 20 dBm, label quantity is set to 1, test scene is set to a speed of 20 m/min, then click start measurement button. At the same time, rotating the control button, RFID tags move forward to antenna. When RFID reader reads RFID tags, the results are displayed on the screen of control box, as well as on the screen of computer.

As shown in Fig. 6b, the test result is 3.961 m, which indicates that when the UHF RFID tag enters the electromagnetic field emitted by the antenna, the RFID tag will receive the energy of the electromagnetic field. Next, it will transmit the information to RFID antenna by reflecting the electromagnetic field energy. When the RFID antenna receives information, the distance of 3.961 m is tested by the laser ranging sensor. The test simulates the vehicle through the roadside antenna, which provides a valuable reference for controlling the speed of vehicles and the arrangement of RFID antennas and readers in the road network.

4.3.2 Multi-tag Performance Test Scenario

As shown in Fig. 7a, antenna power is set to 30 dBm, label quantity is set to 10, and the test scene is set to a speed of 20 m/min. We click the start measurement button; at the same time, RFID tags are moved forward to antenna. When RFID reader reads all signals of the 10 RFID tags simultaneously, the results are displayed on the screen of the control box, as well as on the screen of the control computer.

As shown in Fig. 7b, when the 10 tags are read simultaneously, the distance between the tags and the RFID reader is 1.779 m, and the read rate is 100%. The test simulates the situation of typical traffic jam. When multiple vehicles go through the RFID antennas simultaneously, in order to avoid the multi-tag collision effectively, the test provides a technical support for the arrangement of road network in the case of multi-vehicle parallel.

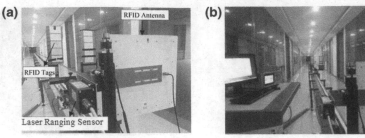

Fig. 7 Multi-tag test scenario

5 Conclusions

In this paper, we propose a semi-physical verification system, which combines a testing platform with computer simulation. In this system, computer simulation technology is used to simulate the real environment. Reading range and anti-collision performance are tested and analyzed by range sensors.

The ultimate goal of dynamic testing of EVI is to establish an automated testing platform. The semi-physical verification technology can not only simulate the situation that a vehicle goes through the RFID antenna, but also simulate multiple vehicles going through the RFID antenna simultaneously. This work plays an important part in testing of vehicle electronic products.

Acknowledgements This work was supported by the China Postdoctoral Science Foundation under Grant NO. 2015M580422 & NO. 2016T90452, the Jiangsu Province Natural Science Foundation for Youths under Grant NO. BK20141032 as well as the 352 Talent Project of Jiangsu Bureau of Quality and Technical Supervision.

References

1. Da Xu L, He W, Li S. Internet of things in industries: A survey. IEEE Transactions on industrial informatics, 2014, 10(4): 2233–2243.
2. Mallik S. Intelligent transportation system. International Journal of Civil Engineering Research, 2014, 5(4): 367–372.
3. Marais H, Grobler M J, Holm J E W. Modelling of an RFID-based electronic vehicle identification system. Africon, Pointe aux Piments, Mauritius, 9–12 Sept. 2013: 303-307.
4. Colella R, Catarinucci L, Coppola P, et al. Measurement platform for electromagnetic characterization and performance evaluation of UHF RFID tags. IEEE Transactions on Instrumentation and Measurement, 2016, 65(4): 905–914.

5. Hu, H, et al. Adaptability evaluation of electronic vehicle identification in urban traffic: a case study of Beijing. Tehicki Vjesnik, 2016, 23(1): 171–179.
6. Cheng JJ, Cheng J L, Zhou M C, et al. Routing in internet of vehicles: A review. IEEE Transactions on Intelligent Transportation Systems, 2015, 16(5): 2339–2352.
7. ISO/IEC18046-3. Information technology-Radio frequency identification device performance test methods-Part 3: Test methods for tag performance. 2007.

Forest Fire Visual Tracking with Mean Shift Method and Gaussian Mixture Model

Bo Cai, Lu Xiong and Jianhui Zhao

Abstract Forest fire region from surveillance video is non-rigid object with varying size and shape. It is complex and thus difficult to be tracked automatically. In this paper, a new tracking algorithm is proposed by mean shift method and Gaussian mixture model. Based on moment features of mean shift algorithm, the size adaptive tracking window is employed to reflect size and shape changes of object in real time. Meanwhile, the Gaussian mixture model is utilized to obtain the probability value of belonging to fire region for each pixel. Then, the probability value is used to update the weighting parameter of each pixel in mean shift algorithm, which reduces the weight of non-fire pixels and increases the weight of fire pixels. With these techniques, the mean shift algorithm can converge to forest fire region faster and accurately. The presented algorithm has been tested on real monitoring video clips, and the experimental results prove the efficiency of our new method.

Keywords Non-rigid object · Tracking algorithm · Mean shift method
Gaussian mixture model

1 Introduction

Forest fire not only causes many casualties as well as property losses, but also damages forest resources and ecological balance. Thus, many countries in the world are committed to research and development of fire detection methods and equipment that can forecast fire disasters in early time. Since color video has the ability to monitor large and open space, the visual-based approach for fire detection and

B. Cai · L. Xiong · J. Zhao (✉)
School of Computer, Wuhan University, Wuhan 430072, Hubei, China
e-mail: jianhuizhao@whu.edu.cn

© Springer Nature Singapore Pte Ltd. 2019
B. K. Panigrahi et al. (eds.), *Smart Innovations in Communication and Computational Sciences*, Advances in Intelligent Systems and Computing 670,
https://doi.org/10.1007/978-981-10-8971-8_30

prediction has received more and more attention from the researchers [1–3]. During which, the successful tracking of forest fire region from monocular video is a very important step.

Most objects of the current video tracking research are rigid objects which meet affine change, but the research for non-rigid object tracking is rare. As a typical example of visual pattern of high entropy, forest fire is the object with shape that cannot be pre-defined [4, 5]. Therefore, automatic detection of forest fire region is the greatest obstacle for fire prevention [6–8].

Mean shift theory was first used in cluster analysis of data [9], then was introduced in image processing [10], and then was successfully applied to target tracking [11]. The basic idea of mean shift [12–14] tracking is using kernel histogram of color as the target template, which contains color information and spatial information of pixels at the same time. The template of target to be tracked is obtained by manual selection. Then, the cost function (the Bhattacharyya coefficient) is used to express the degree of similarity between the template of candidate target in current frame and the target template. And then, iterative optimization method is utilized to find local maximum of the cost function. The position of the candidate target, which corresponds to the local maximum value, is taken as the location of target in current frame. In most cases, mean shift algorithm can track target well; meanwhile, it is very simple and rapid. However, it often falls into local optimum [15], which results in a lower tracking precision. So it is prone to the loss of target in tracking, especially for the non-rigid objects such as forest fire.

Because of the high entropy characteristics of forest fire, the traditional mean shift algorithm is difficult to track fire region well. In this paper, an improved mean shift algorithm is proposed with the Gaussian mixture model [16, 17, 18] to obtain higher precision of fire tracking. First, one size adaptive tracking window is adopted from moment features of mean shift algorithm, which can real-timely reflect the changes of fire in size and shape. Second, the Gaussian mixture model is used to calculate probability value of each pixel in the fire color distribution function. Then, the probability value is used to update the weight of each pixel in mean shift algorithm. It can reduce the weight of non-fire pixels and increase the weight of fire pixels, so that mean shift algorithm can converge to the position of fire target faster and accurately. To test the performance of our presented method, experiments are performed on real videos with forest fire.

The rest of our paper is organized as follows: The novel algorithm based on mean shift and Gaussian mixture model is described in Sect. 2 with details, experimental results and performance comparison are presented and discussed in Sect. 3, and then the conclusion is given in Sect. 4.

2 The New Tracking Method for Fire Region

2.1 Gaussian Mixture Model

The Gaussian mixture model is obtained by superimposing several weighted probability distributions. It is the extension of a single Gaussian probability function, so it smoothly approximates the density distribution of arbitrary shape. If a set of points in high-dimensional space approximate an ellipsoidal distribution, the probability density function of these points can be described using a single Gaussian density function:

$$g(x, u, \Sigma) = \frac{1}{\sqrt{(2\pi)^d |\Sigma|}} \exp\left(-\frac{1}{2}(x - u)^T \sum{}^{-1}(x - u)\right) \tag{1}$$

where u indicates the center position of points' density function, while $\sum$ indicates the covariance matrix of the density function.

When the distribution of points in high-dimensional space is not a single ellipsoid, the above single Gaussian probability density function cannot be used to describe the distribution of these points. In order to accurately describe the probability density distribution of the high-dimensional space points in this case, the Gaussian mixture model needs to be constructed, using more than a single Gaussian probability density function. If k Gaussian functions are needed, the following formula is used to indicate the mixture Gaussian density:

$$p(x) = \sum_{i=1}^{k} a_1 g(x, u, \Sigma) \tag{2}$$

The parameters of the probability density functions are $(a_1, a_2, \ldots, a_k, u_1, u_2, \ldots, u_k, \Sigma_1, \Sigma_2, \ldots, \Sigma_k)$, within which $a_1, a_2, \ldots, a_k$ must satisfy the following condition:

$$\sum_{i=1}^{k} a_i = 1 \tag{3}$$

In our method, we get the parameters of above flame Gaussian mixture probability density function by training a lot of flame images. With the trained Gaussian mixture model, the probability value of belonging to fire region for one pixel can be calculated, taking the RGB value of the pixel as the input of function (2).

Then, the weight of pixel in the mean shift algorithm can be updated by the following formula:

$$w_i = w_i' \sum_{i=1}^{k} a_1 g(x_i, u, \Sigma) \tag{4}$$

where w_i' is the original weight of pixel x_i in the mean shift algorithm, w_i is the new weight after adjustment, and $(a_1, a_2, \cdots, a_k, u_1, u_2, \cdots, u_k, \Sigma_1, \Sigma_2, \cdots, \Sigma_k)$ are related parameters of the Gaussian mixture model obtained from the training of multiple flame images.

2.2 Mean Shift with Size Adaptive Window

The mean shift tracking with size adaptive window is employed to deal with forest fire region with varying size and shape. Based on the weighted image of mean shift algorithm, the following second moment features can be obtained by calculation.

$$M_{00} = \sum_{i=1}^{n_k} w_i \tag{5}$$

$$M_{10} = \sum_{i=1}^{n_k} w_i x_{i,1} \tag{6}$$

$$M_{01} = \sum_{i=1}^{n_k} w_i x_{i,2} \tag{7}$$

$$M_{20} = \sum_{i=1}^{n_k} w_i x_{i,1}^2 \tag{8}$$

$$M_{02} = \sum_{i=1}^{n_k} w_i x_{i,2}^2 \tag{9}$$

$$M_{11} = \sum_{i=1}^{n_k} w_i x_{i,1} x_{i,2} \tag{10}$$

where $(x_{i,1}, x_{i,2})$, respectively, indicate (x, y) coordinates of the point i that locates in the candidate region, w_i indicates the weight of the pixel, and n_k is the number of all pixels in the candidate region.

According to the above formulas, the second moment characteristics can be converted into the second central moments as given below:

$$\mu_{20} = M_{20}/M_{00} - \bar{x}_1^2 \tag{11}$$

$$\mu_{11} = M_{11}/M_{00} - \bar{x}_1\bar{x}_2 \tag{12}$$

$$\mu_{02} = M_{02}/M_{00} - \bar{x}_2^2 \tag{13}$$

where $\bar{x}_1$ is the center coordinate of x, and $\bar{x}_2$ is the center coordinate of y. Then, the covariance matrix is computed as:

$$\text{Cov} = \begin{bmatrix} \mu_{20} & \mu_{11} \\ \mu_{11} & \mu_{02} \end{bmatrix} \tag{14}$$

It is known that covariance matrix is positive-definite matrix; thus, the singular value decomposition (SVD) of matrix Cov is:

$$\text{Cov} = U \times S \times U^T$$
$$= \begin{bmatrix} u_{11} & u_{12} \\ u_{21} & u_{22} \end{bmatrix} \times \begin{bmatrix} \lambda_1^2 & 0 \\ 0 & \lambda_2^2 \end{bmatrix} \times \begin{bmatrix} u_{11} & u_{12} \\ u_{21} & u_{22} \end{bmatrix}^T \tag{15}$$

where λ_1^2 and λ_2^2 are the eigenvalues of the covariance matrix Cov.

If an oval is used to indicate the target to be tracked, vectors $(u_{11}, u_{21})^T$ and $(u_{12}, u_{22})^T$, respectively, indicate the two principal axes of the tracking object in the candidate area. Using a and b to, respectively, indicate the long axis and the short axis of the oval, there is:

$$\lambda_1/\lambda_2 \approx a/b \tag{16}$$

Let $a = k\lambda_1$ and $b = k\lambda_2$ where k is the scale factor, and then let A indicate the area of target, there is:

$$\pi ab = \pi(k\lambda_1)(k\lambda_2) = A \tag{17}$$

Therefore, there are:

$$a = \sqrt{\lambda_1 A/(\pi\lambda_2)} \tag{18}$$

$$b = \sqrt{\lambda_2 A(\pi\lambda_1)} \tag{19}$$

where $A = c(s)M_{00}$, $c(s) = \exp\left(\frac{s-1}{\sigma}\right)$, while s is the Bhattacharyya coefficient of the target model and the candidate model, and σ is used to adjust the response of $c(s)$ to s.

In summary, the covariance matrix shown as below reflects the width, the height, and the direction of the target being tracked:

$$\text{Cov} = \begin{bmatrix} u_{11} & u_{12} \\ u_{21} & u_{22} \end{bmatrix} \times \begin{bmatrix} a^2 & 0 \\ 0 & b^2 \end{bmatrix} \times \begin{bmatrix} u_{11} & u_{12} \\ u_{21} & u_{22} \end{bmatrix}^T \tag{20}$$

2.3 Tracking by Mean Shift and Gaussian Mixture Model

Based on above Gaussian mixture model and mean shift algorithm with size adaptive window, the novel method is proposed for forest fire tracking. The main steps of our method are as follows:

(1) Manually select the target area in the first frame image, and then calculate target model as $\hat{q}$ according to the formula

$$\hat{q}_u = C \sum_{i=1}^{n} k\left(\|x_i\|^2\right) \delta[b(x_i) - u] \tag{21}$$

where $k(x)$ is an isotropic kernel function, the function $b(x_i^*)$ maps pixel x_i to its corresponding feature grid box u, the range of u is $\{1 \ldots m\}$, m is the series of kernel color histograms, δ is the Kronecker delta function, n is the number of pixels, C is a standardized constant coefficient, making the sum of probability densities to be 1. Then, let the center of target area be the initial position (y_0) of the candidate region in the next frame.

(2) Initialize the number of iterations k to be zero.

(3) The candidate model is obtained as $\hat{p}(y_0)$ with the formula

$$\hat{p}(y) = C_h \sum_{i=1}^{n_k} k\left(\left\|\frac{y - x_i}{h}\right\|\right) \delta[b(x_i) - u] \tag{22}$$

where h is the window width of the candidate region, n_k is the number of pixels that locate in the candidate region, C_h is standardized constant coefficient, and y is the center coordinate of the candidate region.

(4) Compute the weight of pixel x_i according to the formula

$$w_i' = \sum_{u=1}^{m} \sqrt{\frac{\hat{q}}{\hat{p}_u(y_0)}} \delta[b(x_i - u)] \tag{23}$$

then, update the weights using Formula (4).

(5) The centroid of the target candidate region can be obtained by the following formula

$$y_1 = \frac{\sum_{i=1}^{n_k} x_i w_i g\left(\left\|\frac{y-x_i}{h}\right\|^2\right)}{\sum_{i=1}^{n_k} w_i g\left(\left\|\frac{y-x_i}{h}\right\|^2\right)} \tag{24}$$

where $g(x) = -k'(x)$.

(6) Get the distance (d) between target model and candidate model by the formula

$$d(y) = \sqrt{1 - \rho[\hat{p}(y), \hat{q}]}$$
$$= \sqrt{1 - \sum_{u=1}^{m} \sqrt{\hat{p}_u(\hat{y})\hat{q}_u}} \tag{25}$$

then, let $y_0 \leftarrow y_1$. If d is less than the minimum distance threshold or k is more than the maximum number of iterations, then go to Step (6); otherwise, go to Step (3).

(7) Use Formula (20) to calculate the width, the height, and the direction of current target, and then set the width and the height as the size of target being tracked in current frame.

(8) The region, whose size is computed by Step (7) with the center y_0, is the tracking result obtained from the current frame. If the current frame is the last frame, then end the algorithm. Otherwise, let y_0 be the initial center of candidate region in the next frame, go to Step (2) to continue the tracking.

3 Experimental Results and Analysis

Based on some real video clips with forest fire, the new method has been tested. The above algorithms are implemented in the environment of Visual Studio C++ and OpenCV. As shown in Fig. 1, forest fire regions are tracked with mean shift algorithm with size adaptive window. As shown in Fig. 2, fire areas are detected with the improved mean shift algorithm by the Gaussian mixture model.

From Fig. 1, it can be found that, using only mean shift algorithm with size adaptive window for forest fire tracking, the tracking accuracy is low and the target may be lost after some frames.

From Fig. 2, it is obvious that the tracking accuracy is improved, and the size of target is accurately adjusted according to the changes of fire region.

Fig. 1 Tracking result of mean shift algorithm with size adaptive window

Fig. 2 Tracking result of improved mean shift algorithm with the Gaussian mixture model

4 Conclusion

We have presented an improved mean shift algorithm with the help of the Gaussian mixture model to track forest fire regions from video clips. Our contributions include the following: The size adaptive tracking window is computed from the moment features of mean shift algorithm, and it is used to reflect the size and shape changes of fire region; the Gaussian mixture model is employed to calculate the probability value of each pixel, and it is utilized to adjust the weight of each pixel in mean shift algorithm, which helps the algorithm converge to the position of fire target faster and accurately. The performance of our presented method has been tested on real video clips with forest fire, and the experimental results prove its improvement on tracking precision.

References

1. Liu Z.G., Yang Y., Ji X.H. Flame detection algorithm based on a saliency detection technique and the uniform local binary pattern in the YCbCr color space. International Journal of Signal, Image and Video Processing, 2016, 10, 277–284.
2. Zhang H.J., Zhang N., Xiao N.F. Fire detection and identification method based on visual attention mechanism. Optik, 2015, 126, 5011–5018.
3. Ye W., Zhao J., Wang S., Wang Y., Zhang D., Yuan Z. Dynamic Texture Based Smoke Detection Using Surfacelet Transform and HMT Model. Fire Safety Journal, 2015, 73, 91–101.
4. Roberto R.R. Remote detection of forest fires from video signals with classifiers based on K-SVD learned dictionaries. Engineering Applications of Artificial Intelligence, 2014, 33, 1–11.
5. Yuan F.N., Fang Z.J., Wu S.Q., Yang Y., Fang Y.M. Real-time image smoke detection using staircase searching-based dual threshold AdaBoost and dynamic analysis. IET Image Processing, 2015, 9(10), 849–856.
6. Kosmas D., Panagiotis B., Nikos G. Spatio-Temporal Flame Modeling and Dynamic Texture Analysis for Automatic Video-Based Fire Detection. IEEE Transactions on Circuits and Systems for Video Technology, 2015, 25(2), 339–351.
7. Zhang Z.J., Shen T., Zou J.H. An Improved Probabilistic Approach for Fire Detection in Videos. Fire Technology, 2014, 50, 745–752.
8. Zhao Y.Q., Tang G.Z., Xu M.M. Hierarchical detection of wildfire flame video from pixel level to semantic level. Expert Systems with Applications, 2015, 42, 4097–4014.
9. Miguel A. Carreira-Perpinan. Gaussian Mean-shift is an EM Algorithm. IEEE Transactions on Pattern Analysis and Machine Intelligence, 2007, 29(5), 767–776.
10. Ido Leichter, Michael Lindenbaum. Mean Shift tracking with multiple reference color histograms. Computer Vision and Image Understanding, 2010, 114, 400–408.
11. Chunhua Shen. Fast Global Kernel Density Mode Seeking: Applications to Localization and Tracking. IEEE Transactions on Image Processing, 2007, 16(5), 1457–1469.
12. K. Fukunaga and L.D. Hostetler. The Estimation of the gradient of a density function with applications in pattern recognition. IEEE Transactions on Information Theory, 1975, 21(1), 32–40.
13. Y. Cheng. Mean Shift, Mode seeking, and clustering. IEEE Transactions on Pattern Analysis and Machine Intelligence, 1995, 17(8), 790–799.
14. D. Comaniciu, V. Ramesh and P. Meer. Kernel-based object tracking. IEEE transactions on Pattern Analysis and Machine Intelligence, 2003, 25(4), 564–575.
15. D. Comaniciu, V. Ramesh and P. Meer. Real-time tracking of non-rigid objects using mean shift. In Proc. of IEEE Conf. on Computer Vision and Pattern Recognition, Hilton Head, SC, Volume II, June 2000, 142–149.
16. C. Biemacki, G. Celeux, G. Govaert. Assessing a Mixture Model for Clustering with the Integrated Completed Likelihood. IEEE Transactions on Pattern Analysis and Machine Intelligence, 2000, 22, 719–725.
17. Heng-Chao Yan, Jun-Hong Zhou, Chee Khiang Pang. Gaussian Mixture Model Using Semisupervised Learning for Probabilistic Fault Diagnosis Under New Data Categories. IEEE Transactions on Instrumentation and Measurement, 2017, 66(4), 723–733.

A Novel Construction
of Correlation-Based Image CAPTCHA
with Random Walk

Qian-qian Wu, Jian-jun Lang, Song-jie Wei, Mi-lin Ren
and Erik Seidel

Abstract CAPTCHA has been widely adopted throughout the World Wide Web to achieve network security by preventing malicious interruption or abuse of server resources. Existing text-based and image-based CAPTCHA techniques are not robust enough to resist sophisticated attacks using pattern recognition and machine learning. To overcome this challenge, we designed a new approach to construct an image-based CAPTCHA by using a random walk on image with correlated contents, which capitalizes on human knowledge on the relevance of images. The usability and robustness of the proposed scheme have been evaluated by both numerical analysis and empirical evidence. Early testing has shown it to be a promising approach to enhancing and replacing the existing Web CAPTCHA techniques when fighting against bots.

Keywords CAPTCHA · Image correlation · Random walk

Q. Wu · S. Wei (✉) · E. Seidel
CSE School, Nanjing University of Science and Technology, Nanjing, China
e-mail: swei@njust.edu.cn

Q. Wu
e-mail: qqwu@njust.edu.cn

E. Seidel
e-mail: seidel@njust.edu.cn

J. Lang
Affiliated High School, Nanjing Normal University, Nanjing, China
e-mail: 1668304800@qq.com

M. Ren
Beijing Engineering Research Center of NGI & Its Major Application Technologies Co. Ltd.,
East Zhongguancun Rd, Beijing 100084, China
e-mail: rmilin@163.com

© Springer Nature Singapore Pte Ltd. 2019   339
B. K. Panigrahi et al. (eds.), *Smart Innovations in Communication and
Computational Sciences*, Advances in Intelligent Systems and Computing 670,
https://doi.org/10.1007/978-981-10-8971-8_31

1 Introduction

Completely Automated Public Turing Test to Tell Computers and Humans Apart (CAPTCHA) is an interactive test that can distinguish between computer programs and human users [1]. CAPTCHA is widely used on the Internet to protect Web sites from being attacked by malicious programs such as password crackers. CAPTCHA plays an important role in Internet security.

As an alternative to text media [2], image CAPTCHA is more complex and robust, which is based on the difficulty of image recognition and fuzziness of content understanding. However, there are some limitations in the existing image-based CAPTCHA construction techniques. It depends on an annotated image library with numerous images which are not automatically generated [3]. Recently, the Artificial Neural Networks (ANN) technique has achieved significant effectiveness in object recognition and image classification [4], which poses a serious threat to image-based CAPTCHAs.

In this paper, we propose a new way to construct image CAPTCHA based on the correlation and dependency [5] of words annotating images from real-time online search engines. The correlation degree of word tags is measured based on their frequency of occurrence in news articles. Word tags are visualized as images in the CAPTCHA construction. These images are instantly retrieved from the Internet with tags as keywords to generate a dynamic resource library, which is an effective solution to the above-mentioned vulnerability. The newly designed CAPTCHA is implemented with a text word challenge and a lays out eight image choices, which are retrieved by randomly walking [6] through the image correlation graph to avoid regular patterns.

The following section introduces the related works. Section 3 describes the details of the proposed CAPTCHA construction methodology, and Sect. 4 presents the experimental evaluation of the proposed solution. Section 5 highlights the paper contributions and discusses some future work.

2 Related Works

In this section, we discuss the related work on state-of-the-art CAPTCHA generation methods.

BOT is one of the most malicious threats to Internet security. It runs automated tasks over the Internet and increases the scale of malicious online activities. Usually, CAPTCHAs are based on the three principles proposed by Chew and Tygar [7]: (1) CAPTCHAs should be easy for humans to solve. (2) CAPTCHAs should be hard for computers to solve. (3) CAPTCHAs should be easy to generate and verify. For example, Rahman et al. [8] proposed a dynamic image-based three-tier CAPTCHA system to deter against BOT. The proposed system makes a random guess at all levels; therefore, it is difficult for BOT to attack. The proposed system makes cracking a difficult task for BOT, but at the same time it is easier for human to overcome the challenge.

In 2014, Google unveiled a significant revision of reCAPTCHA. ReCAPTCHA requires users to click a checkbox, or to solve a challenge by identifying images with similar content. The system is widely used. However, Suphannee et al. [9] designed a novel attack for image-based CAPTCHAs which extract semantic information from images. By using image annotation services and libraries, the attacker is able to identify the image content and select response images depicting similar objects.

To counter the above method, we propose a highly dynamic CAPTCHA system that randomly walks through images with correlated contents to create tasks. The scheme capitalizes on human reasoning ability on image correlation instead of recognition ability on similar images.

3   Design and Implementation

A traditional CAPTCHA challenge is typically composed of questions and answer texts or choices. Users are required to choose or type in the correct answer to pass a test, such as Fig. 1a presenting some contorted words that users have to recognize. No matter what text or image used, CAPTCHA answers are usually another representation of the question or challenges, i.e., the correct answers are "equal" to the semantic meaning of the question content. Recent advances in pattern recognition and semantic analysis techniques have made such CAPTCHA techniques vulnerable to big data based on searching and cracking. We propose to construct a more sophisticated image CAPTCHA method by using semantic correlation to connect question keywords with answer choices. Such correlations are only noticeable and interpretable by human beings with an ability for abstract thinking and cultural understanding. Figure 1b shows an example that people have to choose in order to pass the test.

(a)   **(b)**

Equivalence-based CAPTCHA   Correlation-based CAPTCHA

Fig. 1 CAPTCHA examples in different schemes

The process of generating such correlation-based CAPTCHAs in Fig. 1b can be divided into two stages: semantic graph generation and random image selection. The first stage builds and measures the semantic correlations of keyword contents, while the second stage assembles keywords to produce image challenges.

3.1 Semantic Graph Generation

We first build a thesaurus of keyword seeds. The thesaurus is a collection of common nouns, which are periodically expanded and updated with tagged Web content. To expand, we use a random function to select a word from the thesaurus denoted as W then search related news articles using a news search engine, with W as the keyword to grab N results and get their news headlines. We then parse the news headlines into words and tag the words with parts of speech, and select common nouns to form data sets. The Apriori algorithm is applied to extract association rules from the data sets. Words with higher-than-threshold correlation degrees are filtered out to augment the thesaurus.

In the thesaurus, the correlation degree of two words is measured by mutual information (MI) between them. With two words key W_i and key W_j, we first use each to get online statistics from search engine denoted as $c(\text{key } W_i)$ and $c(\text{key } W_j)$, search key $W_i + \text{key } W_j$ in order to get search results denoted as $c(\text{key } W_{i,j})$, search key $W_j + \text{key } W_i$ in order to get search results denoted as $c(\text{key } W_{j,i})$. Then, we infer

$$c(\text{key } W_i + \text{key } W_j) = \left(c(\text{key } W_{i,j}) + c(\text{key } W_{j,i})\right) \div 2 \tag{1}$$

The mutual information is represented by $MI(\text{key } W_i, \text{key } W_j)$. We compute the mutual information of key W_i and keyW_j as

$$\text{MI}(\text{key } W_i, \text{key } W_j) = \log_2\left(c(\text{key } W_i, \text{key } W_j) \times N\right) / \left(c(\text{key } W_i) \times c(\text{key } W_j)\right) \tag{2}$$

where N represents the maximum of the search results. $MI(\text{key } W_i, \text{key } W_j)$ is the degree of correlation of key W_i and key W_j. Specifically, MI is 0 when they are same.

By drawing all the keywords and their relations together, the correlation of words in the thesaurus can be presented and recorded as a semantic graph [10] like that in Fig. 2. The semantic graph consists of nodes representing words and undirected edges representing correlation. The graph is stored in three tuples, as (node1, node2, correlation-degree-between-node1-node2).

If the degree of correlation is greater than a predefined threshold, the two nodes are connected by an edge. The weight on each edge is just the degree of correlation of a word pair.

Fig. 2 Example of a
semantic graph

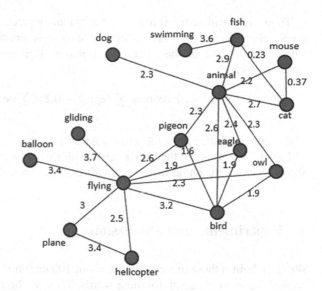

3.2   Random Image Selection

The image selection process follows the Markov chain model [11]. Selecting N images implies randomly walking N steps. We first establish the adjacency matrix denoted as A of the semantic graph. A_{ij} represents the edge weight between node i and node j. Obviously, A_{ij} is identical to A_{ij}. D_i represents the vertex degree of the node i. Next, we denote by $M = (p_{ij})$ the matrix of transition probabilities of this Markov chain. p_{ij} is the probability of from i reaching j in one step and is given by

$$p_{ij} = A_{ij} \div D_i \tag{3}$$

If at the t-th step, the probability from i to j is given by the ij-entry of the matrix M^t. Assume we start at a node v_0 and need to walk n steps. For each step, we select the next node by the matrix of transition probabilities. When finishing n steps, we get a sequence of random nodes, which is a list of words. Then, we search images corresponding to the list of words from the Internet to generate images for CAPTCHAs.

We use a result judgment mechanism as the standard to decide whether users can pass the test or not. We first calculate the sum of the correlation degrees of the right images as an optimal value:

$$\text{optimal value} = \sum_i \text{degree}_i, i \in \{\text{good answers}\} \tag{4}$$

Then, we calculate the sum of the correlation degrees of the images selected by users following the rule that if the correlation is positive. Then we add it, or otherwise add 20% of the negative correlation degree. Then, we get user's correctness score:

$$\text{user score} = \sum_i \text{degree}_i + 0.2 \times \sum_j \text{degree}_j \tag{5}$$

where $i \in \{\text{good answers}\}, j \in \{\text{error answers}\}$.

Finally, we compare the user's score with the optimal response. If the score is above a certain threshold of the optimal response, then this user passes the test.

4 Experiments and Discussion

We have built a thesaurus containing about 100 common words, and we have also created the semantic graph for these words. Then, we built a Web site which would present users with the novel CAPTCHA challenge. Graduate students in NJUST were invited to evaluate this novel image-based CAPTCHA. There were nearly one thousand tests in total conducted with ten students. An automated program was used to do one thousand tests by selecting answers randomly as well. We set different pass thresholds of the optimal response to find the best one for result judgment. The students needed to select the images that they thought most relevant to the question keywords and tried to pass the test.

Figure 3 shows the relation between different pass thresholds for the optimal response and pass rate. It is obviously that the pass rate of the automated program is very low compared to humans. We observed that when the pass threshold of the optimal response is greater than 60%, the variation range of pass rate becomes large. In Fig. 3, the proper threshold should be greater than 60%. In Fig. 4, the pass rates differ greatly when the threshold is set as 65, 75, 85, and 95%, respectively. If the pass threshold of the optimal response is 75%, which is represented by the solid line in Fig. 4, it shows that about 80% of the students can pass the test in one attempt, and nearly all of the volunteers can pass the test within four trials. So, a pass threshold 75% of the optimal response has a good balance between human-bot discrimination and appropriate level of difficulty.

Fig. 3 Pass threshold and pass rate

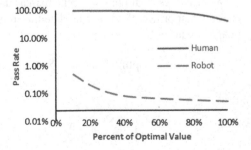

Fig. 4 Pass rate with various pass thresholds

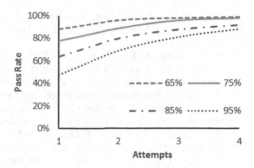

5 Conclusion

User experience has been carefully considered in our design. In order to improve user experience, we created a thesaurus which covered only common topics. Furthermore, people in different fields may have different understandings of the same topics. To balance the potential difference, a certain degree of the tolerance is necessary. The method allows tolerance for small mistakes arising from different knowledge levels of users.

The thesaurus database will be updated regularly, which forces the thesaurus to stay up-to-date and reduces the risk of cracking the database. The images used by the novel CAPTCHAs are retrieved from the Internet in real time and are selected by random walk procedure to avoid regular patterns being found. The high level of randomization of the images can better protect Web sites from brute force-based and exhaustive attacks.

For future work, we intent to improve our CAPTCHA method by expanding our thesaurus via adopting new methods of extracting related words, such as mining associative terms from tagged resources in social networks. We will also fine tune the threshold selection for optimal value.

Acknowledgements This material is based upon work supported by the China NSF grant No. 61472189, the CERNET Innovation Project No. NGII20160601, and the Innovation Projects of Beijing Engineering Research Center of Next Generation Internet and Applications.

The authors confirm that an ethic approval for this particular type of study is not required in accordance with the policy of the involved institutes.

References

1. Yan, J., & Ahmad, A. S. E. (2011). Captcha robustness: a security engineering perspective. Computer, 44(2), 54–60.
2. Bursztein, E., Martin, M., & Mitchell, J. (2011). Text-based CAPTCHA strengths and weaknesses. ACM Conference on Computer and Communications Security, CCS 2011, Chicago, Illinois, Usa, October (pp. 125–138). DBLP.

3. Datta, R., Li, J., & Wang, J. Z. (2009). Exploiting the human–machine gap in image recognition for designing captchas. IEEE Transactions on Information Forensics & Security, 4(3), 504–518.
4. He, K., Zhang, X., Ren, S., & Sun, J. (2016). Deep Residual Learning for Image Recognition. Computer Vision and Pattern Recognition (pp. 770–778). IEEE.
5. Lopezpaz, D., Hennig, P., & Schölkopf, B. (2013). The randomized dependence coefficient. Advances in Neural Information Processing Systems, 1–9.
6. Fouss, F., Pirotte, A., Renders, J. M., & Saerens, M. (2007). Random-walk computation of similarities between nodes of a graph with application to collaborative recommendation. IEEE Transactions on Knowledge & Data Engineering, 19(3), 355–369.
7. Chew, M., Tygar, J.D. (2004) Image recognition captchas. In: Zhang, K., Zheng, Y. (eds.) ISC 2004. LNCS, vol. 3225, pp. 268–279. Springer, Heidelberg.
8. Rahman, R. U., Tomar, D. S., & Das, S. (2012). Dynamic Image Based CAPTCHA. *International Conference on Communication Systems and Network Technologies* (pp. 90–94). IEEE.
9. Sivakorn, S., Polakis, I., & Keromytis, A. D. (2016). I am Robot: (Deep) Learning to Break Semantic Image CAPTCHAs. IEEE European Symposium on Security and Privacy (pp. 388–403). IEEE.
10. Moawad, I. F., & Aref, M. (2013). Semantic graph reduction approach for abstractive Text Summarization. Seventh International Conference on Computer Engineering & Systems (pp. 132–138). IEEE.
11. Lee, J., Cho, M., & Lee, K. M. (2010). A Graph Matching Algorithm Using Data-Driven Markov Chain Monte Carlo Sampling. International Conference on Pattern Recognition, ICPR 2010, Istanbul, Turkey, 23–26 August (pp. 2816–2819). DBLP.

Matching Algorithm and Parallax Extraction Based on Binocular Stereo Vision

Gang Li, Hansheng Song and Chan Li

Abstract By using binocular stereoscopic vision and planar images, this paper details the process of obtaining 3D information for interested objects and obtains the world and pixel coordinates of any point on the object. The main contents of this article are focus on camera calibration, image correction, stereo matching, and parallax extraction. Furthermore, various algorithms and implementation methods are studied and analyzed. Finally, by comparing correction and stereo matching algorithms, more effective correction algorithm and matching algorithm are achieved.

Keywords Binocular stereo vision · Calibration algorithm · Correction algorithm Stereo matching algorithm · Disparity map

1 Introduction

Binocular stereoscopic vision is an important branch of computer vision, that is, from two different cameras at the same location to obtain the same scene of the two views, by calculating the spatial point in the two images in the parallax, get the point of the three-dimensional coordinates [1]. This article is based on this principle set up two cameras to collect information and the corresponding treatment. Binocular stereo vision can be very effective access to the measured object of the three-dimensional information, and its system structure is simple, high efficiency, high accuracy, and fit.

Binocular stereoscopic vision [2] is mainly the use of parallax principle to obtain three-dimensional information, the target point in the formation of two different imaging equipment in the horizontal, there are differences in coordinates (i.e., parallax $d = x^l - x^r$), and Z is defined the distance from the point of detection to the imaging plane; there is a reverse relationship between them: $Z = f * T/d$. Figure 1 shows

G. Li (✉) · H. Song · C. Li
School of Information Engineering, Chang'an University, Xi'an 710064,
Shanxi, China
e-mail: gangli@chd.edu.cn

© Springer Nature Singapore Pte Ltd. 2019
B. K. Panigrahi et al. (eds.), *Smart Innovations in Communication and Computational Sciences*, Advances in Intelligent Systems and Computing 670,
https://doi.org/10.1007/978-981-10-8971-8_32

binocular stereo vision flowchart. Figure 2 shows the imaging schematic. The left and right images are the imaging plane of the binocular camera, respectively. The distance between the two projection centers in the binocular camera is L (called the baseline). The center of the two lenses is the origin of the camera. The coordinate system is shown in Fig. 2 where X_c-Y_c-Z_c is the camera coordinate system, U-O-V is the image coordinate system, and the camera's optical axis and the left and right imaging plane focus are O^1, O^2. The X and Y are coordinate systems, respectively. Assuming that the space is p, its coordinates of the world coordinate system are (X_c, Y_c, Z_c), the coordinates of the corresponding points in the left image are P_1 (U_1, V_1), and the coordinates of the corresponding points in the right image are P_2 (U_2, V_2).

Assuming that the image formed by the binocular camera is on a horizontal plane, then the point is the same as the vertical coordinate system of the left image coordinate system and the right image coordinate system, that is, $V_1 = V_2$. Derived from the triangular geometric relationship:

$$u_1 = fx_c/z_c \quad u_2 = f(x_c - b)/z_c \quad v_1 = v_2 = fy_c/z_c \tag{1}$$

f in formula (1) is the focal length of two cameras. By the known parallax for the same point in the left and right two images abscissa difference between, that is, $d = u_1 - u_2$:

$$d = u_1 - u_2 = f * b/z_c \tag{2}$$

The calculated parallax d into u_1, v_1, u_2, v_2 can be calculated x_c, y_c, z_c, so as to get the coordinates of the world coordinate system:

$$x_c = b * u_1/d \quad y_c = b * v/d \quad z_c = b * f/d \tag{3}$$

Therefore, as long as we can accurately find the point P in the left and right images in the point P_1, P_2, and through the binocular calibration to obtain the camera's internal and external parameters, you can determine the point P in the world coordinate system in the three-dimensional coordinates.

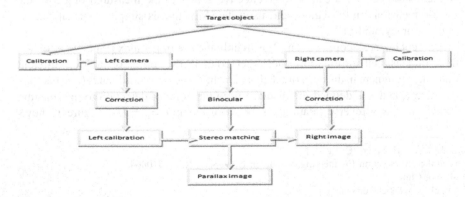

Fig. 1 Binocular stereoscopic flowchart

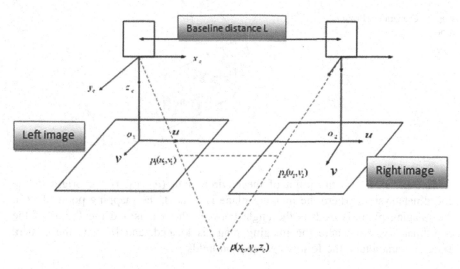

Fig. 2 Schematic diagram of binocular stereoscopic imaging

2 Camera Calibration

2.1 Basic Principles of Calibration

The basic principle [5] is to establish the mathematical relationship between the position of any point in the space and the plane position of the camera. The path is based on the pinhole imaging model, by the need to observe the point of the image coordinates and the world coordinates to solve the camera parameters. The parameters include the camera's internal parameters (e.g., focal length) and external parameters (rotation matrix and translation vector).

2.2 Calibration Method

The basic method of traditional camera calibration [3] is that under certain camera model, based on the specific experimental conditions such as shape and size of the reference object, after its image processing, using a series of mathematical transformation and calculation method, take the camera model internal parameters and external parameters. This article uses Zhang Zhengyou calibration method [4] (Fig. 3).

The method assumes that the target used in the calibration of the Z-axis in the world coordinate system is 0, the optimal solution of the parameters within the camera is solved linearly, and then, the maximum likelihood estimation is used for the nonlinear solution. Brief description of the principle:

Fig. 3 Camera calibration
model

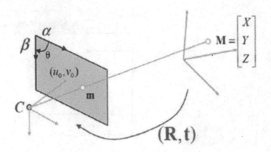

$M = [X, Y, Z]^T$ is the point of the environment, (u_0, v_0) is the origin of the
coordinate system where the imaging plane is located, the mapping point of M in
the imaging plane is m, θ is the angle between the α axis and the β axis of the
coordinate system where the imaging plane is located, and $[R\ t]$ is the camera
external parameters; the following equation holds:

$$sm = A[R\ t]M \tag{4}$$

where A is the camera's internal parameters, is $A = \begin{bmatrix} \alpha & \gamma & u_0 \\ 0 & \beta & v_0 \\ 0 & 0 & 1 \end{bmatrix}$ make

$H = A[r_1\quad r_2\quad t]$, there is

$$sm = HM \tag{5}$$

Since M and m are known at the time of correction, we can find the H matrix.

After the above steps are completed, the calibration is not nearing the end. The next
step is to decompose the internal parameter matrix A and the outer parameters (rotation
matrix and translation vector) of the binocular camera $[h_1\ h_2\ h_3] = \lambda[r_1\ r_2\ r_3]$, ($\lambda$ is the
reciprocal of the scaling factor) r_1 and r_2 are orthogonal; there are

$$h_1^T A^{-T} A^{-1} h_z = 0 \tag{6}$$

Thus, we can find $[R\ t]$, that is, the camera's internal parameters and external
parameters.

2.3 Experimental Results After Calibration of the Camera

The experiment with the board calibration plate, the number of checkers X for the 6,
Y direction for the 8, the size of the board is 26 mm; the results obtained by
calibration is shown Fig. 4.

Fig. 4 Calibration process image

3 Binocular Correction

3.1 Binocular Correction Algorithm

This paper uses the Bouguer correction algorithm. The number of reprojection of any one of the images used for matching is as small as possible, and the area of interest that we can observe is maximized after the above values are minimized. The core is that the distortion of the projection is also as small as possible, and the rotation matrix R is divided into R_l and R_r. R_l is the synthetic rotation matrix of the left camera, and R_r is the synthetic rotation matrix of the right camera. Since R is split, the left and right cameras only need to rotate halfway, and the main ray is parallel to the main ray formed by only one rotation matrix. Because only half the rotation, so the two cameras in the same plane, but in the horizontal direction is not aligned. Then, create a new direction of the e_1 direction of the beginning of the rotation matrix, e_1 direction for the two camera projection center between the translation vector directions:

$$e_1 = T/\|T\| \tag{7}$$

$$T = [\,T_X \quad T_Y \quad T_Z\,]^{\mathrm{T}}   \tag{8}$$

Vector e_2 must be orthogonal to e_1. The direction of e_2 can be obtained by multiplying the product of e_1 with the main ray and then dividing the vector into the unit vector:

$$e_2 = [\,-T_Y \quad T_X \quad 0\,]/\sqrt{T_X^2 + T_Y^2}   \tag{9}$$

After obtaining e_1 and e_2, e_3 is orthogonal to e_1 and e_2, and e_3 is the cross product of both of them:

$$e_3 = e_1 \times e_2 \tag{10}$$

You can map the outer poles of the left camera to infinity through matrix R_{rect}, which is converted as follows:

$$R_{rect} = \left[(e_1)^T \quad (e_2)^T \quad (e_3)^T \right]^T \tag{11}$$

The R_l and R_r are multiplied by R_{rect} to obtain the overall rotation matrices R'_l and R'_r, and the left and right camera coordinates are multiplied by the respective R'_l and R'_r so that the main optical axis of the left camera is parallel to the right camera and the image plane is parallel to the baseline.

$$R'_l = R_{rect} * R_l \quad R'_r = R_{rect} * R_r \tag{12}$$

This completes the image correction.

4 Stereo Matching

4.1 Matching Principle

The principle of binocular stereo matching is based on the selected matching primitives, along the correction of all the three-dimensional image of the horizontal line of the matching search, to find the matching image in the pixel-one correspondence, which can calculate the corresponding parallax figure. The essence of the stereo matching algorithm is to rationally use the basic assumptions and constraints to eliminate the incorrect stereo matching, and the stereo matching problem is transformed into solving the energy function optimization problem.

4.2 Matching Steps

Regardless of the matching algorithm used, it typically includes three basic steps: calculating the matching cost, superimulating the matching cost, and calculating and optimizing the parallax.

(1) To calculate the cost of matching: This step is the basis for the need to test the gray similarity. The most commonly used method is to find the absolute value AD (Absolute Intensity Differences) or the squared SD (Squared Intensity Differences) of the grayscale difference. When calculating the original matching complexity, you can set a threshold so that the errors that are caused by improper matching are impaired when subsequent overlays are complex.

(2) Overlay matching cost: Under normal circumstances, the use of the first step is to calculate the cost of the study to further study the global algorithm, and the use of the window will match the cost of stacking so that the reliability is enhanced in the regional algorithm.

(3) Calculate and optimize parallax: For the global algorithm, there is a known energy evaluation function before the matching complexity is superimposed, and then, the function is taken to the minimum by constantly changing the value of the parameter so that the parallax. It is easy to calculate. And the regional algorithm, after the window through the stack so that the reliability of enhanced, access to parallax becomes easy.

4.3 Matching Algorithm

The main algorithm used in this paper is BM and SGM algorithms [6].

BM (bidirectional matching): Bidirectional matching is a widely used three-dimensional (local and global) detection of outliers and is a fast matching algorithm. The corresponding problem was solved twice. For the first time, assume that the left image is the reference DLR (x, y), and the second assumes the right image as the reference DRL (x, y). The disparity difference between the two maps is classified as an outlier, and the following formula is executed:

$$\mathrm{DLR}(x, y) - \mathrm{DLR}(x + (\mathrm{DLR}(X, Y), y) < T| \tag{13}$$

T is usually set to 1, also known as left and right consistency checks.

SGM algorithm [7, 8]: Improvements to the bidirectional matching algorithm. The core steps of the SGM algorithm are as follows: select the matching primitives; construct the cost energy and function of the scanning lines based on multiple directions; find the optimal solution of the energy cost and function. The basic idea is to calculate the parallax of each pixel in the image to form a parallax. Set a global energy function whose parameter list contains the previously calculated parallax value, which allows the function to be minimized by optimization so that the parallax value of each pixel reaches the optimal value. Compared with the BM algorithm, this algorithm makes the edge of the object more obvious, so the resulting parallax effect is more obvious.

Compared with the BM matching algorithm, the parallax graph result is poor, the contour is not clear, the effect is seriously deformed, and the probability of false matching is very high, but the advantage is that the matching speed is faster, and the SGM algorithm will be slightly smaller than the BM algorithm. Well, the level is more clear, but there are some wrong match, but the matching speed is slower. It is concluded that the BM algorithm is suitable for situations where the processing effect is not demanding but requires real-time conditions such as industrial real-time monitoring. SGM algorithm is suitable for situations where processing effect is

Fig. 5 Extraction effect of disparity map by using BM algorithm

high, such as medical image analysis. In this paper, the following conclusions are obtained: The SAD window size and parameter minimum parallax are relatively large for the parallax results.

4.4 Simulation Results

After analysis, simulation by using BM-based binocular stereo matching algorithm and disparity extraction is shown. The single picture experiment simulation is given in Fig. 5.

Figure 5 shows the left view, right view, and disparity map of object. It shows that using BM algorithm has a clearer disparity map and a faster matching speed. Specifically, the color matching time of a 320 * 240 color frame is about 59 ms.

5 Conclusion

(1) This paper introduced the whole process of binocular stereo vision. This system has advantages of high overall recognition rate, high efficiency, simple system structure, and accurate accuracy. Compared with the traditional monocular, the recognition rate of objects is dramatically improved.
(2) This system is suitable for more application scenarios such as measurement, target location, 3D reconstruction, and other fields, especially for non-contact product detection and quality control.

Acknowledgements The work described in this paper was funded by The Project of Shaanxi Provincial Science and Technology Program (2014JM8351). And it was also funded by Fundamental Research Funds for the Central Universities of China (2013G1241109).

References

1. DING Hong-wei, CHAI Ying, LI Kui-fang A fast binocular vision stereo matching algorithm [J]. Acta Optia Sinica, 2009, 29 (8).
2. Yue Ming, Qiuqi Ruan. Face Stereo Matching and Disparity Calculation in Binocular Vision System [C]. 2010 2nd International Conference on Industrial and Information Systems, 2010.
3. YANG Xing-fang, YOU Mei, GAO Feng, YANG Xin-gang, HAN Xu-sha.A new algorithm for corner detection of checkerboard image for camera calibration [J]. Chinese Journal of Scientific Instrument, 2011.
4. Zhang Zheng You. A Flexible Camera Calibration by Viewing a Plane from Unknown Orientations [A], ICCV99 [C], 1999.
5. Yang Xiaomei, three-dimensional image of the correction algorithm [D], Xi'an University of Technology, March 2010.
6. Zhang Huan, Amway, Zhang Qiang, etc. SGBM algorithm and BM algorithm analysis [J]. Surveying and mapping and spatial geographic information, 2016, 39 (10).
7. Jaco Hofmann, Jens Korinth, and Andreas Koch. A Scalable High-Performance Hardware Architecture for Real-Time Stereo Vision by Semi-Global Matching [C], 2016 IEEE Conference on Computer Vision and Pattern Recognition Workshops, 2016.
8. Heiko Hirschmuller. Accurate and Efficient Stereo Processing by Semi-Global Matching and mutual Information [M]. IEEE, 2005.

Wild Flame Detection Using Weight Adaptive Particle Filter from Monocular Video

Bo Cai, Lu Xiong and Jianhui Zhao

Abstract Wild flame detection from monocular video is an important step for monitoring of fire disaster. Flame region is complex and keeps varying, thus difficult to be tracked automatically. A weight adaptive particle filter algorithm is proposed in this paper to obtain flame detection with higher accuracy. The particle filter method considers color feature model, edge feature model, and texture feature model and then fuses them into a multi-feature model. During which related adaptive weighting parameters are defined and used for the features. For each particle corresponding to target region being tracked, the proportion of fire pixels in the area is computed with Gaussian mixture model, and then it is used as an additional adaptive parameter for the related particle. The presented algorithm has been tested with real video clips, and experimental results have proved the efficiency of the novel detection method.

Keywords Wildfire · Detection algorithm · Particle filter method
Weight adaptive

1 Introduction

Since monocular video has the ability to monitor large and open space, visual-based approach for fire detection and prediction has thus received more attentions from the researchers [1–3]. During which, the successful tracking of wild flame regions is a very important step. For the existing visual detection methods, most objects are rigid which have affine changes, while research for non-rigid object tracking is comparatively rare. As an example of visual pattern with high entropy, wild flame is

B. Cai · L. Xiong · J. Zhao (✉)
School of Computer, Wuhan University, Wuhan 430072, Hubei, China
e-mail: jianhuizhao@whu.edu.cn

B. Cai
Information Center, Wuhan University, Wuhan 430072, Hubei, China

© Springer Nature Singapore Pte Ltd. 2019
B. K. Panigrahi et al. (eds.), *Smart Innovations in Communication and Computational Sciences*, Advances in Intelligent Systems and Computing 670,
https://doi.org/10.1007/978-981-10-8971-8_33

the object with shape that is hard to be pre-defined [4, 5]. Therefore, automatic detection of wild flame region is the greatest obstacle for fire prevention [6–8].

Particle filter [9–11], also known as condensation, is a practical algorithm to solve Bayesian optimal estimation. Because it uses nonparametric Monte Carlo method for recursive Bayesian filter, particle filter is thus applicable to any nonlinear system and any case of non-Gaussian noise, and the results of particle filter can approximate the optimal estimate. Basic idea of particle filter tracking algorithm is using a series of samples with associated weights that obtained through random sampling and the estimates based on these samples to calculate posterior probability distribution of state, while the estimate can approximate posterior probability distribution when the size of sample is large enough.

The particle filter algorithm has high accuracy and robustness in tracking, and it works for the situation that the information of target is of nonlinear system and with non-Gaussian noise, so its application range is very wide. However, the detection accuracy is closely related with the number of particles, i.e., the greater the number of particles, the higher the tracking precision, and the lower detection speed at the same time. Meanwhile, there may be the phenomenon of particle degradation. Therefore, improvements have to be made for particle filter in wild flame detection. To improve the tracking accuracy and robustness, weight adaptive techniques are considered in our work. The rest of this paper is organized as follows: the novel algorithm using weight adaptive particle filter is described in Sect. 2 in details, experimental results and performance analysis are presented in Sect. 3, and then the conclusion is given in Sect. 4.

2 The Weight Adaptive Particle Filter

2.1 Color Feature Model

Since HSV color model is more consistent than RGB model in the way human eyes perceive color, and the hue component of HSV color model has a light and rotation invariant. The hue kernel histogram is used as the color characteristic of wild flame target.

Select the target area in current frame and then calculate its target model as $\hat{q}$ according to the formula:

$$\hat{q}_u = C \sum_{i=1}^{n} k\left(\|x_i\|^2\right) \delta[b(x_i) - u] \tag{1}$$

where $k(x)$ is an isotropic kernel function, the function $b(x_i^*)$ maps pixel x_i to its corresponding feature grid box u, the range of u is $\{1\ldots m\}$, m is the series of kernel color histograms, δ is Kronecker delta function, n is the number of pixels, C is a standardized constant coefficient, making the sum of probability densities to be 1.

Then let the center of target area as the initial position (y_0) of the candidate region in the next frame.

Correspondingly, candidate model of candidate region is obtained as $\hat{p}(y_0)$ with the formula:

$$\hat{p}(y) = C_h \sum_{i=1}^{n_k} k\left(\left\|\frac{y - x_i}{h}\right\|\right) \delta[b(x_i) - u] \qquad (2)$$

where h is the window width of the candidate region, n_k is the number of pixels that locate in the candidate region, C_h is standardized constant coefficient, y is the center coordinate of the candidate region.

Then the color feature model is defined as the distance between target model and candidate model:

$$D_c = \sqrt{1 - \sum_{u=1..m} \sqrt{p^{(u)} q^{(u)}}} \qquad (3)$$

2.2 Edge Feature Model

Compared with general edge characteristics, gradient direction histogram cannot only effectively describe the features of edges and contours, but also highlight the motion information of target area. The gradient direction histogram can be extracted by the following steps: firstly get the motion features through calculating the absolute difference from two neighboring video frames, then obtain the gradient of the frame difference image which exactly express the motion edge features, and then compute the gradient direction histogram.

Let I_k and I_{k-1} represent the k-frame and the $k - 1$-frame of the video, respectively; then the frame difference image diff_k can be obtained according to the following formula:

$$\text{diff}_k = |I_k - I_{k-1}| \qquad (4)$$

Obtain the motion edge image E_k through calculating the gradient of image diff_k:

$$E_k = \nabla \text{diff}_k = \left[\frac{\partial \text{diff}_k}{\partial x} \quad \frac{\partial \text{diff}_k}{\partial y}\right] \qquad (5)$$

The direction angle θ is:

$$\theta(x, y) = \arctan\left[\frac{\partial \text{diff}_k}{\partial y} \middle/ \frac{\partial \text{diff}_k}{\partial x}\right] \qquad (6)$$

Obviously, the range of angle θ is $[0, 2\pi]$. Let $\Delta\theta$ be the quantized distance between the direction angle, and then the direction code can be calculated by:

$$c_{ij} = \begin{cases} [\theta_{ij}/\Delta\theta] & |\partial f/\partial y| + |\partial f/\partial x| > T \\ m & \text{otherwise} \end{cases} \tag{7}$$

Assume that the direction code is quantized into m, the range of c_{ij} is $\{0, 1, 2, \ldots, m - 1\}$. Threshold T can be determined based on experience, while 5 is used in our work as the value of T. The quantified distance between the direction angle is taken as $\pi/8$, i.e., $m = 16$.

Obtain the direction histogram by counting statistics, the frequency of each direction code appeared in the image. So the frequency of direction code $u_{0,1,2,\ldots,m}$ is:

$$f(u) = \sum \delta(u - c_{ij}) \tag{8}$$

where δ is the Kronecker delta function, and then it is normalized. Distance of motion edge features between two video images can now be obtained according to the following formula:

$$D_e = \sqrt{1 - \sum_{u=1\ldots m} \sqrt{f_p^{(u)} f_q^{(u)}}} \tag{9}$$

Texture feature model.
Because wavelet transformation [12] can reflect both time information and frequency information in the same time, it is selected as the texture extraction tool in our work. After a three-layer wavelet transform, the image is decomposed into ten frequency sub-bands, and texture information of one sub-band is expressed as:

$$e_m = \frac{1}{MP} \sum_{i=1}^{M} \sum_{j=1}^{P} |x(i,j)| \tag{10}$$

where M and P respectively indicate the length and the width of each sub-band image, $x(i,j)$ is the wavelet coefficient of position (i,j). The texture feature of the whole image can then be represented by feature vector t which contains 10 elements:

$$t = [e_1 \; e_2 \; e_3 \; e_4 \; e_5 \; e_6 \; e_7 \; e_8 \; e_9 \; e_{10}]^{\mathrm{T}} \tag{11}$$

After normalization of vector t, the distance of texture features between two video images can be obtained according to the following formula:

$$D_t = \sqrt{1 - \sum_{u=1...m} \sqrt{e_p^{(u)} e_q^{(u)}}} \tag{12}$$

2.3 Multi-feature Fusion Model

Based on the color feature, motion edge feature, and texture feature models mentioned above, a multi-feature fusion model is constructed. Compared with each of the traditional single feature model, it improves the accuracy and robustness of tracking. To fuse the multiple features together, adaptive weighting parameters are utilized for relevant elements. The multi-feature fusion model can be obtained by the following formula:

$$
\begin{aligned}
P_{\text{fused}}(z_k | x_k^i) &= \omega_1 P_{\text{color}}(z_{\text{color},k} | x_k^i) \\
&\quad + \omega_2 P_{\text{edge}}(z_{\text{edge},k} | x_k^i) \\
&\quad + \omega_3 P_{\text{texture}}(z_{\text{texture},k} | x_k^i)
\end{aligned}
\tag{13}
$$

where P_{color}, P_{edge}, and P_{texture} indicate color, edge, and texture feature models, respectively; ω_i is the adaptive normalized weight corresponding to each feature, which is calculated as the following steps.

(1) Compute the averaged values of the Bhattacharyya coefficients of color feature, edge feature, and texture feature for all particles and express them as Bha_ave$_1$, Bha_ave$_2$ and Bha_ave$_3$.

(2) Calculate the Bhattacharyya coefficient sum of the important relevant elements as:

$$\text{Bha\_sum}_i = \sum_{j=1}^{M} \left(\text{Bha}_{i,j} | \text{Bha}_{i,j} > \text{Bha\_ave}_i \right) \tag{14}$$

where M is the number of elements corresponding to feature i, Bha$_{i,j}$ is the Bhattacharyya coefficient of the element j of feature i, Bha_ave$_i$ is the averaged Bhattacharyya coefficient of all elements of feature i, while Bha_sum$_i$ is the Bhattacharyya coefficient sum of relevant elements, whose Bhattacharyya coefficient is more than the average value.

(3) Generate the averaged Bhattacharyya coefficient for the important relevant elements by:

$$w_i' = \text{Bha\_sum}_i / \text{num}_i \tag{15}$$

where num$_i$ is the number of important elements of feature i, which has greater Bhattacharyya coefficient than the average value, while the range of i is [1, 2, 3], corresponding to the 3 weighting parameters.

(4) Obtain the normalized w'_i, i.e., the weight of each feature in the multi-feature fusion model with:

$$w_i = w'_i / \sum_{j=1}^{N} w'_j \tag{16}$$

where w'_i is the weight of feature i before normalization, w_i is the normalized value, while N is the number of particles in particle filter algorithm.

2.4 Proportion of Flame Pixels

In our work, another adaptive weighting parameter is considered, i.e., the proportion of flame pixels in the target area to be tracked. In particle filter, each particle has a weight which indicates the contribution degree for the location of the target region in current frame. The more similar between particle characteristic model and the target model, the higher is the weight. In order to make particle filter track wild flame objects better, the proportion of fire pixels for each particle is obtained by Gaussian mixture model [13, 14]. If the proportion of fire pixels is greater, there is higher similarity between the flame region and the particle; thus, the position of fire target in the current frame is more likely to be located in this particle. Therefore, the particle with more proportion of fire pixels should be provided with greater weight.

To further improve the tracking accuracy of particle filter, the additional adaptive weighting parameter has been considered as follows.

For particle k, calculate the fire proportion of particle area using the following formula:

$$\text{firepixel\_scale}_k = \sum_{i=1}^{m} \sum_{j=1}^{n} \text{is\_fire}_{ij} / mn \tag{17}$$

where m and n respectively indicate the values of the height and the width of the particle area, is_fire$_{ij}$ is used to determine whether pixel (i, j) is fire pixel, which is defined as follows:

$$\text{is\_fire}_{ij} = \begin{cases} 1 & \text{gmm\_pro}_{ij} > \text{gmm\_prothres} \\ 0 & \text{otherwise} \end{cases} \tag{18}$$

where gmm_pro$_{ij}$ is the value of GMM probability for pixel (i, j), and gmm_prothres is the pre-defined fire pixel probability threshold.

Based on the proportion of fire pixels, the weight of particle k is adjusted as:

$$w_k = \text{firepixel\_scale}_k w_k' \tag{19}$$

where w_k' is the original weight of particle k in particle algorithm, w_k is the adjusted weight using the adaptive proportion value of flame pixels.

2.5 Detection Algorithm for Wild Flame

The proposed weight adaptive particle filter tracking algorithm is described as follows.

Step 1. Initialize particles
In the initial time t_0, extract an flame target template $f(ab)$ with size $m \times n$, $(a1 \ldots mb1 \ldots n)$, and the initial motion parameters $T^{init}(TX^{init} TY^{init})$ of the fire target, where TX^{init} and TY^{init} indicate the X and Y coordinates of the target center. To obtain the initial target tracking template, calculate the color feature model, edge feature model, and texture feature model of the tracking target, respectively.
Step 2. Resample
Define two threshold values of M and Z. For those particles whose weight is less than Z, when the number is more than M, resampling is applied to generate a new set of particles.
Step 3. State prediction
Use the following system transition formula to predict the state of each particle, i.e., to propagate the particles.

$$x_t A x_{t1} B v_{t1} \tag{20}$$

where A and B are constant parameters, v_{t1} is the random value within the range of [11].
Step 4. Weight updating
Each particle is observed after propagation, i.e., the similarity between the particle and the target model is measured by the multi-feature fusion model, while the weighting parameters of feature models and particles are adaptively adjusted.
Step 5. Calculate posterior probability
The posterior estimate of tracking target is calculated as follows:

$$E(x_t) \sum_{i0}^{N1} w_t^i x_t^i \tag{21}$$

Then go to Step 2 to continue tracking until all frames of the video are tracked.

3 Experimental Results and Analysis

Using some real video clips with wildfire, the new method has been tested. The proposed particle filter is implemented in the environment of Visual Studio C++ and OpenCV. As shown in Fig. 1, wild flame regions are tracked with traditional particle filter. As shown in Fig. 2, the fire areas are detected with the new weight adaptive particle filter method.

From Fig. 1, it can be found that, using traditional particle filter, the detection accuracy is low and there are many non-fire pixels in the detected result.

From Fig. 2, it can be found that detection accuracy has been improved obviously, i.e., most pixels in the detected result are flame pixels.

Fig. 1 Detection result of traditional particle filter

Fig. 2 Detection result of proposed weight adaptive particle filter algorithm

4   Conclusion

In this paper, an improved particle filter algorithm with weight adaptive method is presented to detect wild flame from monocular video. Our contributions include: fire pixel proportion of each particle region is used to adjust the weights of particles, and the averaged Bhattacharyya coefficients for important elements are used to adjust the weights for features in the multi-feature fusion model. Performance of our novel algorithm has been tested on real video clips with wild flame, and the experimental results prove that the new method has better stability and accuracy for wild flame detection.

References

1. Ye W., Zhao J., Wang S., Wang Y., Zhang D., Yuan Z. Dynamic Texture Based Smoke Detection Using Surfacelet Transform and HMT Model. Fire Safety Journal, 2015, 73, 91–101.
2. Liu Z.G., Yang Y., Ji X.H. Flame detection algorithm based on a saliency detection technique and the uniform local binary pattern in the YCbCr color space. International Journal of Signal, Image and Video Processing, 2016, 10, 277–284.
3. Zhang H.J., Zhang N., Xiao N.F. Fire detection and identification method based on visual attention mechanism. Optik, 2015, 126, 5011–5018.
4. Yuan F.N., Fang Z.J., Wu S.Q., Yang Y., Fang Y.M. Real-time image smoke detection using staircase searching-based dual threshold AdaBoost and dynamic analysis. IET Image Processing, 2015, 9(10), 849–856.
5. Roberto R.R. Remote detection of forest fires from video signals with classifiers based on K-SVD learned dictionaries. Engineering Applications of Artificial Intelligence, 2014, 33, 1–11.
6. Zhao Y.Q., Tang G.Z., Xu M.M. Hierarchical detection of wildfire flame video from pixel level to semantic level. Expert Systems with Applications, 2015, 42, 4097–4014.
7. Kosmas D., Panagiotis B., Nikos G. Spatio-Temporal Flame Modeling and Dynamic Texture Analysis for Automatic Video-Based Fire Detection. IEEE Transactions on Circuits and Systems for Video Technology, 2015, 25(2), 339–351.
8. Zhang Z.J., Shen T., Zou J.H. An Improved Probabilistic Approach for Fire Detection in Videos. Fire Technology, 2014, 50, 745–752.
9. Sanjeev Arulampalam M., Maskell S., Gordon N., Clapp T. A Tutorial on Particle Filters for Online Nonlinear/Non-Gaussian Bayesian Tracking. IEEE Transactions on Signal Processing, 2002, 50(2), 723–737.
10. Van der Heijden F. Consistency Checks for Particle Filters. IEEE Transactions on Pattern Analysis and Machine Intelligence, 2006, 28(1), 140–145.
11. Wang H. Adaptive Object Tracking Based on an Effective Appearance Filter. IEEE Transactions on Pattern Analysis and Machine Intelligence, 2007, 29(9), 1661–1667.
12. Chen X. Real Wavelet Transform-Based Phase Information Extraction Method: Theory and Demonstrations. IEEE Transactions on Industrial Electronics. 2009, 56(3), 891–899.
13. Biemacki C., Celeux G., Govaert G. Assessing a Mixture Model for Clustering with the Integrated Completed Likelihood. IEEE Transactions on Pattern Analysis and Machine Intelligence, 2000, 22, 719–725.
14. Yan H., Zhou J., Pang C. Gaussian Mixture Model Using Semisupervised Learning for Probabilistic Fault Diagnosis Under New Data Categories. IEEE Transactions on Instrumentation and Measurement, 2017, 66(4), 723–733.

Author Index

© Springer Nature Singapore Pte Ltd. 2019
B. K. Panigrahi et al. (eds.), *Smart Innovations in Communication and Computational Sciences*, Advances in Intelligent Systems and Computing 670,
https://doi.org/10.1007/978-981-10-8971-8

Printed in the United States
By Bookmasters